'A SUBTLE AND MYSTERIOUS MACHINE'

STUDIES IN HISTORY AND PHILOSOPHY OF SCIENCE

VOLUME 18

The titles published in this series are listed at the end of this volume.

EMILY BOOTH

'A SUBTLE AND MYSTERIOUS MACHINE'

The Medical World of Walter Charleton (1619-1707)

Springer

A C.I.P. Catalogue record for this book is available from the Library of Congress.

ISBN-10 1-4020-3377-X (HB)
ISBN-10 1-4020-3378-8 (e-book)
ISBN-13 978-1-4020-3377-3 (HB)
ISBN-13 978-1-4020-3378-0 (e-book)

Published by Springer,
P.O. Box 17, 3300 AA Dordrecht, The Netherlands.

www.springeronline.com

Printed on acid-free paper

Printed in The Netherlands.

TABLE OF CONTENTS

	Acknowledgements	vii
I.	Introduction	1
II.	Rewriting Walter Charleton: Physick and Natural Philosophy	32
III.	'The Alembic of our Pen': Charleton's Identity as a Physician	52
IV.	'The Animal Oeconomy': *Natural History* (1659) in the Context of English Physiology	81
V.	'The Republick of Letters': Charleton's Identity in the Royal Society & College of Physicians	109
VI.	*Enquiries into Human Nature* (1680): Charleton's Anatomy and Physiology after the Royal Society	137
VII.	*Three Anatomic Lectures* (1683): Ways of Knowing and the Anatomical Body	178
VIII.	Conclusions	216
	Annotated bibliography of the works of Walter Charleton	223
	Primary Sources	243
	Secondary Sources	254

ACKNOWLEDGEMENTS

I am grateful for the assistance of the Wellcome Institute for the History of Medicine, London, in the research toward this volume. Thanks also to Cathy Fowler at the Library of the Royal College of Physicians of London for her kind help.

This book would never have eventuated if it were not for the guidance (and endless patience) of Lotte Mulligan—a tireless reader, rigorous critic and true friend. My profound thanks to Inga Clendinnen for reminding me why it was all worth it. Thanks also to John Cashmere for his kindness and enthusiasm, and to the History Department at La Trobe University, the support of whom I have been lucky enough to enjoy.

I'm immensely grateful to Stephen Gaukroger and John Schuster, for their patient and detailed editorial suggestions. And to John Henry and Conal Condren, for their thoughtful insights into the work's strengths and weaknesses. Needless to say, all failings are my own.

I would long ago have faltered were it not for the wisdom, clarity and excellent company of Lisa MacKinney, Jo Wallwork and Marina Bollinger. All of them have helped me in ways more significant than they'll ever realise. My wonderful family has been a source of unquestioning support and love throughout this project, and I could not have finished this book were it not for them.

My most humble debt of gratitude is to Darren James, whose intelligence, love and humour are a continual inspiration, and whose effortless insights into Walter Charleton have never ceased to infuriate me. Thankyou.

CHAPTER I

INTRODUCTION

Walter Charleton is an intriguing character—he flits through the diaries of Pepys and Evelyn, the correspondence of Margaret Cavendish, and his texts appear in the libraries of better-known contemporaries. We catch sight of him conversing with Pepys about teeth,[1] arguing with Inigo Jones about the origin of Stonehenge, being lampooned in contemporary satire,[2] stealing from the Royal Society, and embarrassing himself in anatomical procedures. While extremely active in a broad range of Royal Society investigations, his main discovery there seems to have been that tadpoles turned into frogs.

As a practising physician of limited means, Walter Charleton was reliant for his living upon patrons and his medical practice—in addition he had the misfortune to live in an era of dramatic political change, and consequently of unpredictable fortune. His achievements were known on the Continent. Despite his embarrassments in Royal Society anatomical investigation he was offered the prestigious chair of anatomy at the University of Padua. He turned down this extraordinary opportunity, only to die destitute in his native country a couple of decades later. The lugubrious doctor is without doubt an enigma. Charleton's Anglicanism and staunch Royalism were unwavering throughout his career. The latter caused difficulties for him when he attempted to gain membership of the College of Physicians during the interregnum. His religious views were a source of concern when he was offered the position at Padua.

At the forefront of contemporary thought in his translation of the continental philosophies first of Van Helmont and then of Gassendi, Charleton seems to have swerved publicly from hermetic to atomistic philosophy in the 1650s. Partly for this reason, his writings have been annexed by historians over the years almost uniformly to perpetuate the idea of 'scientific revolution'. He has been depicted as a 'barometer' of contemporary thought.[3] This book aims to present

[1] Pepys records that on 28 July 1666 he dined with Charleton, Lord Brouncker and his mistress and Sir William Warren, at the Pope's Head, where Charleton delivered a 'very pretty discourse . . . concerning Nature's fashioning every creature's teeth according to the food she intends them.' *The Diary of Samuel Pepys*, ed. R. Latham and W. Matthews, London, Bell, 1970-1983, vol. 7, pp. 223-4. Nicholas Dew discusses this occasion in detail in 'The Politics of the Body in Restoration England: Anatomy and Theology in the Work of Walter Charleton', MSc Dissertation, Oxford, 1995, p. 1.

[2] See Samuel Butler, 'An occasional reflection on Dr Charleton's feeling a dog's pulse at Gresham College by R[obert] B[oyle]', in *The Genuine Remains in Verse and Prose of Mr Samuel Butler*, 1749, vol. 1, pp. 404-10.

[3] C. Webster, *The Great Instauration: Science, Medicine and Reform, 1626-1660*, London, Duckworth, 1975, p. 278.

him as a historically contextualised figure, against a background that he himself would have recognised, rather than as a symbol of a theoretical perspective derived from our own era.

Those who claim for Charleton a transition from 'ancient' to 'modern' argue that this took place in the 1650s, when he allegedly rejected hermeticism for atomism. Examinations of his writings are thus often restricted to the texts surrounding this alleged transition. This characterisation is not particularly helpful, or accurate, in relation to his medical thought, the study of which does not support the view of him promoted by scholars who look at him solely as a natural philosopher. Definitions of Charleton's career as contemporaneous and commensurate with the process of scientific revolution have promoted a reading which omits his medical writings. His publications ranged over numerous disciplines, from his first triad of Helmontian texts to his translation from Pierre Gassendi; several 'physico-theologicall' treatises; translations from classical authors; a history of the passions; a history of Stonehenge, and finally a series of physiological and anatomical medical tracts. Although his life was founded on the practice of physic he is depicted by scholars as a natural philosopher in both identity and aim. His medical status and his medical texts are neglected aspects of his career.

This book considers the works he published in the discipline within which he made his living, as they illuminate aspects of his self-presentation not widely recognised. His medical works are the centrepiece of my study, which explores Charleton's construction of an authoritative identity. I also consider the relationship between identity and epistemology, primarily in Charleton's medical writings. Just as these writings have been neglected in existing historiography, so has their author's status as a professional physician. The restricted view of Charleton as a man preoccupied with natural philosophy has led us to neglect some of the major insights that his works offer into other areas of his career, and into the professional and intellectual landscape of the seventeenth century physician.

One of the most important and fascinating insights offered by his medical works is that he operated within a truly eclectic mode. Charleton tended to repeat rather than generate original explanations, and deliberately presented himself as a compiler, rather than an innovator. His openness to a variety of theories, and willingness to explore them without adopting any single systematic framework, illustrates important features of the genres within which he wrote. Some recent historiography of epistemology emphasises the significance of eclecticism. While I concur with the emphasis, I disagree with the tendency of that school to link eclecticism solely with the skepticism of the virtuosi, and therefore with empirical practice. This study highlights the importance of eclecticism as an influence upon Charleton's self-presentation as a physician. I argue that eclecticism is manifest in his works both as a method and a philosophical approach. Far more than simply a sign of uncertainty (as it is sometimes depicted), eclecticism was integral to Charleton's ability to assimilate and maintain ancient authority and to craft an appropriate medical persona. A detailed account of the origins of eclecticism as a philosophical tradition is a topic on which more research is required, and is beyond the scope of this volume. This study illustrates how important further research into this subject might be for a closer understanding of seventeenth century thinkers such as Charleton.

While experimental experience has been ascribed primary epistemological significance by historians of the period, it is crucial to look at the value Charleton himself placed upon empirical evidence in his anatomical and physiological writings. Those who perceive practices as the centrepiece of identity fail to recognise the important discrepancies that can emerge between practices and textual presentation. I examine not only the experimental practices with which Charleton was involved, but also how he treated experiment, and whether it constituted a persuasive and frequently-invoked aspect of his physiological explanation. I believe that there is an important discrepancy between his experimental activities and his self-presentation.

Identified by many scholars as a 'virtuoso' experimental philosopher, Charleton is generally assumed to have complied with the practices associated with the natural philosophical community. Prominent historians Shapin, Dear and Henry see this as entailing participation in the collaborative empirical generation of units of knowledge, known as 'matters of fact', and an emphasis upon experimental identity and upon the laboratory as the authoritative context for the construction of knowledge. In contrast, I argue that Charleton's medical works illustrate the continued importance of many activities ignored by this model. These include an emphasis on textual authority, an eclectic acceptance of a range of alternative theories, and an emphasis upon the meditative and solitary creation of knowledge. While these characteristics would have been shared by many in the natural philosophical community, they are not recognised in the works of many historians, and it is with this important omission that I take issue.

To understand Charleton's medical writings, and his role in the late seventeenth century intellectual landscape, we must place him within the context in which he lived and worked: that of a professional physician. This, I argue, sheds new light upon his life, and offers an understanding not afforded by his portrayal as a natural philosopher.

This study examines some of the reasons why Charleton's self-construction differed from that of the 'virtuosi', as well as exploring the ways in which it did differ. In doing this, epistemological issues central to our understanding of seventeenth century thought about the study of the natural order are augmented.

WHO WAS WALTER CHARLETON?

The following is a sketch of some of the main phases in Charleton's life, and is designed to illuminate features relevant to the context within which he wrote and practised. The content of specific works is left for later chapters.

The present examination of Charleton's personal and professional circumstances aims to assist our understanding of how he constructed himself as a physician, and looks at the sources from which he drew authority. This provides the background to the subsequent chapters, which focus on the later medical texts to illuminate aspects of his thought not previously discussed by historians.

Two main types of sources offer evidence concerning Charleton's life. One of these consists of the records of the activities of the institutions to which he belonged. Birch's *History of the Royal Society of London* records that organisation's activities in great detail.[4] Birch describes the physician's experimental contributions and involvements in the Royal Society. The College of Physicians' *Annals* document the discussions and developments there, but are less detailed about investigations, and are primarily a document of professional regulatory matters.[5] Charleton's appearances and contributions are noted in both institutions' records.

The other chief source of evidence is the physician's own published and unpublished writings. The former represent the author's public presentation, and the significance that he himself attributed to his experimental life at the Royal Society. There exists a small quantity of manuscript material, including some letters, copies of some presentations at the Royal Society, and medical (though not clinical) notes. These manuscripts are at times referred to through the text, but I have also attached an annotated bibliography in which all of Charleton's writings are documented.

Oxford years

Walter Charleton was born on 2 February 1619. He received his early education from his father, Walter Charleton, who was Vicar of Ilminster and later Rector of Shepton Mallett, Somerset. The nature of his early tuition perhaps placed the young man at a slight disadvantage when, in 1635, he entered Magdalen Hall, Oxford. Many of his peers had attended Westminster in London, and a network of school associates provided a powerful substructure of allegiances and favours within the university.[6] However his entry into Oxford at an opportune historical moment assured that he was not on the outside for long. At Magdalen Charleton studied under John Wilkins (1614-1672), 'by whose instruction he profitted much beyond his years, in logic and philosophy.'[7] Prominent in mechanics, mathematics, astronomy and microscopy, Wilkins was Warden at

[4] Thomas Birch, *The History of the Royal Society of London for the Improving of Natural Knowledge, from its First Rise*, 4 vols, London, 1756-57, New York, Johnson Reprint Co., 1968.

[5] See *Annals of the Royal College of Physicians of London* (hereafter *Annals*), typescript translation. My thanks to the Royal College of Physicians for permission to cite their records.

[6] See R. G. Frank, *Harvey & the Oxford Physiologists: Scientific Ideas and Social Interaction*, Los Angeles, University of California Press, 1980, p. 59.

[7] A. Wood, *Atheniæ Oxonienses . . . to which are added the fasti or annals of the said university etc.*, London, Johnson Reprint Co., 1967, p. 752.

Wadham College during the Commonwealth, in which period the college became active in the investigation of natural philosophical questions.[8] The basic undergraduate education involved four years' reading and disputation in the traditional categories of grammar, rhetoric, logic, mathematics, as well as moral philosophy.[9] The MA was achieved by a further three years devoted to philosophical studies. This background prepared the successful graduate for one of the higher faculties: theology, law or medicine.

As he was not wealthy, Charleton was obliged to gain a profession. His choice was to study physic.[10] It was possible to achieve a BM after three years, and DM after a further four. Medicine at Oxford during the early seventeenth century had been at low ebb, with inadequate staff and no anatomical or botanical teaching facilities. Relatively few passed through the seven-year process. Charleton, like a small number of his peers, received his DM by 'creation', a process by which academic merit was rewarded, and one through which favours were bestowed.[11]

English universities emphasised thorough knowledge of philosophy, rather than clinical experience. Graduate physicians were to be proficient in Latin and thoroughly familiar with authoritative texts. This contrasted with education at some Continental universities (such as Leiden) at which more time was spent on chemistry and clinical cases. The university medical curriculum combined three disparate sources of medical knowledge: Aristotelian natural philosophy, Galenic anatomy and Hippocratic practice.[12] The professional requirements of physicians demanded practical medical understanding, which could only derive from Galen.[13] As aspiring physicians began their education in Arts faculties, a command of logic and dialectics was essential to their training.[14] Medical

[8] R. Westfall, *Science and Religion in Seventeenth century England*, Hamden, Archon, 1970, p. 18. On Wilkins, see Barbara Shapiro, *John Wilkins, 1614-1672; An Intellectual Biography*, London, University of California Press, 1969. Wilkins was later Bishop of Chester and a founding member of the Royal Society.

[9] R. G. Frank, 'Medicine', in N. Tyacke (ed.), *The History of The University of Oxford*, vol. IV, 'Seventeenth Century', Oxford, Clarendon Press, 1997, p. 508.

[10] *Biographia Britannica: Or the lives of the most eminent persons who have flourished in Great Britain and Ireland from the earliest ages down to the present times*, 1747-1766, Hildesheim, Georg Olms Verlag, 1969, vol. II, p. 1286.

[11] Frank, 'Medicine', pp. 506-9.

[12] R. French, *The History of the Heart: Thoracic Physiology from Ancient to Modern Times*, Aberdeen, Equipress, 1979, p. 66.

[13] Although medicine was a natural culmination of an Aristotelian education, Galen was the cornerstone of both theory and practice. University education, by definition, perpetuated Galenic tradition. French, *History of the Heart*, pp. 64-6.

[14] According to the sixteenth century commentary of John Securis [1566] physicians' training should include 'grammer, Logick, musicke, Astronomie and geometrie, and also Philosophie' ('grammer' here refers to Latin grammar.) John Securis, *A Detection and Querimonie of the daily enormities and abuses committed in Physick*, sig. Avi^{v}. See also Harold J. Cook, *The Decline of the Old Medical Regime in Stuart London*, Ithaca, Cornell University Press, 1986, p. 61, on the substance of physicians' education.

education included astronomy and geometry among those arts which defined learned physicians in contrast with 'empirics'.[15] University medical faculties had used as their templates the disciplines of theology and law, which relied upon the study of ancient texts, and the practice of disputation before peer groups.[16] Central to training in physic was command of ancient medical texts in Latin and Greek, in conjunction with scholastic discursive reasoning, through which the student learned to generate conclusions based upon textual propositions.[17] Medical students attended the lectures given by the Regius professor in the discipline, and participated in disputations.[18]

Cook claims that the English medical education encouraged students to present themselves as scholars and gentlemen, according to aristocratic principles of honour. The kind of education that Charleton received at Oxford inculcated traditional scholastic notions of virtue and the hierarchies of natural knowledge. Medical practice was founded upon the development of 'sound judgement'. Medical education was intended therefore not simply to impart particular knowledge, but to develop within aspiring physicians a certain character, from which the necessary powers of judgement would be consequent.[19] Physicians' emphasis upon their university training was a consequence of their need to distinguish themselves from the range of unlicensed practitioners, quacks and apothecaries as well as Barber Surgeons. The basis of a physician's training was the assumption that the greatest benefits of learning were derived from the study of classical texts and ancient authorities. The creation of a distinctive identity for physicians is a key focus of this study, and the role of medical training in the construction of that identity is important in gaining an understanding of Walter Charleton's later writings. The importance of these determinants has been ignored by others, who are intent on portraying him as a natural philosopher, rather than as a physician.

Charleton was scarcely twenty, when the 'flames of . . . intestine commotions first brake forth into open hostility.'[20] In October 1642 Charles I and his court retired to the university town, where they remained until 1646. Oxford life was affected not only by the immediate presence of the king, but by the upheavals in

[15] J. Henry, 'Doctors and Healers: Popular Culture and the Medical Profession', pp. 191-221 in S. Pumfrey, P. L. Rossi and M. Slawinski (eds), *Science, Culture and Popular Belief in Renaissance Europe*, New York, Manchester University Press, 1991, p. 207. The empirics were constrained by the limitations of their own experience.

[16] It was held that since the Fall knowledge had been declining gradually, and that 'the older a text was the closer it was to true wisdom'. Complete comprehension of the ancients was therefore the path to recapture of the lost wisdom of man. Henry, 'Doctors and Healers', p. 194.

[17] Cook, *Decline*, p. 50.

[18] On education at Cambridge at this time, see W. T. Costello, *The Scholastic Curriculum at Early Seventeenth-Century Cambridge*, Massachusetts, Harvard University Press, 1958, pp. 130-35.

[19] H. Cook, 'Good advice and little medicine: The professional authority of early modern English physicians', *Journal of British Studies*, vol. 33, no. 1, 1994, p. 15.

[20] W. Charleton, *The Immortality of the Human Soul*, London, 1657, p. 31.

political and theological orthodoxy. The civil war had prevented numerous clergymen and professionals from performing their regular duties, and many turned their attention instead to the new philosophical enquiries.[21] Charleton cited this as a cause of the revival in medical education: 'our late Warrs and Schisms, having almost totally discouraged men from the study of Theologie; and brought the Civil law into contempt: The major part of young schollers in our Universities addict themselves to Physick.'[22] Oxford at this time was a vibrant atmosphere for the development of new ideas, and a meeting place for royalists. Charleton's position among the Oxford fraternity brought him into contact with various eminent supporters of the king.[23] He came into contact with men from a broad range of disciplines and perspectives. In such extraordinary political circumstances, a young man of his means could flourish, as these conditions allowed an uncharacteristic openness of discourse and social mobility.[24]

Charleton's years at Oxford initiated relationships with men who were later to be co-experimentalists in the Royal Society (Thomas Wharton, Nathaniel Highmore, Thomas Willis, Christopher Merrett), patrons (William Brouncker, Henry Pierrepont), and friends (John Aubrey[25] and John Evelyn[26]). Many of these men were possessed of means that far outstripped Charleton's modest background. For this reason these university years were crucial in establishing the basis for his later patronage network. He encountered many whose interests were close to his own, and whose ideas he was to share. The extent and significance of these influences upon his thought is discussed in later chapters, which deal with the contexts of specific works. In addition, the extraordinary political circumstances of the 1640s led to Charleton's contact with the king, giving his career an early impetus.

Amongst those who played a significant role in Charleton's development were some who eventually became his patrons: Henry Pierrepont, later Marquis of Dorchester; John Prideaux, later Vice Chancellor of the University and Bishop of Worcester; and William, later Viscount Brouncker of Castle Lyons. These contacts were crucial, since patrons were vital to the success of physicians. Brouncker, who was close to both Charles and the Prince,[27] was extremely useful to Charleton (though they later fell out spectacularly). Brouncker gained

[21] Shapiro, *John Wilkins*, p. 130.

[22] Charleton, *Immortality*, p. 50.

[23] L. Sharp, 'Walter Charleton's early life, 1620-1659, and relationship to natural philosophy in mid-seventeenth century England', *Annals of Science*, vol. 30, 1973, p. 317.

[24] Sharp, 'Early life', p. 315.

[25] Charleton was later to propose Aubrey's entry into the Royal Society. The two men corresponded well into the 1670s, and Charleton appears in J. Aubrey, *Brief Lives, Chiefly of Contemporaries, set down by John Aubrey between the years 1669 and 1696*, ed. Andrew Clark, Oxford, Clarendon Press, 1898.

[26] Evelyn (1620-1706) was influential throughout his life and work. Charleton based the character Lucretius in his *Immortality* upon Evelyn. The two men probably met when Evelyn attended Balliol between 1637 and 1640.

[27] Sharp, 'Early life', p. 317.

Charleton the attention of the king, by whose favour he was created doctor at the relatively early age of 24.[28] In early 1642 Charleton was sworn King's Physician at Oxford. This was an unsalaried position as physician-in-ordinary to the ill-fated monarch.[29] As Charles was already attended by William Harvey (1578-1657), the younger physician's title probably entailed little practical medical assistance, but was merely a reward for his loyal support.[30] In any case the position would have assured contact with Harvey, one of the most respected figures in anatomy and physiology at the time.

Charleton left the University with a strongly traditional approach to literary composition. His early publications demonstrate both a familiarity with the ancients and a knowledge of recent Continental philosophy. The physician later indicated the stimulating intellectual nature of his Oxford years, portraying himself and Evelyn as 'fellow Collegiates in Oxford . . . discoursing freely and calmely of some argument or other in philosophy.'[31] Once again, it should be noted that his entire medical education, and the circumstances of his need to attract a patron, would have inclined Charleton to emphasise his classical and scholarly attributes. These highly influential aspects of his early career remained pivotal elements of Chareton's public self-construction as his career progressed.

Physiological experimentation flourished at Oxford in the early 1650s. It has been argued that the main innovations within medical training at the University occurred only after the Civil War. Interested converts to Harvey's method, meeting in Wilkins' room at Wadham, had begun their own discussions and research on questions of physiology.[32] Robert Frank places Charleton's education within this context.[33] However, though a product of the Oxford education system, the young physician was not closely attached to this 'Oxford group' and its concerns. By 1651 when the group was at its peak, he had already left for

[28] Wood's *Athenæ Oxonienses*, p. 752, has him 'created doctor . . . in Feb. 1642', whereas his *Fasti* records this as on 16 Jan. Munk echoes January, see W. Munk, *The Roll of the Royal College of Physicians*, London, 1878, vol. 1, pp. 390-3. According to *Biographia Britannica* this was in Feb 1642, at age 22. *Biographia Britannica*, 1747-1767, vol. ii, pp. 1286-1292.

[29] Wood, *Athenæ Oxonienses*, p. 752.

[30] Sharp, 'Early life', p. 317. Also P. Rattansi 'Paracelsus and the Puritan Revolution', *Ambix*, vol. 11, 1963, p. 26, and H. Rolleston, 'Walter Charleton, DM, FRCP, FRS', *Bulletin of the History of Medicine*, vol. 8, 1940, p. 404.

[31] Charleton, *Immortality*, p. 3. However, this depiction of university life must be viewed in the context of Charleton's rather unsubtle comparison between rational and amicable English intellectual life and its hot-headed and uncivil French counterpart! The depiction is discussed in more detail in my Chapter III.

[32] T. Brown, *The Mechanical Philosophy and the 'Animal Oeconomy'*, New York, Arno, 1981, p. 58. This 'Experimentall Philosophicall Clubbe' included Seth Ward, William Petty, Thomas Willis, John Locke, Robert Boyle, and other future Royal Society fellows.

[33] Frank, *Oxford Physiologists*, pp. 54 and 56.

London.[34] Removed from the physiological and anatomical activities at Oxford, he missed these years of scientific experimentation, and the clubs in which it occurred. Charleton also missed the Gresham group, arriving in London at about the time of its decline around 1648-9.[35] He did not begin his main experimental career until about a decade after the laboratory boom at Oxford.[36] Consequently his background differed from that of some of his later medical colleagues at the Society, who were involved with this Oxford experimental group.[37]

While many of his Oxford peers performed anatomical and physiological demonstrations among the early scientific communities, Charleton devoted his attention in the 1650s to more theoretical writings. His early texts were concerned not with medicine, but with philosophical, theological and moral issues. Having missed the experimental phase at Oxford, he caught up with the more abstract speculative philosophical writings from the Continent, and with hermeticism. The absence of medical works in this early period perhaps accounts for historians' comprehensive neglect of his status as a professional physician.

Charleton married at some stage during this decade, possibly as early as 1642, and certainly prior to 1650.[38] His wife, whose name is unknown, was the daughter of the Somerset divine Bartholomew Parsons, Oriel graduate and rector at Ludgateshall, Wiltshire, from 1620 to 1642. Wood records Parsons as a 'most laborious and frequent preacher', much admired for his hospitality and piety.[39] Charleton later recalled that soon after the King's return from the battle of Edgehill (at which he was possibly called to attend the royal party) 'I went thence to visit my wife at Ludgateshall', only to find his father-in-law dying. Charleton dissected the rector's body to deduce the cause of death.[40]

[34] In subsequent decades the number of Oxford medical graduates drastically increased, and curriculum changes led to the development of new subject areas. Although he missed the main period of innovation at the university, Charleton's works contributed to the body of new literature that shaped subsequent medical education in England. This curriculum included Cornelius Hogehlande, Johann Vesling, Thomas Bartholin, George Ent, Thomas Wharton, Nathaniel Highmore, Thomas Willis and others. Frank, *Oxford Physiologists*, pp. 48-9. Charleton was read by Newton and Locke, amongst others in the later decades, as their library inventories reveal.

[35] This assembly had included John Wallis, Jonathan Goddard, George Ent, Francis Glisson, John Wilkins and others. Had he been in London, this 'Gresham group' would have been a logical place for Charleton to appear.

[36] Though a member of the College of Physicians by 1650, Charleton attended few meetings. Sharp, 'Early Life', p. 324.

[37] For an instructive look at the composition and role of early seventeenth century natural philosophical communities, see J. Henry, 'The Scientific Revolution in England', in Roy Porter and Mikulas Teich (eds), *Scientific Revolution in National Context*, New York, Cambridge University Press, 1992.

[38] On 5 January 1650 Charleton 'was entrusted with £100 to give to his mother-in-law'. *Calendar of the Proceedings of the Committee for Advance of Money* [1642-1656], ed. M. A. Green, London, 1888, part 2, p. 1141.

[39] Wood, *Athenæ Oxonienses*, p. 25.

[40] See Charleton's Letter to a friend of Wood, 20 January 1671, Bodleian Library, MS Wood F 40, vol. B-C.

Unfortunately little information is available about Charleton's wife, or about their lives together.

As the royal cause became hopeless, Charleton retired to London to establish himself as a practising physician,[41] but still lacked the requisite qualifications to set up his own practice. The universities' emphasis on classical learning meant that physical training was not a part of the curriculum.[42] Knowledge of the practical elements of medicine was to be acquired after the MD, through assisting a more experienced physician.[43] By 1650 Charleton still needed to gain some clinical experience. The only existing evidence of his movements between 1646 and 1649 is a letter from the French physician Theodore Turquet de Mayerne to Lord Conway, regarding Charleton's treatment of the latter's complaint.[44] Mayerne, who arrived in England in 1606, had been appointed as first physician to Charles I, and in this capacity he probably encountered Charleton. A personal friend to the royal family, Mayerne was later appointed physician to Charles II.[45] Lindsay Sharp reasonably suggests that the letter indicates that Charleton gained his practical training as assistant to Mayerne.[46] As Mayerne's junior assistant, he would have been capable of diagnoses, but still dependent upon training in more complex instances.[47] The association between the two men would have lasted no longer than a few years, but long enough for Charleton to receive the requisite experience to launch his own clinical career.[48]

Charleton set up his own practice around 1650, in Covent Garden.[49] His clinical career proceeded, though not without incident. On 6 December 1650 he complained about Dr Cassell who had, in a patient's presence, condemned Charleton's treatment as 'dangerous and ill-advised.' On 5 March 1651 the cure was defended by the College of Physicians, against criticism by the husband of a woman who died after the abortion of twins under Charleton's treatment. Members of the College generally received the benefit of the doubt under these

[41] Wood, *Athenæ Oxonienses*, p. 751. Frank, *Oxford Physiologists*, p. 30. However, little evidence exists to place Charleton in either London or Oxford between the latter's fall to Parliamentarian forces and 1649 when he corresponded with the College of Physicians.

[42] Henry, 'Doctors and Healers', p. 195.

[43] Cook, *Decline*, p. 49. See also Henry, 'Doctors and Healers', p. 195.

[44] The letter, dated 19 September 1648, stated that 'Charleton has been to see me and tells me that the mischief in your ear is internal, so he will not undertake it, in which I think he is right but we will discuss the matter when we meet . . .' See *Calendar of State Papers, Domestic, Addenda* 1625-1649, ed. W. D. Hamilton, London, 1897, p. 717.

[45] In the same year he retired to Chelsea, where he died in 1655.

[46] Sharp, 'Early Life', p. 319.

[47] Sharp, 'Early Life', p. 318.

[48] Charleton later wrote the 'Præloquium' and an epitaph for a book of Mayerne's medical writings, *Praxis Medica*, which was published in London in 1690, and Geneva in 1692. Copies of both editions are held in the British Library.

[49] Rolleston refers to his address in Russell Street. Rolleston, 'Walter Charleton', p. 404. In a letter of 1671 Charleton referred to 'my house in York St, Covent Garden', suggesting that he lived nearby.

circumstances, and thus support like this was not unusual.[50] Having collected the first of many professional criticisms, Charleton was a practicing London physician by the early 1650s.

Helmontian translations

As we have seen, historians have frequently divided Charleton's intellectual development into phases, following the chronology of his publications. During 1650 he released the works that have been seen to define the hermetic stage of his career. Although we should be cautious about attributing to Charleton a comprehensive Helmontian philosophy, he did show great interest in Helmont's ideas, as manifest in his publication of two texts. One of Charleton's first publications was *Spiritus Gorgonicus* (1650). The title described the subject as the 'Gorgonic Spirit deprived of its stone-producing power, or a discourse of the cause, symptoms and cure of the stone'. The text discussed 'the universal stone-forming spirit', depicted as the origin of urinary concretions or 'stones' in man, and of rock deposits in the material world.[51] The macrocosm-microcosm analogy implicit in this approach is evident throughout the work, and *Spiritus* clearly illustrated a strong interest in, and knowledge of, the hermetic tradition. The text covered the traditional subject matter: material and remote causes, symptoms, prognosis and diagnosis, varieties of the affliction, and therapeutic recommendations. The book's publication in Leiden, and not in England, demonstrated that the young physician was already well connected on the Continent.[52] Containing cabbalistic and neo-Platonic material, the text drew reference to hermetic authors, such as Hartmann, Severinus and especially Paracelsus. *De Lithiasis* (1644), the treatise on urinary calculus by Johannes Baptista van Helmont, was perhaps the single greatest influence. Charleton's fascination with the formation of stones seems to have continued into his later years, and he contributed findings on the subject at the Royal Society in June 1663.[53] He showed an interest in viewing the same topic through different approaches.

Spiritus was followed by two volumes of expanded translations from Helmont. Hermeticism underwent a vogue during the 1650s, as suspicion of classical authorities generated both the desire for reform and the acceptance of newer medical approaches. The breakdown of various traditional intellectual, social and political structures in England led to a burgeoning of religious, political and medical alternatives to orthodoxy.[54] Webster ascribes Charleton's translations to

[50] H. Cook, *Trials of an Ordinary Doctor: Joannes Groenevelt in seventeenth-century London*, Baltimore, Johns Hopkins University Press, 1994, p. 1.

[51] W. Charleton, *Spiritus Gorgonicus, vi sua saxipara exutus; sive De causis, signis &sanatione lithiaseωs, diatriba,* Ludg. Batav. Ex Officina Elseviriorum, 1650.

[52] On the issue of where works were published, see A. Johns, *The Nature of the Book: Print and Knowledge in the Making*, Chicago, University of Chicago press, 1998.

[53] 3 June, and 10 June 1663. Birch, *History*, vol. 1, p. 251.

[54] R. Kargon, 'Walter Charleton and the Acceptance of Atomism', in *Atomism in England from Hariot to Newton*, Oxford, Clarendon Press, 1966.

a prevailing trend in the medical community.[55] While these publications may indeed have been instigated by this trend, we should remember that Charleton's interest always extended to the exploration of different perspectives.

The physician's first translation was of Helmont's writings on sympathetic medicine, deposits in wine, and magnetism. Entitled *A Ternary of Paradoxes: Magnetick Cure of Wounds, Nativity of Tartar in Wine, and the Image of God in Man*, this was the first English translation from Latin of the influential Flemish hermetic. Rattansi cites *Ternary* as a demonstration of Charleton's 'adherence to the doctrines of Helmont'[56], assuming his complete acquiescence with the translated text. If we attribute to Charleton a magical philosophy in 1650, his 1654 rejection of sympathetic cures appears as a radical transition in thought.[57] However, Charleton stated his purpose in *Ternary* as being 'to clear the prospect, by the necessary remove of such *Doubts*, as seemed very much to obscure the resplendent lustre of *Magnetisme*, and render the Excellencies of *Sympatheticall Remedies* imperceptible'.[58] Thus he did not assert the truth of the theory, but rather its plausibility. He did not explicitly argue against the Helmontian spiritualist perspective, but claimed to leave the illumination of spiritual matters to those writers best equipped to illustrate them.[59] He cited the preceding works of Baptista della Porta, Severinus, Hortmann, Kircher, Cabeus, Robert Fludd, and Digby as a point of reference for those interested to know more about 'how *Sympathetically magnetick* Agents transmit their *Spirituall Energy*, unto determinate Patients, at vast and intermediate distance'.[60]

Charleton's highly derivative use of sources did not necessarily signal his agreement with the theories outlined. He presented the writings of Helmont not as a doctrine, but as a suggestive framework within an eclectic tradition. To understand anything substantial about Charleton we need to dispense with the idea that he adhered to a sequence of discrete doctrines throughout his intellectual career. Mulligan observes that although Charleton chose to translate three Helmontian works, his 'Prolegomena' to *Ternary* contains subtle refutations of Helmont's epistemology. Although supportive of the Helmont's discoveries, and of a great many of his cures, his acceptance of the idea of magnetic cure at a distance 'never relied on the existence of an *anima mundi* but on a mechanical theory of atoms'.[61] In recounting Helmont's argument about the action of magnetism, Charleton redescribed the operation of magnetism through an account which specifically excluded the spiritual aspects so pivotal to Helmont.

[55] Webster, *Great Instauration*, p. 278.
[56] Rattansi, 'Puritan Revolution', p. 26.
[57] Rattansi, 'Puritan Revolution', p. 30.
[58] Charleton, 'Prolegomena', *Ternary*, p. XVIII.
[59] Charleton, 'Prolegomena', *Ternary*, pp. VIII-XIX.
[60] Charleton, 'Prolegomena', *Ternary*, pp. XVIII-XIX.
[61] L. Mulligan, '"Reason", "right reason" and "revelation" in mid-seventeenth-century England', in B. Vickers (ed.), *Occult and Scientific Mentalities in the Renaissance*, New York, Cambridge University Press, 1984, p. 381.

Gelbart also emphasises the continuities which characterised his philosophical approach. She observes that the finer details of Charleton's weapon salve theory diverge from the mystical explanation of Helmont.[62]

The *Ternary* was dedicated to William, Viscount Brouncker. The dedicatory epistle provides biographical material on the relationship between Brouncker and Charleton, which suggests that the two men were close.[63] The dedication deliberately reinforced the friendship between the physician and his mentor, while asserting Brouncker's superiority in intellect and learning. The dedication contained a standard avowal of the physician's lack of interest in publication. Charleton protested that he had been 'intempensively drawne upon the Stage', by obedience to his patrons.[64] It was under Brouncker's 'command' that he had devoted himself to the project. These generic excuses defended the book with self-effacing claims of inadequacy: the author had devoted 'two months onely' to the work's composition, and he himself considered it unworthy of publication.

Charleton's final publication of 1650 was *Deliramenta Catarrhi; or the incongruities, Impossibilities and Absurdities couched under the Vulgar opinion of Defluxions*. As its title suggests, this volume questioned accepted explanations of catarrhal defluxions, in concurrence with Helmont's belief that traditional explanations were in need of reform. A neglected but fundamental aspect of Charleton's *Deliramenta Catarrhi* is the author's tirade against dogmatism. Some historians have interpreted this argument as a rejection of Helmontianism, but I would argue that his words constituted a statement of proper method.[65] He claimed that doctrines with which we were first taught could rarely be displaced, even if inaccurate, as they 'hold our credulities enslaved to an implicite conformity, by the tyrannous title of Præscription.' Charleton borrowed Hobbes' argument that, once registered, false ideas were almost impossible to correct and prejudiced the reader against even the most self-evident correction:

> We judge of the truth or falshood not onely of things subject to the apprehension of sence, but also of Philosophicall and Religious opinions, as we have been accustomed from the minority of our Understandings: and although many times we are greatly deluded, yet cannot the arm of the strongest reason bend us from our accustomed judgement.[66]

62 N. R. Gelbart, 'The Intellectual Development of Walter Charleton', *Ambix,* vol. 18, no. 3, 1971, pp. 149-68.

63 Charleton wrote of the 'happy honour of admission to your closest lectures', and of his patron's 'careful influence bestowed upon each sheet of this work, successively before the Ink could be dry.' Charleton, 'Epistle Dedicatory', *Ternary*, p. xi.

64 Charleton, 'Epistle Dedicatory', *Ternary*, p. vi.

65 He claimed to find Helmont's method 'stronger at Demolishing the Doctrines of the Antient Pillars of our Art, then Erecting a more substantial and durable Structure of his own, his Witt more acute and active at Contradiction, then his judgement profound and authentick at Probation; yet shall I usurp the liberty to say, that many of the Grounds of his quarrell against the Schools, in this particular of Catarrhs, are sufficiently justifiable'. Charleton, 'To the Reader', *Deliramenta Catarrhi,* p. v.

66 Charleton, 'To the Reader', *Deliramenta*, p. ix.

Purely content-based interpretations of Charleton's Helmontianism, such as that of Rattansi, inevitably miss the significance of his epistemology. To argue that this work was Helmontian simply because it was a translation from Helmont is to oversimplify Charleton's reasons for translation. I see these reasons as a reflection of his interest in exploring a variety of philosophies without committing himself to any one of them. By translating he was able to present a set of theories in which he was interested without necessarily showing his support for them. This eclectic mode benefited him as a physician—he could show his familiarity with innovation, while reaffirming the classical traditions upon which his professional authority was founded.

During the early 1650s Charleton's publications illustrate an interest in the atomistic philosophy. His writings of this period feature prominently in current historiography, and are often seen to have defined his agenda. This atomist orientation has been associated by some scholars with the possibility that Charleton visited France himself, and met many of the Continental philosophers whose work he admired.[67] Charleton was travelling physician to Charles II, and physician in ordinary to the king while in exile.[68] However, whether or not he went to France, his work in this period was strongly influenced by various strands of Continental speculative philosophy. Charleton's interest in atomism may have been sparked by the interest of contemporaries, such as Thomas Browne, in the Gassendian philosophical perspective. He discovered the writings of 'the glorious Marinus Mersennus' and 'that heroicall Wit, Renatus De's Cartes',[69] in addition to Hogehlande and Gassendi in the early 1650s. Some of the other important works on Charleton's new horizon were those of Margaret Cavendish, the Duke of Cavendish, Kenelm Digby and 'our eminent Mr Hobbs'.[70]

One of the significant works of this period is Charleton's *Darknes of Atheism*,[71] in which he presented a materialist account of the universe. As such it has been described as the prelude to *Physiologia*. In *Darknes* Epicurean atomism was presented as an argument for God's existence. The book thus exemplifies Charleton's aim of reconciling ancient with modern concerns. Osler sees the

[67] Charleton's activities between December 1650 and March 1652 are not fully recorded. His attendance at meetings at the College of Physicians was irregular between his enrolment and March 1652. Sharp, 'Early Life', p. 324; Frank, *Oxford Physiologists*, p. 30. Frank, Webster and Sharp claim that the young physician visited France at this time. Frank, *Oxford Physiologists*, p. 93; Webster, *Great Instauration*; Sharp, 'Early Life'.

[68] This position was shared with his Oxford peer Edward Greaves. Rolleston, 'Walter Charleton', p. 405.

[69] These men in particular were lauded in Charleton's preface to *Darknes of Atheism*, pp. xv-xvi.

[70] M. Feingold, 'Mathematical sciences and New Philosophy', in N. Tyacke (ed.), *University of Oxford*, p. 406. Hobbes seems to have shown interest in the book, and was asked by Abraham du Prat to send a copy to Paris in 1655. Thomas Hobbes, *The Correspondence*, ed. Noel Malcolm, Vol. 1: 1622-1659, Oxford, Clarendon Press, 1994, pp. 213-14, 246-7, 342-3.

[71] W. Charleton, *The Darknes of Atheism dispelled by the Light of Nature*, London, 1651. See my annotated bibliography for more information.

crucial thrust of the book as consisting in Charleton's claim for the 'absolutely free and unhindered exercise of God's will in his dominion over nature', which she sees as 'essential to understanding Charleton's system of nature and natural knowledge.'[72] The book was dedicated to Francis Prujean, whom the author thanks for his assistance in gaining admission into the College of Physicians, and for his personal assistance in the physician's health.

Charleton published a translation and amplification of Gassendi's *Animadversiones in decimum librum Diogenes Laertii* (1649), entitled *Physiologia Epicuro-Gassendo-Charltoniana.*[73] This work was one of the first of the century seriously to confront the atheistic taint associated with atomism. It was also the first systematic presentation of Gassendian views in the vernacular. *Physiologia* is seen by many to signal Charleton's complete rejection of Helmontianism. However, while his recantation of Helmont is interesting, it is certainly not evidence of a comprehensive shift in his philosophy.[74] *Physiologia* presented an argument for the primacy of non-purposive, physical factors in guiding the action of the internal body. Actions were explicable through reference to function, rather than sentience. His discussion of occult phenomena repudiated Helmont as 'Hairbrain'd and Contentious', and his writings as a 'Delirium'.[75] *Physiologia* thus seems, to many historians, to reject exactly that philosophy which its author had embraced in 1650. His own rejection of the weapon salve theory, he claimed, was founded upon its failure to be demonstrated by experiment, revealing 'the lightness and invalidity of my own and other mens Reasons'. He was now 'fully convinced' of, and 'wholly Converted' from his 'former Error'.[76]

Charleton used the recanting of Helmont's view as an opportunity to reaffirm his own philosophical nature as 'strictly obliged, to præfer the interest of Truth, infinitely above that of Opinion, how plausible and splendid soever, and by whomsoever conceived and asserted'.[77] I argue that this refutation of Helmont's doctrine did not indicate a fundamental shift in Charleton's personal philosophy; rather, he used the refutation to reaffirm his own status as an eclectic. In this I agree with Eric Lewis, who argues that Charleton does not defend his earlier explanation, but instead rejects the phenomenon itself.[78]

Physiologia rejected ideas of occult qualities, attraction, faculties, operations at a distance, and indeed all invisible processes. These were, he argued, not

[72] Osler, 'Descartes and Charleton', p. 452.

[73] Walter Charleton, *Physiologia Epicuro-Gassendo-Charltoniana, or, A fabrick of science natural, upon the hypothesis of atoms*, London, 1654.

[74] This perspective on Charleton's apparent reversal is shared by Lewis, 'Early modern eclecticism', p. 664.

[75] Charleton, *Physiologia*, p. 58.

[76] Charleton, *Physiologia*, p. 382.

[77] Charleton, *Physiologia*, p. 382.

[78] E. Lewis, 'Walter Charleton and early modern eclecticism', *Journal of the History of Ideas*, vol. 62 no. 4, 2001, p. 664.

immaterial, but effected by 'Corporeal, though both impalpable and invisible Organs.'[79] The operations of these organs were deduced, through analogy, from actions in the corporeal sphere. The reader was urged to consider that the invisible might operate in exactly the same ways as the visible. Man could explore the uncertainties of the invisible through his reasoned knowledge of the visible. The mere fact that such instruments were inaccessible to human sense was no disproof of their operations. This echoed Cartesian notions of the similarity between invisible and visible entities.[80] Interestingly, it also echoed the kind of refiguring of Helmont's argument in which he had been engaged in the *Ternary*.

Physiologia was composed under the roof of Sir Robert and Mrs Elizabeth Villiers. The book was dedicated to Elizabeth Villiers, daughter of a prominent parliamentarian.[81] Sharp suggests that Charleton's choice of her as patron was politically astute, as protection from a prestigious person of quiescent status would have advanced his career.[82] However, as a gambit for protection in troubled times, it does not seem to have been particularly successful.

During the early 1650s Charleton encountered the Newcastle circle. Sharp and Feingold believe that their writings 'changed the young author's mode of thought almost entirely'.[83] Focused around the Cavendish brothers, Sir Charles and William, William's wife Margaret, and Thomas Hobbes, the collective included in its periphery William Petty, John Evelyn, Kenelm Digby and other prominent thinkers. This group was active in the exploration of the atomist philosophy. In the late 1660s the Cavendishes were patrons of Charleton, and the physician struck up a friendship with Margaret Cavendish which continued for some years.[84] In 1667 Margaret Cavendish became the first woman to visit the

[79] Charleton, *Physiologia*, p. 346.

[80] Descartes claimed that the difference between the operations of machines and those of natural objects was that, 'the workings of machines are mostly carried out by apparatus large enough to be readily perceptible by the senses (as is required to make their manufacture humanly possible), whereas natural processes almost always depend on parts so small that they utterly elude our senses.' R. Descartes, 'Principia', part II, chap. Xxxvi, in *Descartes: philosophical writings*, ed. and trans. E. Anscombe and P. T. Geach, London, Nelson, 1964, p. 215.

[81] Robert Villiers had served in the royal army, and Charleton may have met Villiers in this capacity at Oxford. However, by 1654 he had switched his allegiances and married Elizabeth, thereby gaining 'influence and protection' after the king's defeat. Sharp, 'Early Life', p. 332.

[82] Sharp, 'Early Life', pp. 331-3. The dedications preceding his 1650s works praised prominent parliamentarians, and politically expedient patrons. After 1657 his situation seems to have been eased, through the patronage of Henry Pierrepont, Marquis of Dorchester.

[83] Sharp, 'Early Life', p. 324. Feingold, 'Mathematical sciences', p. 406.

[84] A 1655 letter in her collection is the earliest example of their correspondence. See *Letters and Poems in Honour of the Incomparable Princess Margaret, Duchess of Newcastle*, London, 1676, pp. 142-9. Visiting Margaret Cavendish, in 1667, Mrs Evelyn 'found Dr Charlton with her, complimenting her wit and learning in a high manner.' Mrs Evelyn to Mr Bohun, from John Evelyn, *Diary and Correspondence*, ed. W. Bray, London, Routledge, 1952, pp. 731-2.

Royal Society, a visit facilitated by Charleton.[85] In 1668 his translation into Latin of Margaret's biography of her husband was published, for the European market.[86] Charleton and this eminent lady exchanged ideas on philosophical matters, and in May 1669 she made him a gift of her recent writings on moral philosophy.[87]

The new influences upon Charleton during the 1650s are seen to define the second, atomistic, period of his intellectual career. This was marked by his publication of three works: *Darknes of Atheism, Physiologia, Epicurus' Morals* and *Immortality*. These have been the major focus of interest in Charleton, concerning his ethics and natural philosophy. While I cite these works at various points in this study, they are not the focus of my discussion. It is noteworthy that all of these atomist works were published in the vernacular, suggesting that the author's aim was dissemination of continental ideas to England, rather than the reverse.

The Immortality of the Human Soul consists of two dialogues between three speakers, in which proper conduct, scientific progress and philosophical method are discussed. The speakers are Lucretius (Evelyn), Athanasius (Charleton) and Isodicastes (Henry Pierrepont, to whom the text was dedicated). Armistead notes the eclecticism manifest in this work—while the most revered thinkers are Epicurus, Bacon, Descartes and Digby,[88] *Immortality* contains also scattered references to Hermes Trismegistus, Plato, Aristotle, Cicero, and the Church Fathers.[89] As such the text exemplifies Charleton's eclectic vision, and his dedication praised this same attribute in his patron, the Marquis of Dorchester. The preface to *Immortality* thanked the latter profusely, declaring that 'from you alone I have received more both of Encouragement and Assistance in my

[85] The visit, which she made on Thursday 30 May 1667, is recorded in Birch, *History*, vol. 2, pp. 175-8. Charleton, 'A Short Harangue designed to be made to the President of the Royal Society, at their entertainment of the Duchess of Newcastle with sight of some select Experiments', Bodleian MS Smith 13, document 12, pp. 21-3. The occasion is discussed by Samuel Mintz, 'The Duchess of Newcastle's visit to the Royal Society', *Journal of English and German Philology*, vol. 51, 1952, pp. 168-76. See also L. T. Sarasohn, 'A Science turned upside-down: Feminism and the natural philosophy of Margaret Cavendish', *Huntingdon Library Quarterly*, vol. 47, 1984, pp. 299-307. Londa Schiebinger notes the event in 'Margaret Cavendish: Natural Philosopher', in M. E. White (ed.), *A History of Women Philosophers*, Dordrecht, Kluwer, 1988. Pepys also noted the spectacle in his *Diary* for 30 May.

[86] It was published as *Guilielmi Ducis Novocastrensis Vita*, London, 1668.

[87] Letter from Charleton to Margaret Cavendish, Bodleian MS Smith 13.

[88] Armistead suggests that the structure of Charleton's *Immortality* was perhaps borrowed from Digby's *Two Treatises*. J. M. Armistead, 'Introduction' to Walter Charleton, *The Immortality of the Human Soul*, New York, A.M.S. Press, 1985, p. viii. Digby's first 'Treatise' set out an atomistic theory of matter, on the basis of which the second proved the soul's immortality through the fact that it was unaffected by the laws governing bodies. See Kenelm Digby, *Two Treatises: Of the Nature of Bodies; Of the Nature of Mans Soule,* Paris, 1644.

[89] Armistead, 'Introduction', p. vii.

studies, than from the whole World besides.'[90] Pierrepont remained a prominent patron, upon whom Charleton relied.[91]

The two 'stages' of his 1650 publications—hermetic and atomistic—have been accepted by historians as providing a complete characterisation of Walter Charleton. However these groups of texts represent only his moral philosophical writings, and some of his natural philosophy. They include no medical writings, and therefore do not allow extensive insight into his life as a physician. An overview of Charleton's career suggests that the subjects of his publications did not indicate his adherence to the broader philosophies upon which they touched. Rather he seems to have published according to trends in the marketplace, and to have explored willingly the categories of thought in vogue at any time. His references indicate a wide range of authorities of whom he approved and his adherence to these was, at best, qualified.

Translation was not equivalent to persuasion, but reflected a more complex relationship. Charleton wished to maintain his authoritative classical identity as a physician, while simultaneously demonstrating a firm knowledge of recent innovations. Thus he preserved the scholarly prestige of his profession, without succumbing to the dangers of dogmatism.

Royal Society

1662 saw Charleton's election as a fellow in the early Royal Society. This period of his life is discussed in depth in chapter five, which explores his institutional involvements. Though this part of Charleton's career has been mentioned within studies on the activities of the Society as a whole, no previous research has focused on the physician's trajectory from his early Royal Society involvement onward, or the relationship between his philosophy and his activities in the Society. I consider the exact nature of his investigations at this time, and the relationship between his activities and writings. I also analyse Charleton's self-identification as a natural philosopher within his written works.

His involvement across an extremely broad range of activities was characteristic of the Royal Society's declared ideals, but his writings often made no reference to the experiments that he performed. This suggests that he did not wish to present himself as an experimenter. I examine possible reasons why experimental activities might not have constituted an appropriate subject for medical texts. Specifically, I argue that Royal Society experimentalism could, in a medical context, be

[90] Charleton, 'Dedication', *Immortality*.

[91] Henry Pierrepont (1606-1680) joined the court at Oxford during king's retreat there between 1642 and 1646. Here he would have met Charleton and also Harvey, to whom he became a friend and a patient. The university conferred on him an MA in 1642, and Charles created him Marquis of Dorchester (on 25 March 1645) and admitted him to the privy council. Pierrepont was the first elected Honorary Fellow of the College of Physicians, in 1647. In 1663, he joined the Royal Society. When Dorchester died, Charleton was put in charge of the Marquis' books that were given to the College. *Annals*, 9 April 1688.

identified with the category of an 'empiric'. This would carry negative implications for Charleton, who instead desired the status of a learned physician.

In the early 1660s he composed an antiquarian work, *Chorea Gigantum; Or the Most famous Antiquity of Great Britain, Vulgarly Called Stone-Heng.*[92] In it Charleton argued that the ancient monument was a Danish creation. This initiated considerable controversy, and was generally ill-received in London. Having attacked Inigo Jones's argument on the Roman origin of the monoliths, *Chorea* earned a severe retaliation from Jones's son-in-law.[93]

The exception to Charleton's otherwise non-medical contributions to the Society was his presentation, on 13 June 1661, of *Exercitationes pathologicae*.[94] This introduction to the study of pathological 'Physick' examined the nature, generation and causes of most known diseases. *Exercitationes pathologicae* followed the tradition of medical compilation, providing a nosology, a summary of contemporary arguments, and progressed through the origins, signs and causes of diseases. This volume was composed before Charleton's entry into the Society. A fuller discussion of the text is given later in this book.

Dedicated to Charles II, *Exercitationes pathologicae* preceded closely Charleton's publication of an *Imperfect Pourtraicture of His Sacred Majesty Charles the II* (London, 1661). This 'character' of the newly restored monarch depicted religion as one of Charles' greatest concerns. Charleton had been keen to gain the approval of the new king, and seems to have been successful. In July 1660 he was appointed Physician in Ordinary to the restored monarch, with a salary of £100 a year.[95] This represented a substantial improvement in Charleton's fortunes.

Cheshire

In the late 1660s his participation at the Society lapsed, and Charleton retired to Cheshire, where he remained for some time in the hospitality of the Crewe

[92] W. Charleton, *Chorea Gigantum; Or the Most famous Antiquity of Great Britain, Vulgarly Called Stone-Heng*, London, 1663.

[93] Wood claims Charleton's text was 'exploded by most persons when t'was published', Wood, *Athenæ Oxonienses*, p. 754. However, it found support with Sir William Dugdale, and some interest from the king, to whom it was dedicated. Charleton in turn recommended Dugdale's *History of imbanking and draining of divers fens and marshes, both in foreign parts and in this kingdom* (London, 1662) at the Society in 1664. In July, when the physician presented his observations on Stonehenge to the Society, Aubrey was asked to look into the matter [Birch, vol. 1, p. 272]. He indicated that the King was quite taken with Charleton's theory about Stonehenge. Charleton and Aubrey attended the king the Duke and Duchess of York when they visited the area in 1663. A. Powell, *John Aubrey and his Friends*, London, Hogarth, 1988, pp. 106-8.

[94] W. Charleton, *Exercitationes pathologicae, in quibus morborum pene omnium natura, generation, & causae, ex novis anatomicorum inventis sedulo inquiruntur*, London, printed 28 January 1661.

[95] *Calendar of State Papers, Domestic, 1660-61*, ed. W. D. Hamilton, London, 1897, pp. 134, 208.

family.[96] At Crewe Hall Charleton practised as a doctor to the local gentry. This period of his life has left us the greatest existing collection of his correspondence.[97] It was common for medical practitioners to spend periods out of the city, as they did in summer, following the habit of the gentry.[98] However, this was a considerably longer period than the normal seasonal peregrinations of a practitioner. Charleton's exact reasons for the retirement are unclear, but a range of possibilities can be suggested. As we see in chapter five, he may have found his combined embarrassments at the Society were too well known for his comfort throughout London.[99] The controversial *Chorea Gigantum* had recently been published, and very little of the ensuing dispute was in Charleton's favour.[100] Brouncker, previously his friend and patron, but now an enemy, was president at the Society. In addition, the Great Fire and the onset of the plague would surely have contributed to his hardship. He may have struggled to maintain a clinical profile, for these reasons and others discussed later.

Charleton published several works, between his participation at the Society and 1676, possibly to improve his profile and ensure his College admission. In 1672, he published *De Scorbuto liber singularis; cui accessit Epiphenomena in Medicatros*.[101] The text discussed the nature, origin and cause of scurvy. The physician himself was out of London when the volume was published (he received his own copy in early 1671, in Cheshire, from his daughter in London).[102]

In 1674 Charleton published, anonymously, *Natural History of the Passions*.[103] In this text he engaged with Thomas Willis's *De Anima Brutorum*,[104]

[96] In 1671 Nathaniel Crewe, for whom Charleton wrote an epitaph, was elected Bishop of Oxford, and given the living of Whitney and the Rectorship of Lincoln College. It seems therefore that Charleton was more likely to have been in the company of Nathaniel's uncle, Sir Randolph Crewe, who is recorded to have settled in Cheshire. *Biographia Britannica*, p. 1520.

[97] Charleton's letters later came into the possession of Thomas Hearne.

[98] Cook, *Trials*, p. 4.

[99] Charleton's theft of some poison, given to the Royal Society by the King of Macassar, for trial on a dog at his own home was publicised by Samuel Butler in his satirical 'An Occasional Reflection', p. 407.

[100] The poet Dryden famously praised the work in a poem prefacing the printed version. The works of the period on this subject have been published in a collected edition: Inigo Jones', *Stone-heng*, Walter Charleton's, *Chorea gigantum* and John Webb's *A vindication*, with a new introduction by Stuart Piggot, Farnborough, Gregg, 1971.

[101] W. Charleton, *De Scorbuto Liber Singularis; cui accessit Epiphenomena in Medicatros*, Londini, Typis E. Tyler, & R. Holt, prostant apud Guliel. Wells & Rob. Scot, 1672. Charleton's notes on this subject exist in the British Library, MS Sloane 1532, ff. 41-61.

[102] Charleton, 'Letter to a friend of Wood', 20 January 1671, Bodleian Library, MS Wood 40.

[103] W. Charleton, *Natural History of the Passions*, London, Printed by T. N. for James Magnes, 1674. The work has been incorrectly assumed to be a translation from Senault's *De l'usage des passions* [Paris, 1641]. R. A. Hunter and E. Cuttler, 'Walter Charleton's *Natural History of the Passions* (1674) and J. F. Senault's *The Use of the Passions* (1649): A case of mistaken identity', *Journal of the History of Medicine and the Allied Sciences*, vol. 13, 1958. The book's dedication describes it as 'the product of my late ten weeks' solitude in the country remote from my library', and it appears he composed the book in his retreat at Cheshire.

[104] Thomas Willis, *De Anima Brutorum*, Oxford, 1672.

as well as the writings of Gassendi, Descartes, Hobbes and Digby.[105] The dedication, to an unnamed person of honour, states that *Passions* was written 'because my accumulated Misfortunes had . . . reduced me to a necessity of consulting that part of Philosophy, about the most effectual Remedies against Discontent.' Charleton posited that 'all the Good and Evil of this life depends upon the various Passions incident to the Mind of man', the conclusiveness of which he claimed to know from 'my own dearly bought experience'. Like afflictions of the body, those of the mind might be more easily cured through understanding of their 'nature, causes, motions [effects] &c.' Charleton's treatise, in explicating these features of the passions, aimed to discover possible 'Remedies against their Excesses.' *Natural History of the Passions* delineated the interaction of body and soul, focusing on the conflict between reason and emotion. Charleton returned to this theme, by several different approaches, throughout his *oeuvre*.

Away from busy London life, he exercised one of his continuing interests—translation. In 1675 the physician published, anonymously, a translation from Greek which he entitled *Plato, His Apology of Socrates*.[106]

College of Physicians

Trained in both theoretical physic and practice, Charleton presented himself at the College of Physicians on 6 July 1649. A vote was taken as to whether the usual examination process could be omitted because he was in the king's employ.[107] The College members decided that he should pass through the normal procedures.[108] On 8 April 1650 Charleton was made a candidate.[109] This was a probationary stage, prior to admission as an elected fellow. Sharp attributes this decision, to enforce full examination procedures, to an attempt to placate the republican government.

[105] In particular, Digby, *Two Treatises*. Thorpe sees Charleton's volume on the passions as inspired by Hobbes' work on the same subject, by whom the text is strongly influenced. Large sections of Charleton's work were drawn almost directly from Hobbes' *Elements of Law*, and the impact of the *Leviathan* is also evident. The text also demonstrates Charleton's familiarity with Aristotle, the Scholastics and the Cambridge Platonists. C. D. W. Thorpe, *The Aesthetic Theory of Thomas Hobbes*, Ann Arbor, University of Michigan Press, 1940, p. 181.

[106] *Plato, His apology of Socrates, and Phædo, or Dialogue concerning the Immortality of Mans Soul, and Manner of Socrates his Death*, London, 1675. See annotated bibliography.

[107] Cook, *Decline*, p. 115. See *Annals* vol. 4, 20b, 22a-24b, 26b.

[108] L. Sharp, 'The RCP and Interregnum Politics', *Medical History*, vol. 19, no. 2, 1975, p. 114. Sharp emphasises the intellectual, rather than political, motivations for Charleton's acceptance, and believes the Helmontian texts Charleton had published influenced his eventual election. However, he provides no proof for this assertion.

[109] Cook sees Charleton's admission as the result of an increasingly conservative turn in the College in 1649-50. Growing numbers of royalists, he argues, were allowed to join the College, in contrast to their exclusion in the 1640s.This conservatism in admission was accompanied by the expulsion of radical critics of the College. Cook, *Decline*, p. 115. See *Annals* vol. 4, pp. 20b, 22a-24b, 26b. See also Rolleston, 'Walter Charleton', pp. 406-7.

On 3 May 1655, Charleton was proposed as a Fellow, but his character was deemed unworthy of that status.[110] The College committee heard 'certain things' against the aspiring Fellow, 'less than worthy in a future Fellow'. It therefore chose to defer the matter, 'and meanwhile to investigate the whole affair with the evidence of witnesses.'[111] At the following week's meeting, Charleton and a Dr Wedderbourne were summoned. The latter accused Charleton 'not only of harmful practices against himself, but also against the Society itself and good sense.' Charleton was 'renounced for it, wholly without hope of obtaining favour with us for the highest rank.'[112] The issue was again raised on 14 July 1655, and a secret ballot was held. Twelve out of seventeen voters found against Charleton, and he was denied the right to proceed to full fellowship of the College.[113] He did not achieve full fellowship status until 1676.

The reasons for this refusal and the subsequent lengthy postponement have been the focus of some debate among historians.[114] Many scholars claim that the physician's difficulties were the consequence of his overt royalism.[115] Cook argues that the vote against Charleton's incorporation as a fellow was a victory for the 'Cromwellian' faction, and an indication of conflict within the College over political allegiances.[116] Webster, likewise, explains the delay in full admission as the consequence of his 'obdurate adherence to the royal party'.[117] As this loyalty had been publicly declared in the dedicatory prefaces to his medical publications it was unlikely to go unnoticed. His slow acceptance in London has thus been attributed to the very same royalist sympathies which had worked in his favour at Oxford. Frank blames Charleton's 'staunchly, not to say aggressively, Royalist sentiments' for his remaining a mere candidate throughout the 1650s.[118] Although none of these scholars cites textual evidence that Charleton's royalist sympathies were significant in his exclusion from College fellowship, the importance of his royalism is widely agreed upon.[119]

110 As discussed below, the position of physician was seen to entail not only a certain amount of learning, but also a particular character and discipline, which qualified the individual to counsel patients.

111 *Annals*, vol. 4, p. 65.

112 *Annals*, vol. 4, p. 65.

113 Sharp, 'Early Life', p. 332. It is worth noting that during this century as many as a third of London's practising physicians were not members of the College. Cook, *Decline*, p. 79.

114 See Cook, *Decline*, p. 115. Sharp, 'Interregnum Politics', pp. 114-8. Dew, 'Politics of the Body', pp. 12-13; Webster, *Great Instauration*, p. 309.

115 *Biographia Brittanica* notes that Charleton's entry into the College would have been extremely difficult if he had not had the support of some principal members, especially Francis Prujean, who was President between 1650 and 1655. *Biographia Britannica*, p. 1287.

116 Cook, *Decline*, pp. 115, 127; *Annals*, vol. 4, pp. 55b, 56a-b, 57a, 57b, 59b, 63a-b.

117 Webster, *Great Instauration*, p. 309.

118 R. G. Frank, 'The Physician as Virtuoso in seventeenth-century England', in R. G. Frank and B. Shapiro (eds), *English Scientific Virtuosi in the Sixteenth and Seventeenth Centuries*, Los Angeles, University of California Press, 1979, p. 90.

119 Historian of the College, George Clark, claims that even ignoring Charleton's later acceptance and rise to presidency under the restored monarchy, 'there would be sufficient grounds for believing that

Charleton's own declaration, upon his eventual acceptance to the College in 1676, casts a slightly different light on his previous exclusion. In his speech upon acceptance he declared that he had at last been given a place of honour in the College, 'whence 21 years before, for execrating impious matters and with yet late consequences, I may have fallen down because of Cromwell's despotic rule'.[120] He appears to claim that his rejection was on the basis of Cromwell's government, and that his stand against the former's impiety was the cause.[121]

An alternative explanation, less heavily reliant on a questionable dichotomy between Cromwellian and royalist loyalties within the College, might rest in Charleton's controversial publications at the time. It seems that his potentially impious materialism, as represented in *Physiologia*, may have earned his some disrepute as a physician. Sharp believes that Charleton's poor profile at the College might have been a consequence of his perceived atheism.[122] While this may be closer to the truth than the argument over his royalism, Sharpe misses the aspects which I believe crucial to an understanding of the situation—he pays attention to the physician's philosophy, but ignores the issues of self-presentation and identity.

In 1657 Charleton complained of the critical reception of his Gassendian text three years previously. He had been censured for 'negligence in the duties in my profession, and invading the certainty of all its rules and Maxims, while I wholly addicted my selfe to the Innovation of its Fundamentalls.'[123] His presentation of Gassendian matter theory undercut the epistemological basis of medical practice. It seems some critics believed he should learn the art and practice of medicine, instead of attempting to dissect the basis of the profession itself. Such censure may have emanated from the College itself, and may have influenced Charleton's rejection in 1655. This response might explain the physician's wheedling praise of College principles and activities in the 1657 *Immortality*, as he attempted to regain favour through his pious defence of the soul's immortality.

Thus it seems that while he did not necessarily adhere to each of the philosophies he presented in his published works, he did suffer through association with some of them. He was not always successful in his bids to negotiate an authoritative identity as a practising physician with broader intellectual interests.

Charleton's differences with the College suggest that conflict may have arisen over the proper public identity for physicians. The exploration of newer theories had controversial implications for the foundations of physic and the profile of professional physicians. Charleton was perhaps seen to have transgressed the

---

this . . . doctor was kept out by his royalism'. G. Clark, *A History of the Royal College of Physicians of London*, 2 volumes, Oxford, Clarendon Press, 1964-6, p. 282.

120 23 January 1676. See Charleton's papers in the Bodleian Library, MS Smith 13.

121 Interestingly, this evidence is not cited by any who discuss his attempts to enter the College.

122 Sharp, 'Interregnum Politics', p. 118.

123 Charleton, *Immortality*, p. 10.

loyalty to authority demanded by this professional institution. His questioning of some of the 'fundamentals' upon which academic physic rested would have compromised his character, according to the College criteria of a learned and sombre professional capable of good judgement. Knowledge and character together constituted the determinants of a physician's authority. Therefore Charleton's presentation of unorthodox learning would have compromised his character to the extent that the College would not support him. Medical education, and the distinctive status of physicians by contrast with other practitioners, relied on classical modes of authority.

His actions revealed a character engaged in conflict with the public profile of an authoritative physician. It was perhaps not even Charleton's expressed beliefs which landed him in disrepute with the College, but his association with the mortalist heresy of Hobbes and the Cavendish circle.

Charleton returned to London in the mid 1670s, and was at last accepted into the College of Physicians, where he rose rapidly in status, and eventually achieved the position of President. Information on his practical involvements at this time is scarce, in comparison with his years at the Royal Society (the College's role was not research but professional regulation).[124] However, several of his anatomical lectures were printed, and these illustrate his continued association with colleagues Francis Glisson, George Ent and Thomas Wharton. This period of Charleton's activities and medical publications has been discussed infrequently, and his medical lectures under the College aegis scarcely at all. It is with the texts published at this time that this book is primarily concerned.

Acceptance into the College of Physicians seems to have given an international profile to Charleton's career. His involvement in the licensing of practitioners broadened his contacts immensely. Charles Goodall noted in 1684 that 'Charleton's very learned and laborious Works . . . have given him a very high and deserved Reputation in our own as well as foreign Universities'.[125] In 1678, shortly after his acceptance into the College, Charleton was offered the Principal seat of Superordinary Professor of Practical Medicine at the Paduan academy. His books *Spiritus Gorgonicus*, *Deliramenta Catarrhi*, *Oeconomia Animalis*, *Exercitationes Pathologicæ*, *Inquisitiones Duæ Anatomico Physicæ* and *De Scorbuto* could be obtained by readers on the Continent, and might therefore have established his reputation. However the offer came before his major anatomical works, which were *Enquiries into Human Nature* and *Three Anatomic Lectures*. It is likely that his medical prowess was promoted through international correspondence (as we have seen, Charleton was mentioned in correspondence emanating from Royal Society Fellows).

[124] Whereas Birch's *History* provides detailed observations of the activities of the fellows at each meeting, there is no comparable record for the College. Though crucial for administrative purposes, the *Annals* made no record of experimental or theoretical contributions from College members. The dearth of source material on Charleton's College activities perhaps explains why his later years are scarcely touched upon by historians.

[125] Charles Goodall, *The Royal College of Physicians of London*, London, 1684, p. xix.

Charleton seems to have intended to take up the role, as he prepared an introductory speech for his inauguration.[126] However his subsequent letters to Paul Sarotti (Governor of Venice) indicate that he was unable to accept the position.[127] He had communicated a list of conditions under which he would take up the offer,[128] which included a request for one thousand gold crowns as travelling money. He confirmed his acceptance of Sarotti's offer of a salary of one thousand five hundred florins *per annum*, to increase by 300 florins after five years if he was still in service.

Charleton requested that he be excused from any of the Catholic religious rituals that were necessary for the 'gathering of Favour'. He also asked that he might be allowed to read from his own notes, rather than recite from memory. He claimed that this was how he lectured in England, for he was 'little tenacious of memory'. The physician hoped that he would be excused from having to occupy his mind with recitations, so that he could instead devote himself to the search for truthfulness in relation to the Hippocratic art. Now in his sixties, he made clear his desire to be able to retire from a teaching role in the School of Physicians, if he found himself unable to continue 'with decorum and dignity'. He asked to be released from service after four years, and returned to his country of birth along with his possessions (notably his library). Given his request for an increase in salary after five years, this may have been a reference to a paid visit to England. Sarotti's letter has not survived, and sadly and we know nothing of how this matter was resolved.

We can infer from Charleton's letter that at this time he was financially precarious, and did not trust his own abilities. His caveats regarding memory and 'decorum', and the fact that he was still anxious not to damage his chance of further patronage by participating in any Catholic rites, indicate that he was anxious to safeguard his interests.

Later life and activities

As in his early years of practice, Charleton's later professional life was not without incident. In March 1688-9 the *Annals* of the College of Physicians recorded that Charleton, then Vice President, was summoned, 'to answere an accusation from Dr. Blackmore for disparaging his Practice'.[129] This incident allows us a glimpse of the kind of competitive marketplace in which Charleton practiced.

[126] Charleton, 'Oratio Inauguralis, in Gymnasii Patavini sede Primaria solemniter habenda', 1678. This is held in the manuscript collection of Charleton's effects in the Bodleian Library, MS Smith 13, no. 44.

[127] The letters are held in the Bodleian Library, MS Smith 13, documents no. 40, 41, 42 and 43 (to Baptista Nannius).

[128] See 'Charltoni Postulata Epistola ad Illustris D. D. Paulus Sarotti', Bodleian Library, MS Smith 13, document no. 39.

[129] Charleton was accused of criticising Dr Blackmore's method to a former patient of Blackmore's. *Annals*, 22 March 1688-9.

In this capacity, his Royal Society experiments and his publications would have won him no substantial advantages.

He seems to have been a popular President at the College. In 1690, when his term as president was due, the *Annals* record that 'Dr. Walter Charleton lost no time in resigning his office of President, though not before he had received the Fellows standing on all sides with a most elegant speech. When this was finished and the Elects had withdrawn a little (as was their habitual custom) he was briefly, unanimously and readily re-elected as President for the following year.'[130]

In 1692 Charleton retired to Nantwich, and later spent some time on the Isle of Jersey, his financial circumstances 'becoming narrow'.[131] Some sources attribute this to the gradual attrition of his old royalist friends and patrons.[132] His contemporary Wood, however, made no mention of poverty as a cause. The College *Annals* suggest that Charleton was in the King's service at this time. On 4 August 1693, a committee of 'the President, Consiliary, and Censors' convened, to consider whether Charleton's position as Elect 'was void by his absence from London'. They were informed that he was acting as the King's physician, which was 'testified by Sr Tho. Millington from my L. Chamberlaine himselfe', as well as 'Mr Swift the College Atturney.' This fact 'was also affirm'd by Dr Briggs from some officers in Guernsey that he was in the Kings service in that Island'.[133] Being in the king's service meant no great financial advance. The main importance of such a role was the honour and status it accorded to the individual. It offered no subsequent economic security. On 29 December 1693 his place as Elect was forfeited by his absence.[134] In 1695 Wood recorded that Charleton was still in Jersey, where he 'hath been some years'.[135]

Charleton reappeared in London at the College on 22 September 1696, from which time he was present intermittently at meetings until 1704. In 1698 he regained his position as Senior Censor at the College, which he held until 1706. He was re-appointed to the position of Elect on 5 December 1701, when the first vacancy appeared, and was Consiliarius from 1702 to 1706 inclusive.[136] Charleton thrice delivered the Harveian Oration.[137]

The ageing doctor seems to have been keen to redeem any moneys owed him, and on 21 May 1704 he asked to be paid in retrospect 'the Salary of the Plate usually allow'd the President', for the year of his presidency which fell outside the period during which all College salaries had been ceased. His request

[130] *Annals*, 22 September 1690.

[131] *Biographia Britannica*, p. 1290. Munk concurs with 'Dr. Charleton's circumstances being straitened.' *Roll*, vol. 1, p. 391. Details of his movements at this time are extremely limited.

[132] *Biographia Britannica*, p. 1290. See also *Dictionary of National Biography*.

[133] *Annals*. Charleton's Elect status was judged valid at that meeting, but only a few months later, on 29 December 1693 the place was turned over to Dr John Downes instead.

[134] Munk, *Roll*, vol. 1, p. 391.

[135] Wood, *Athenæ Oxonienses*, p. 752.

[136] Munk, *Roll*, vol. 1, p. 391.

[137] 5 August 1680, again in 1702 and on 16 August 1705. Munk, *Roll*, vol. 1, p. 391.

illustrates the dire state of his finances. Without sons to support him, Charleton had no security once his practice dwindled. His increasing age may have deterred potential clients.[138] On 6 December 1706 he was appointed Harveian Librarian, and the College, aware of his financial plight, offered a stipend of £20, which Charleton was able to use as a pension.[139] On 24 April 1707, shortly after his appointment, he died 'after a long and tedious disease'.[140]

Charleton died poor despite a long career in medicine, throughout which he tried assiduously through publication and flattery to attract patrons. After his falling out with Brouncker, he may have lost one of his most substantial benefactors. Ultimately it was the College that came to his aid.

Publication was one way in which writers could attract patrons, and it would have to be said that in his case it was not terribly successful. Charleton's career spanned a turbulent period in England's fortunes. As the balance swung between the monarchy and parliament, finding a patron who would remain in a position of power was fraught with difficulties. Charleton was long-lived by the standards of the time, and it seems possible that he out-lasted his major patrons. None of his major supporters were still alive during his later years: John Prideaux (1578—1650); Francis Prujean (1593—1666); Thomas Belayse, Viscount Fauconberg (1623—1670); Margaret Cavendish (1623—1673); Henry Pierrepont (1606—1680); William Brouncker (1620—1684) and Clement Barksdale (1609—1687).[141] Publication would have been a major expense for as prolific an author as Charleton. Writing and publishing did not provide any appreciable income, and authors often received copies of the book rather than money from a publisher.[142] It is possible that the expense of publication was one of the reasons for his poor financial status at the end of his life. His caveats in relation to the job at Padua suggest a man of uncertain health, not confident of his own abilities, and far from comfortable in financial terms.

AN OUTLINE OF THE BOOK

This study focuses on three medical works that are continuous and related, and span the period of Charleton's involvement in the Royal Society. These offer a test case for my central hypothesis: that this physician did not demonstrate adherence to the principles of identity and epistemology upon which natural philosophical authority was based according to prominent historians of

[138] Cook, *Trials*, p. 196.

[139] This library had been provided for the College in 1651-1654 by Harvey, out of his own resources, and comprised 'a magnificent 'Museum', consisting of library above and meeting room below. He furnished it 'sumptuously and gave it many of his books and dissecting instruments.' Frank, *Oxford Physiologists*, p. 25

[140] Munk, *Roll*, vol. 1, p. 391. See also *Annals*, 5 December 1706.

[141] The only one remaining was John Crewe (1633—1722).

[142] Cook, *Trials*, p. 113.

scientific knowledge such as Steven Shapin. Instead his medical writings reveal a quite different set of determinants of identity, peculiar to his status as a professional physician, within which eclecticism was central.

The second chapter reviews the literature on Charleton. Across generations of shifting historiographical emphasis, the basic characterisation of him, as a barometer of contemporary thought, has remained unaltered. My discussion highlights how understandings of his work have been constrained within a 'scientific revolution' narrative. On the other hand, medical historiography has by and large ignored Charleton, and I argue that the emphases of both historiographies have been to the detriment of our understanding.

The recent emphasis on the discursive construction of identity has not yet revealed new insights into Charleton. Research has been limited by a focus upon natural philosophical authority, at the expense of other forms of identity. I aim to direct attention away from familiar questions of knowledge construction, towards a quieter narrative[143] about the relationship of a physician to experimental authority.

Chapter III concerns Charleton's self-presentation—and considers how a physician might present himself in print. Given the integral relationship between identity and epistemology, I analyse the physician's self-presentation and range of methods across moral, natural philosophical, medical and theological writings. Issues relating to generic self-presentation are outlined, including the shaping of the appropriate character for physicians by their education, textual traditions and modes of practice. I consider the contemporary trends toward eclecticism and probabilism that influenced Charleton's medical writings, and examine how his identity as a publishing physician might have been influenced by questions of etiquette in relation to authorship and audience.

Chapter IV, 'The Animal Oeconomy' analyses the first of three specific medical works, in terms of the epistemological and historiographical issues highlighted in my introductory chapters. *Natural History* was Charleton's first medical work in English. It shows a departure from the matter theory and moral philosophy on which he had previously published, and also signals the point in his career at which historians have tended to lose interest in this author.

Natural History was highly successful, in both Latin and English, and was the first of Charleton's works to enjoy such success. My discussion illustrates the links between the determinants of identity (outlined in previous chapters) and the epistemological and methodological devices upon which the physician drew. I examine direct influences upon Charleton's work, and the range of methodologies that he invoked, in relation to the kinds of professional circumstances identified in my first three chapters.

This first of his original medical texts makes an important point of comparison with Charleton's later medical lectures. *Natural History* was written prior to his practical involvement in medical and anatomical experiment. Between it and those

[143] I am grateful to Eric James for this wonderful expression.

later texts his practical experience was transformed. It is crucial to examine the later medical lectures (thought by some to have been re-writings of this original physiological text), which were composed *after* his involvement in the Royal Society and his acceptance into high office within the College of Physicians.

In order to explore further the relationship between experience and presentation, the next chapter examines Charleton's activities between the 1660s and 1680s—that is, between the publication of *Natural History* in 1659 and that of *Enquiries* in 1680 and *Three Anatomic Lectures* in 1683. Chapter V, 'The Republick of Letters' thus offers an account of Charleton's involvement in the Royal Society and College of Physicians. It contextualises the subsequent chapters, in order to gain a more acute sense of how the three works did, and did not, differ in presentation. This chapter considers the public identities of the institutions with which Charleton was involved. It returns to some of the historiographical themes raised in my second and third chapters, and examines them in the light of the activities with which Charleton was involved at the Royal Society.

As the Royal Society records indicate the activities in which he was involved, we can compare them with his own explanations in later works. The contrast allows us an insight into how his activities in the Royal Society affected the method, epistemology and emphasis of the works he published subsequently. The subsequent two chapters examine specific texts, and investigate Charleton's self-presentation in the light of this discussion.

My textual analyses consider the evidence that Charleton underwent an epistemological shift toward what historians have seen as the innovative values of the Royal Society virtuosi, or toward the experimental and mechanistic emphasis of the College of Physicians. The key question here will be the relationship between experience and textual presentation. Many historians argue that the greatest possible intellectual authority could be derived from experimental reference and claims of impartial witnessing.[144] I examine the works that Charleton published before and after he performed extensive anatomical experimentation, to see how he associated himself with the manipulations he had performed. If a physician's status was directly linked to participation in the experimental community, we might expect him to refer to his activities. In this sense we would expect a contrast between his presentation in pre-Royal Society and post-Royal Society works.

Chapter VI, 'Charleton's Anatomy and Physiology after the Royal Society', is thus an examination of epistemology, method and self-presentation in Charleton's *Enquiries into Human Nature*, published in 1680. This work echoed much of the subject matter of *Natural History*, and as such offers a useful comparison. Through it we can explore the extent to which the author integrated the extensive practical experience he had gained between the two works.

Changes in his presentation are also made more complex by the different genre in which he wrote in 1680, since the later work is a set of lectures. I compare

[144] See the claims of Steven Shapin and Peter Dear in particular.

Enquiries with the earlier *Natural History*, and consider the ways in which Charleton demonstrated and verified his arguments in 1680 in relation to a number of key questions. I examine the genre of the lectures, the likely audience, and discuss change and continuity in his self-presentation between the early and later works. Theodore Brown has claimed that *Enquiries* represented Charleton's declaration of his own and the College of Physicians' adherence to iatromechanical philosophy. This chapter considers the extent to which Charleton followed a single consistent philosophy in this work (I argue that this question should be considered separately from the claim that he represented the College). I examine the value that the author ascribed to a range of epistemologies in the light of his status as a professional physician. My discussion illustrates that Brown's argument cannot be supported when the text is examined in detail. A close analysis of the work reveals that Charleton placed many forms of demonstration above experiment, and other approaches before iatromechanism.

The final chapter, 'Ways of Knowing and the Anatomical Body', also pursues the question of how Charleton demonstrated the theories he presented. The chapter focuses on *Three Anatomic Lectures*, delivered by the author in 1683 at the College of Physicians. It was unusual for lectures to be printed. That Charleton published not one but two books of lectures with the blessing of the College of Physicians suggests that these compositions presented an authority and identity that the College wished to foster.

Three Anatomic Lectures was more specialised in subject than either *Natural History* or *Enquiries*. Consequently it was more consistent in its central hypothesis than were the earlier works. This could be read as evidence of Charleton's shift to a strongly mechanist position (as Brown suggests). Brown bases this view in part on the fact that these lectures took both content and style from Alfonso Borelli's *De Motu Animalium*. However this fact, properly understood, actually highlights Charleton's eclecticism. Far from Charleton personally endorsing a thoroughgoing mechanism, these lectures exemplify his ability to present and examine a philosophy, without attempting to generalise it into an explanatory system. Given his own research on the anatomy of the heart, it is significant that the physician chose to copy a continental text, rather than present his own findings. He was reliant upon the Italian for the majority of the subject matter, but occasionally omitted demonstrations, and added new elements. My final chapter explores some of these divergences, as indicators of how Charleton adapted Borelli's text to satisfy the expectations of an English audience.

Much recent interest has focused on the construction of the authority of 'virtuoso' natural philosophers. A model of such individuals has been generated by Steven Shapin (and formerly also Simon Schaffer), and has been taken up by Peter Dear, John Henry and others, as a way of pinpointing the nature of natural philosophical identity in this period. Where referred to by these historians, Charleton is depicted as adhering to a natural philosophical model whose attributes included active experimentalism, collaborative investigation, the rejection of ancient authority and promotion of innovation. I argue that in order to understand Charleton we must step outside such a characterisation, and look more closely at how he himself made sense of his activities. In so doing we open our minds to a broader set of possible interpretations of natural philosophy and medicine than those permitted within Shapin's framework. While the Shapinian characterisation of natural philosophical discourse is not of course the only strand of historiography relevant to Charleton, it is perhaps one of the most prominent in recent years. Despite its undoubted elegance as a theory, and it persuasive neatness, it nevertheless fails to accommodate the complexities and nuances of actual practice.

This study therefore offers a more detailed examination of the physician's medical writings, and reveals the complexity and diversity of his appeals to authority. Although he was actively involved in many empirical investigations, the physician's medical publications did not draw reference to these practices, nor did they give epistemic primacy to experimental findings in general. His medical works emphasised the continuity of traditional authority and scholarship, and presented their author as scholarly, meditative and solitary, rather than as an active empiricist. I argue that the circumstances of this practising physician resulted in the construction of an identity at variance with that associated with natural philosophers as constructed by Shapin *et al.*

The second characterisation of Charleton that I refute in this work is that which treats him as a personification of the rise of mechanist thought. This approach, championed by Theodore Brown, has been accepted by many scholars, who treat Charleton's alleged mechanism as his central contribution to seventeenth-century thought. While he illustrated familiarity with a range of contemporary theories, Charleton did not embrace specific explanations, such as the mechanical explanation of bodily functions, in any systematic manner. As I demonstrate, his eclectic manner was beneficial to his professional status. It allowed him to maintain the links with the traditional bases upon which medical practice was founded, and also to demonstrate an awareness of recent innovations and discoveries, protecting himself against criticisms of 'dogmatism'.

CHAPTER II

REWRITING WALTER CHARLETON: PHYSICK AND NATURAL PHILOSOPHY

In order to reexamine Charleton, we need to scrutinise some assumptions implicit in existing scholarship on the subject. Specific studies of Charleton are rare. Interest in him seems to have flowered in the 1950s, and is evident in sporadic articles through the 1960s and 1970s. Few historians have studied his full career. Lindsay Sharp's[1] article on his early life was, for decades, the only detailed study, though facsimile editions of several of Charleton's works have been published, bringing with them further surveys.[2] A striking similarity among historical accounts of Charleton is the role attributed to the process of 'scientific revolution'. Many accounts cast him as an indicator of the very process of intellectual transition, due to his public rejection in the 1650s of the hermetic philosophy that he had seemed earlier to embrace. Definitions of Charleton's career as simultaneous and commensurate with the process of scientific revolution reflect and reinforce particular views of that revolution, and restrict our ability to understand him as he might have understood himself.

Many now argue that the notion of a 'scientific revolution' arises from questionable beliefs about the nature of science.[3] These assumptions underwrote histories that glorified the trajectory of science throughout the centuries since the seventeenth.[4] The status of modern science was supported by the idea of a 'scientific revolution' as the point at which dogma, superstition and religion were discarded in favour of rationality, experiment and free inquiry.[5] But as this view

[1] Sharp, 'Early life'. This has now been remedied in two doctoral theses. See Sabina Fleitmann's *Walter Charleton, 1620-1707, 'Virtuoso', Leben und Werke*, Frankfurt, Lang, 1985. Also, Anna Maria Oller í Adam, 'Walter Charleton; Filosofia Natural, Teologia Natural i Etica', PhD dissertation, University of Barcelona, 1995. See also Nicholas Dew, 'The Politics of the Body'.

[2] During the twentieth century some of his works have been reproduced in facsimile editions: *Physiologia Epicuro-Gassendo Charltoniana*, London, Introduced by Robert Kargon, Johnson Reprint Co., 1966; *Epicurus's Morals*, with an 'Introduction' by Frederic Manning, London, Peter Davies, 1926; *Immortality of the Human Soul*, Introduced by J. M. Armistead, New York, AMS Press, 1985; and *The Ephesian Matron*, introduced by Achsah Guibbory, Los Angeles, William Andrews Clark Memorial Library, 1975. Charleton's *Chorea gigantum* has been reproduced in a collection with Inigo Jones' *Stone-heng*, and John Webb's *A vindication*, ed. Stuart Piggot, Farnborough, Gregg, 1971.

[3] For example, A. Cunningham & P. Williams, 'De-centring the "big picture": *The Origins of Modern Science* and the modern origins of science', *British Journal of the History of Science*, vol. 26, 1993, p. 409.

[4] See for example H. Butterfield, *The Origins of Modern Science, 1300-1800*, London, G. Bell, 1957.

[5] Cunningham & Williams, 'De-centring', p. 417.

of science, and its accompanying view of the history of science, have become increasingly difficult to sustain, essential elements of the 'big picture' concept of scientific revolution have been eroded.[6]

I have no wish to argue over whether or not a 'scientific revolution' occurred, or whether that term describes the intellectual environment of the late seventeenth century. Instead the present chapter explores the ways in which the historiography of scientific revolution and the (more recent) 'virtuoso' natural philosopher profile have shaped our understanding of Walter Charleton. The following analysis, of the historical arguments within which he has been framed, aims to illustrate the extent to which the 'scientific revolution' discourse has limited our understanding of him. He has been depicted consistently, across a range of historiographical trends, as an experimental natural philosopher and a symbol of change. This characterisation has perpetuated a disturbing neglect of his role as a physician. A history of Charleton outside the 'scientific revolution' framework is essential because of the striking degree to which he has been defined by association with it. His medical writings illustrate previously unrecognised possibilities concerning the creation of authority, which were perhaps distinctive to the circumstances of professional physicians. What follows is a brief overview of the historiography on Charleton, none of which in my view satisfactorily answers the central questions surrounding his work and identity as a physician.

HISTORIOGRAPHY

Frederic Manning's introductory essay to the 1926 reprint of *Epicurus's Morals* was the first of last century to review Charleton's philosophical contribution.[7] The latter's historical value, according to Manning, is in the insight he allows into contemporary reception of the theories of science's Great Men. As this historian has it, he presented the ideas of his venerated contemporaries 'in direct relation to the age surrounding them, and bare of the additional significance and extension which they have acquired in the course of three centuries.'[8] In such historiography, the study of such minor figures is valid only insofar as it indicates the substance of great men in their own time.[9] Manning believes that Charleton's observations on the writings of Sir Kenelm Digby reveal to him 'the actual process of transition from magic to science'.[10] Charleton historiography has not, essentially, deviated from this path.

[6] See for example S. Shapin, *The Scientific Revolution*, Chicago, Chicago University Press, 1997.

[7] Manning, 'Introduction' to *Epicurus's Morals*. The work was followed in 1940 by Rolleston', 'Walter Charleton'.

[8] Manning, 'Introduction', p. x.

[9] This approach leads Manning to take Charleton's statements at face value, see 'Introduction', p. xi.

[10] Manning, 'Introduction', p. xii.

Robert Kargon charts Charleton's contribution to the development of atomist philosophy in England.[11] Like his older contemporary Douglas McKie, Kargon posits the physician's significance as a proponent of Epicurean-based atomism.[12] In his introduction to the facsimile edition of *Physiologia*, Kargon asserts that the text was significant in the introduction of Continental (Gassendian) atomism to English audiences, as 'a valuable *entrée* to the intricacies of the mechanical philosophy'.[13] Consequently Kargon presents the text as 'an important key to the understanding of the history of atomism' in the seventeenth-century context. Charleton's work, he argues, provides 'a handle for grasping one important aspect of the many-faceted Scientific Revolution.'[14] Thus the physician's modern importance is for Kargon a consequence of his role in the changing face of seventeenth century science. Charleton is here represented as a conduit for Gassendi. Like Manning, Kargon views his subject as worthy of examination due to his illustration of the reception of the works of more significant contemporaries. He is, for Kargon, the embodiment of Restoration science, who 'exemplified the spirit of a new intellectual age. In his works are mirrored all the ferment, controversy, enthusiasm, and excesses of the partisans of the new learning.'[15] While there's nothing wrong with seeing Charleton as a man of his time, I argue that the consistent definition of his era through the lens of 'scientific revolution' has been to the detriment of our understanding of his writings and career. The author's status in these texts is that of a lens onto a much larger phenomenon, 'scientific revolution'.

Richard Westfall also sees Charleton as an indicator of the impact of more enlightened contemporaries. The young doctor is grouped with Digby as a 'would-be philosopher' influenced by Cartesian and Gassendian ideas. Since both Charleton and Digby failed to reach 'great philosophic stature', Westfall sees their systems as 'of little interest in themselves, beyond their illustration of the influence that the mechanical hypothesis exercised over the minds of the virtuosi.' Westfall treats Charleton's references to Descartes' philosophy as definitive of his aspiration to embrace it fully. He attempted, but failed, 'to construct rational systems of nature which explained phenomena by mechanical causes.'[16] Charleton is widely recognised as having made little innovative contribution to natural philosophy and, until quite recently, this was assumed to have

[11] See R. H. Kargon, 'Walter Charleton, Robert Boyle, and the Acceptance of Epicurean Atomism in England', *Isis* vol. 55, no. 2, 1964, pp. 184-92; Kargon, 'Acceptance of Atomism'; and R. H. Kargon, 'Introduction', to Johnson Reprint Co. facsimile of *Physiologia*, London, 1966.

[12] However, where McKie's ambition is simply to detail atomistic thought prior to 'modern chemical atomic theory', Kargon extends the contextual significance of Charleton's work. Douglas McKie, 'English writers on atomism before Dalton', *Endeavour*, vol. 25, 1966, pp. 13-15.

[13] Kargon, 'Introduction', p. xiii. Not only an effective defence of Epicurean atomism against charges of atheism, it promoted atomism as 'a powerful doctrine in the cause of religion and piety.' Kargon, 'Introduction', p. xiv.

[14] Kargon, 'Introduction', p. xiii.

[15] Kargon, 'Introduction', p. xvii.

[16] Westfall, *Science and religion*, p. 80.

constituted a failing. According to Westfall, Charleton's views on providence, 'were framed with the mechanical hypothesis in mind.' His opinion therefore represented one of seventeenth-century mechanical philosophy's solutions to the obstacle of natural religion.[17] Westfall's approach does not recognise Charleton's eclecticism as integral to his epistemology, but rather fixes on the presence of mechanistic thought in the author's work and generalises it to provide a profile of his world view.

A similarly triumphalist attitude to mechanism's trajectory is present in the writings of Margaret Osler, who deals specifically with mechanist views on 'God's relationship to the world.'[18] Within late seventeenth-century natural philosophy, according to Osler, the appropriateness of mechanism was not under question. The issue was rather which mechanical model was superior: Cartesian or Gassendian. Charleton's role in this narrative is as a translator and interpreter of the writings of Gassendi. Osler sees the physician as a convinced Gassendian, and assumes that *Darknes* and *Physiologia* 'enunciate Charleton's world view.' His significance is defined by his presentation of one of the identifiable 'paradigms' of his era. Osler's comparison between Descartes and Gassendi focuses on Descartes and Charleton, because of the latter's explicit discussion of Providence within a mechanistic framework, which seems to her a perfect synopsis of Gassendi's views.[19] Her assumption of Charleton's unitary 'world view' is difficult to sustain in examination of the range of his writings. Osler sees the two theological positions at the heart of essential differences over early modern epistemology. She represents them as opposing frameworks, with irreconcilable principles and 'scientific methods'.[20] Her argument depicts the scientific revolution as a theologically founded epistemological shift from (Cartesian/Catholic/intellectualist) rationalism to (Gassendian/Protestant/voluntarist) empiricism. Here again Charleton has been linked to a particular philosophy, and his status as an object of study justified according to the significance of that philosophy.

One common assumption of the writers so far considered is the notion of coherent and mutually exclusive 'world views', reinforcing the idea of discrete categories: ancient and modern, hermetic and mechanist, rationalist and empiricist beliefs. Mechanist philosophy is portrayed as self-determining: once adopted, it commanded certain corollaries.[21] Eclecticism, which I believe is central to understanding Charleton, is not recognised in any of these accounts as a viable approach, let alone as a desirable one. It is not treated as a true philosophical position, but rather as a contingent muddle arising from prevailing scepticism.

[17] Westfall, *Science and religion*, p. 80.
[18] M. J. Osler, 'Descartes and Charleton on Nature and God', *Journal of the History of Ideas*, vol. 40, 1979, p. 447.
[19] Osler, 'Descartes and Charleton', pp. 446-7.
[20] Osler, 'Descartes and Charleton', p. 456.
[21] This approach to the history of 'mechanist' thought is opposed by Steven Shapin *et al.*

The binaries of Manning, Westfall and Osler have been revised within the history of ideas by studies that question the notion of scientific revolution as the displacement of one exclusive 'world view' by another. In these revisions Charleton is cast in a different light. Nina Gelbart focuses on Charleton's alleged transition from hermeticism and Helmontianism to atomism and mechanistic philosophy,[22] but emphasises the permeability of the boundaries between these frameworks. To her, Charleton's works demonstrate 'the great complexity of seventeenth-century scientific thought.' Like Mulligan, Gelbart sees Charleton as indicating the difficulties associated with the categorical delineation of world views. Both counter the depiction of Charleton as undergoing an irreversible and symbolic transition from magic to science. Gelbart approaches the subject from a different model of the scientific revolution: 'in tracing the meandering route by which Charleton arrived at his atomism, we are reminded that the development of modern science was a slow and circuitous process.'[23] Once again, the author's status is as a gauge of contemporary thought.

Armistead's introduction to the facsimile edition of Charleton's *Immortality* argues that during his residence in London the author encountered 'the full range of avant-garde thought in his time'.[24] Consequently *Immortality* provides 'revealing clues to early scientific thought'.[25] Like Kargon, he sees Charleton as, for approximately fifty years, a 'defender and archivist of the latest scientific developments'[26], and thus as virtual embodiment of this period of intellectual change. However, he does recognise the fundamentally eclectic nature of Charleton's composition. Though these histories of Charleton, arising from different assumptions about the nature of the scientific revolution, have countered the tendency to enforce anachronistic binaries, they are not without problems. In each case the physician is invoked as a barometer of opinion in natural philosophy. His practices, and their relationship to his writings, are not discussed.

These accounts fall into the category that has in the past been termed 'internalist' historiography, within which the momentum of ideas is treated as a sufficient explanation for intellectual change. Charleton has also been analysed by historians concerned with so-called 'external' (institutional, religious and socio-economic) influences on philosophy. The dialogues in his *Immortality of the Human Soul* have been a particular focus, because of their observations on contemporary institutional activities. Charles Webster invokes this text to challenge assumptions about institutional activities. *Immortality* contains a lengthy dialogue regarding the state of scientific endeavour in the Royal College of Physicians during the 1650s, through which Webster generates a revised picture of the Interregnum activities of the College.[27] In this approach he is followed by

22 Gelbart, 'Intellectual Development', pp. 149-69.
23 Gelbart, 'Intellectual Development', p. 168. See Mulligan, 'Right reason'.
24 Armistead, 'Introduction', p. v.
25 Armistead, 'Introduction', p. viii.
26 Armistead, 'Introduction', xiv.
27 C. Webster, 'The College of Physicians: "Solomon's House" in Commonwealth England', *Bulletin of the History of Medicine*, vol. 41, 1967, pp. 393-412.

Lindsay Sharp and Theodore Brown (though Brown disagrees on the nature of the College of Physicians). Robert J. Frank also probes the nature of College activities through the *Immortality* dialogues.[28] Charles Webster's *Great Instauration* contains the influential characterisation of Charleton as 'the intellectual barometer of the age', and his writings as 'a valuable index to contemporary fashions'.[29] This characterisation has been invoked by many subsequent writers.[30] Webster claims that the physician's beliefs, throughout his apparent transition from hermeticism to atomism, reflected the concerns of his contemporaries. The author's adjustments thus become signals not of the displacement of magic by science, but of a decline in the popularity of magical beliefs. Unlike earlier historians, he makes an explicit link between Charleton's orientations and theological and political expediency. The physician's adherence to, and then rejection of hermeticism, according to Webster, simply followed prevailing opinion. Helmont's theories declined in popularity after the 1640s. Webster links Charleton's rejection of Helmont with the former's recognition of an unwanted association between Helmontian philosophy and 'separatist and anti-monopolistic factions'.[31] Again Charleton is seen to epitomise late seventeenth-century change, though this time political and theological motivation are integrated within the account of his works.

Pyarali Rattansi replicates this picture of Charleton's shift, from natural magic to the new mechanical philosophy, in response to external impetus. He sees the individual's transition as 'a remarkable parallel to the revulsion of moderate opinion from the natural magic tradition' in England, as the latter was increasingly linked with heretic religious and social views.[32] Rattansi, like others before and after him, does not pause to interrogate the existence, nor the comprehensive nature, of a complete transition in Charleton's beliefs.

Lindsay Sharp creates a fuller and more accurate account of the physician's early life, in order 'to construct a detailed and reliable interpretation of his intellectual development.'[33] He views Charleton's activities, travel and contacts as vital to an understanding of his intellectual development and of changes in his natural philosophy:[34] 'Only when this evolution is clearly identified can Charleton's ideas be used as evidence for broad phylogenetic theories about the growth of Helmontianism or atomism in England.'[35] Sharp does not explain why a 'phylogenetic' history is the most appropriate.

[28] Frank, 'Virtuoso', pp. 90-92.
[29] Webster, *Great Instauration*, p. 278.
[30] These include Lindsay Sharp 'Early Life' and Jonathan Sawday, 'The Mint at Segovia: Digby, Hobbes, Charleton and the body as a machine in the seventeenth century', *Prose Studies*, vol. 6, no. 1, 1983, pp. 21-36.
[31] Webster, *Great Instauration*, pp. 278-9.
[32] Rattansi, 'Puritan Revolution', p. 31. For this he is attacked by Mulligan, 'Right reason'.
[33] Sharp, 'Early Life', p. 312.
[34] Sharp, 'Early Life', p. 339.
[35] Sharp, 'Early Life', p. 339.

Theodore Brown argues for the necessity of an institutional examination of seventeenth-century natural philosophical debates, as a means of understanding the relationship between mechanical and non-mechanical modes of thought. Brown argues that the relationship between the beliefs expressed and the groups within which individuals wrote is the crucial focus for investigation. The historian's responsibility is to chart the interaction of institutions through time and under a variety of vicissitudes.[36] Brown is concerned with the relationship between institutional adoption of particular beliefs and the authority claims they convey. Here Charleton's mechanism is important as a signal of the specific knowledge claims of the Royal College of Physicians.[37] Institutional/social significance is integral to understanding the adoption of mechanistic philosophy within England's medical institutions. Intellectual transition, according to Brown, is inseparable from the circumstances and authority claims of the RCP in relation to the Royal Society.

This survey of the literature on Walter Charleton shows how it has tended to invoke the concept of 'scientific revolution'. In all of the above accounts Charleton is treated as a personification of the transformations of his era. The focus of these portrayals has been his apparent transition, in the 1650s, from hermetic to atomist and mechanist. My study suggests and contributes to ways in which new research might be carried out, and the purpose of this overview has been to ascertain the assumptions within which it will work, and those it will discard.

A recent focus on discursively bound knowledge communities has offered an alternative to the preceding historiographical emphasis on disembodied ideas and notions of intellectual 'progress'. Current historical sociology of knowledge and epistemological history examine the generation and legitimation of knowledge, focusing on knowledge-making practices, and particularly those of the fellows of the Royal Society.[38] Knowledge is treated as inseparable from its communicators and their means of communication. Such histories of scientific epistemology are concerned not with the 'birth' of modern scientific method, but 'the practices by which types of scientific knowledge were made and their credibility secured in early modern England.'[39]

John Henry argues that English natural philosophers rejected 'slavish adherence' to any authority, be it contemporary or ancient in origin.[40] He identifies several emphases in seventeenth-century natural philosophy, which had their origin in Royal Society Baconianism. These included an emphasis upon 'collaborative effort', to be sustained over long periods, the urgency of gathering and establishing 'matters of fact' and the refusal to indulge in speculative

[36] Brown, *Animal Oeconomy*, p. v.

[37] Brown's views are dealt with more fully in my final chapters.

[38] Followers of this approach include Steven Shapin, Peter Dear, John Henry and Barbara Shapiro.

[39] S. Shapin, *A Social History of Truth: Civility and Science in Seventeenth Century England*, Chicago, University of Chicago Press, 1994, p. xvi.

[40] Henry, 'Scientific Revolution in England', p. 196.

theorizing'.[41] While cautious about attributing this profile to natural philosophers across the board, Henry rightly suggests that the profile is useful in understanding the distinctive English context of natural philosophy. While recognising the utility of this profile, I suggest that in Charleton's case we witness the presence of additional determinants upon the physician's identity.

Shapin sees natural philosophers as a distinctive and relatively new phenomenon. He claims that they had to establish means by which to prove the value of their endeavours, and to demonstrate their unique ability to resolve conflicts and tackle otherwise volatile subjects. Shapiro charts the emergence of a new category of knowledge—'moral certainty'—that could be ascribed to matters of fact over which 'no dispute would be possible'.[42] Virtuoso self-construction supported the authority of experimental natural philosopher as both relevant and authoritative. The assertion of these basic characteristics was imperative, according to these historians, in the legitimation of natural philosophical activities within the theological and social context of late seventeenth-century England.

Shapiro, Shapin, Schaffer and Dear see the question of natural philosophical epistemology in the context of the larger question of how post-Restoration society coped with the recent dissolution of civil harmony, both on a national political and on a theological scale. They, like Eric Lewis,[43] depict the search for knowledge as framed by the urgent sense of the necessity to develop a mode of knowledge-production free from the pitfalls of dogmatism and released from claims of certain and infallible knowledge. Peter Dear claims that devaluation of the primary epistemic authority of scholastic philosophy was 'the most significant symptom of the so-called Scientific Revolution', necessitating new (experimental) criteria for the definition of authoritative knowledge.[44] Shifts in the definition of authority within theology were linked to natural philosophy. Shapin argues that 'the rejection of authority and testimony in favour of individual sense-experience' underlies our own recognition of seventeenth-century practitioners 'as "moderns," as "like us," and, indeed, as producers of the thing we warrant as "science" ', though he notes that the prominence of experimental philosophy was greater in word than in deed.[45]

Recent literature has seen an increasing emphasis on the construction of a new identity for what is seen to be a precarious late seventeenth-century creation, the 'virtuoso natural philosopher'. These scholars argue that the generation of a new kind of authority for this community required the legitimation of

[41] Henry, 'Scientific Revolution in England', p. 189.

[42] See B. Shapiro, *Probability & Certainty in seventeenth century England: A study of the relations between Natural Science, Religion, History, Law, and Literature*, Princeton, Princeton University Press, 1983; see also Simon Schaffer (review of Shapiro), 'Making certain', *Social Studies of Science*, vol. 14, 1984, p. 141.

[43] Lewis, 'Early modern eclecticism'.

[44] P. Dear, 'From truth to disinterestedness in the seventeenth century', *Social Studies of Science*, vol. 22, 1992, p. 628.

[45] Shapin, *Social History of Truth*, p. 201.

its specific activities. Some of the best known proponents of this view, Dear, Shapin and Schaffer, write about the practices of the Royal Society 'virtuosi'. The influence of Bacon upon the stated ideal of the Royal Society provides a backdrop to this examination. Walter Charleton is depicted as a member of this community, because of his experimental activities. Shapin and Schaffer view the careers of men like Charleton within their model of what it meant to be a natural philosopher in late seventeenth-century England. A distinctive aspect of the era, according to these historians, was the shift from private to public context for the creation and legitimation of knowledge (hence the communal nature of natural philosophical endeavour). In demonstrating the legitimacy of their aims, experimenters located themselves as a community.[46] Shapin argues for the definitively public nature of the new philosophy: knowledge-making, after Bacon, was a social/public rather than solitary/private enterprise.[47] Dear claims that 'The dominant ideal of natural knowledge in the seventeenth century involved the crucial assumption that true knowledge was shared or shareable, that knowledge was common property.'[48] Although Dear recognises that alternatives to this idea existed, he sees such deviations only in terms of the major conflicts which erupted during this era. Thus 'natural knowledge needed, by definition, to be seen as the common property of a nonexclusive group rather than being a private, personal conviction'.[49] The seventeenth century was therefore, in this formulation, the first era of collective scientific endeavour. The experience of the Civil War and Interregnum, these scholars claim, had left the people of England with a distrust of individualistic hermetic and enthusiastic beliefs. The establishment of the sense of a unified scientific community was fundamental to the success of the Royal Society and the virtuosi, as part of the technology by which it promoted its ability to generate useful and impartial knowledge. Shapin and Schaffer state that the virtuosi needed to establish their ability to generate indubitable shared truths, which were distinct from conjectures.[50] Therefore its alternative, solitary research, was widely reviled. Individual testimony, especially in relation to questions such as the behaviour of spirits, 'reeked too much of enthusiasm and dogmatism.'[51] Shapin acknowledges that a

[46] Schaffer, 'Making Certain', p. 141.

[47] S. Shapin, ' "The mind is its own place": Science and solitude in seventeenth-century England', *Science in Context*, vol. 4, no. 1, 1990, p. 201.

[48] P. Dear, 'Miracles, experiments, and the ordinary course of nature', *Isis*, vol. 81, 1990, p. 665.

[49] Dear, 'Miracles, experiments', p. 665.

[50] 'Unless the experimental community could exhibit a broadly based harmony and consensus within its own ranks, it was unreasonable to expect it to secure the legitimacy within Restoration culture that its leaders desired. Moreover that very consensus was vital to the establishment of matters of fact as the foundational category of the new practice.' S. Shapin and S. Schaffer, *Leviathan and the Air-Pump: Hobbes, Boyle and the Experimental Life*, Princeton, Princeton University Press, 1985, p. 73.

[51] S. Schaffer, 'Godly men and mechanical philosophers: Souls and spirits in Restoration natural philosophy', *Science in Context*, vol. 1, no. 1, 1987, p. 59.

tradition of scholarly solitude existed upon which a seventeenth-century natural philosophical identity might draw.[52] Yet he argues that in the case of natural philosophical, as with religious, knowledge, individual experience needed to be further ratified through connection with 'public contexts of justification.'[53] Shapin argues that the new emphasis upon experimentalism demanded that empirical knowledge must be confirmed by eye-witnesses, and claims that Boyle and Sprat believed that the most authoritative witnessing was collective witnessing.[54] As this was difficult to achieve in practise, there developed a 'literary technology' by which close experimental description in natural philosophical texts achieved the aim of allowing the reader to be a 'virtual witness' to the experimental event.[55] Schaffer claims that 'the social technology of collective witnessing allowed the production of secure matters of fact.'[56]

Shapin and Dear posit that part of the identity of natural philosophers relied upon the assertion of the modesty of their aims, which revolved around contemporary notions of civility. 'Matters of fact' were social, as well as intellectual constructs, and crucial to their impact was the construction of the author as a provider of reliable testimony.[57] As Shapin argues, 'The presentation of self as modest, sober, restrained, tolerant, and unconcerned for fame was considered effectively to enhance the credibility of what one claimed.'[58] The importance of a modest self-presentation was echoed in ideas about the relationship between temperament and authority. A consequence of this agenda was the rise of probabilism—Shapin claims that the ideal method for gaining credibility lay in the individual 'confessing his own (excusable) faults, by identifying (limited) troubles in the matters he claimed, by giving readers and auditors (inadequate) grounds freely to withhold their assent'.[59] He notes that the most persuasive voice in the community of English natural philosophers was that of the individual who established his own disinterestedness. Portrayal of the self as 'modest, sober, restrained, tolerant, and unconcerned for fame' secured the credibility of knowledge claims, in an atmosphere suspicious of vested interests and of dogmatism.[60] These were the trademarks of Royal Society virtuoso identity and etiquette. The 'civil order of a knowledge-producing community', Shapin

[52] Shapin, 'Science and solitude', p. 206. See also Schaffer, 'Godly men'.

[53] Shapin, 'Science and solitude', p. 207. He claims that private and publicly generated knowledges formed different parts of the knowledge-making process. The individual aspect of creativity was part of a 'context of discovery' while the public was part of a 'context of justification'. Ibid.

[54] S. Shapin, 'Pump and Circumstance: Robert Boyle's Literary Technology', *Social Studies of Science*, vol. 14, 1984, p. 487.

[55] Shapin, 'Pump and Circumstance' and Peter Dear, 'Totius in Verba: Rhetoric and authority in the early Royal Society', *Isis*, vol. 76, 1985, *passim*.

[56] Schaffer, 'Godly men', p. 59.

[57] Shapin, 'Pump and Circumstance', pp. 494-7.

[58] Shapin, *Social History of Truth*, p. 222.

[59] Shapin, *Social History of Truth*, p. 223.

[60] Shapin, *Social History of Truth*, p. 222.

argues, was protected by 'lowering the standards of certainty, accuracy and exactness legitimately to be expected of claims about the world.'[61]

This shift toward probabilism has generally been linked, by many scholars, to the social and political turmoil within which England was engulfed at the time. The latter created a context of aversion to received authority and dogma. As a consequence, historians such as Shapin and Schaffer have tended to describe 'scepticism' as one of the outcomes of the rejection of doctrines, rather than as a consistent position in itself. Contrary to Shapin's claims for innovation in natural philosophical discourse, I argue for the continued relevance of the tradition of philosophical eclecticism that can be identified in a significant strand of European thought—throughout Renaissance Neoplatonism and back to Cicero amongst others.[62]

As an active experimentalist in the Royal Society, Charleton has been placed against the backdrop of prototypical natural philosophers such as John Locke, Robert Boyle and Robert Hooke. His epistemological views are seen by many historians to reveal a classic 'virtuoso' profile. Charleton is cited by Shapiro and Schaffer[63] as a natural philosopher engaged in the kind of knowledge-production that they identify as generic for that community. I disagree with this characterisation on a significant number of points, on which I elaborate in the chapters to follow. While arguments for characteristics of natural philosophical discourse, such as 'virtual witnessing', are fascinating and indeed compelling in some cases, they imply that the performance of experiment necessarily indicated the desire for an empiricist identity.

A critical interrogation of the Shapin depiction of the 'virtuoso natural philosopher' is one of my central concerns in this study. I examine how Charleton navigated the private/public contexts of justification, and how he wrote about the roles of experiment and meditation. I also consider his espoused modesty and skepticism, and argue that these were not elements of a passing phase in self-presentation, but rather show the continuity of humanist traditions, which allowed Charleton to reconcile competing theories. I examine these claims and other aspects of the Shapin thesis in the chapters to follow, and argue that in Charleton's case these assumptions are not supportable. His depiction within a Shapinian model of a 'virtuoso' discourse of knowledge construction is not particularly accurate nor is it helpful in coming to a deeper understanding of his work.

I examine the degree to which his *medical* writings illustrate this profile, and consider how Charleton presented himself, in these writings, on the two issues of eclecticism and empiricism. Charleton's writings show the presence of a coherent philosophy which is more central than has been recognised—namely eclecticism. I suggest that the eclectic approach offered to someone in

[61] Shapin, *Social History of Truth*, p. 308.

[62] On the significance of eclecticism see Stephen Gaukroger, *Francis Bacon and the Transformation of Early-Modern Philosophy*, New York, Cambridge University Press, 2001, esp. pp. 28-36.

[63] Schaffer, 'Making certain', p. 148.

Charleton's position a more appealing source of identity and authority than did the profile mooted for the natural philosophical 'virtuosi'. Again, my claims recognise continuity, over the central role given to innovation within the Shapinian thesis. I argue that Charleton's eclecticism allowed him to maintain appropriate professional status, while also exploring the philosophical ideas that interested him. I do not claim that all physicians were eclectics, but rather that this philosophical tradition potentially offered a great deal to physicians. The importance of eclecticism as a philosophy has been a largely neglected in relation to the English context.[64] In the final stages of writing up this study an article was released by Eric Lewis, upon exactly this aspect of Charleton's thought. Lewis argues, as I do, that the physician's epistemology was eclectic. However, his central claim is that the latter's eclecticism 'should be viewed as an exhaustive effort to prescribe a remedy to the perceived threat of rampant sectarianism.'[65] Although he recognises that the author was a physician by profession, he nevertheless perceives the latter's writing entirely through the lens of natural philosophy. In this way, I believe, he misses some of the other reasons why eclecticism might appeal to Charleton as a philosophical method.

At the centre of these issues is a problem with the assumption that medicine followed the same trajectory as natural philosophy in terms of it knowledge-making practices and authority gambits. This book argues that many determinants were at work on the epistemology of characters such as Walter Charleton, the role of which has been largely ignored. While the significance of the larger political and theological context should not be neglected, this physician seemed oblivious to many of the factors supposed to have influenced him. Despite Lewis' claim that Charleton presented 'a unique attempt to solve the problems of social, religious and intellectual discord',[66] I argue that the differences between Charleton's and the 'virtuosi' epistemological modes reveal divergent determinants.

Robert Frank echoes some of the virtuoso profile arguments with direct reference to physicians. He claims that the mid-seventeenth century witnessed significant changes in the way English physicians saw themselves and were seen by the community in which they lived and practised.[67] They came, he implies, to

[64] Indeed the majority of literature on the subject has been generated by German scholars. The key example is Michael's Albrecht's *Eklektik. Eine Begriffsgeschichte mit Hinweisen auf die Philosophie-und Wissenschaftsgeschichte*, Stuttgart, 1994. Others who have written on the topic include T. J. Hochstrasse, *Natural Law Theories in the Early Enlightenment*, Cambridge, Cambridge University Press, 2000. U. J. Schneider, 'Eclecticism rediscovered', *Journal of the History of Ideas*, vol. 59, no. 1, 1998; D. R. Kelley, 'Eclecticism and the History of Ideas', *Journal of the History of Ideas*, vol. 62, no. 4, 2001, pp. 577-592; Constance Blackwell, 'Sturm, Morhof and Brucker vs. Aristotle: Three Eclectic Philosophers View the Aristotelian Method', in *Method and order in Renaissance philosophy of nature: The Aristotle commentary tradition* D. A. Di Liscia, E. Kessler and C. Methuen (eds), Aldershot, Ashgate, 1998.

[65] Lewis, 'Early modern eclecticism', p. 652.

[66] Lewis, 'Early modern eclecticism', p. 653.

[67] R. G. Frank, *English scientific virtuosi in the sixteenth and seventeenth centuries*, Los Angeles, William Andrews Clark Memorial Library, 1979, p. 66.

promote empiricist ideals in the production of knowledge. By the 1670s, anatomical and chemical researches were important in the establishment of a clinical reputation.[68] A practice might thus be founded successfully upon an experimenter's achievements, and empirical prowess was part of a physician's professional identity. Frank's claim accords with the popular view that medicine followed the same authoritative construction as did natural philosophy. However, Frank seems to ignore the fact that stress upon experiment could be seen to threaten the scholarly status that necessarily underlay learned physic. It could by implication demote them to the status of mere 'mechanicks' or surgeons. Though Charleton did not comment on this explicitly, its presence as a determining factor can help us to understand why he made no mention of his own extensive experimental practices.

An understanding of Charleton's texts and their contemporary meanings requires consideration of the author's self-presentation. This includes the significance he attributed to textual composition, the audience(s) for which he wrote, the language in which he published, the topics upon which he focused, and the identity he constructed through writing. Physicians have been assumed to be a sub-category of natural philosophers, rather than a professional group with distinctive identity and practices. Though participant in natural philosophical knowledge-making communities, physicians were subject to additional determinants of identity quite outside those communities. Their specific concerns (connected to the status hierarchy within which they competed) have been neglected in historians' mapping of communities. While Charleton's writings doubtless aimed to enhance his authority as a professional physician, I argue that the identity to which he aspired was more strongly defined by scholarly than experimental aims. I explore these aspects of Charleton's self-presentation, and consider the ways in which he diverged from the model outlined to understand the self-construction of natural philosophers. In summary, the historiographical approaches represented here leave some major lacunae in their explanations of Walter Charleton's epistemology and self-presentation.

MEDICAL TEXTS AND MODERN INTERPRETATIONS

The preceding survey illustrates that, despite the extraordinary diversity of his interests and writings, Charleton's thought is widely assumed to be defined by a couple of works. The texts generally treated as representative of his *oeuvre* tend to be those which outline his moral philosophy in relation to Epicurean atomism. *Physiologia* is cited repeatedly, as are *Darknes* and *Immortality*. Osler presents Charleton's works of the 1650s as definitive of his 'world view', and Westfall sees the providential theology expressed in his physico-theological treatises as sufficient to rank the author as relatively unimportant. Few historians

[68] Frank, *English scientific virtuosi*, p. 100.

consider any of Charleton's publications after 1660—that is, after the decade in which many of them see him turning from hermetic to mechanist. The absence of studies into his later career is indicative of the role he has been given in the 'scientific revolution' narrative. However, it was in the latter part of his career that Charleton progressed to prominence in the Royal Society, Presidency in the College of Physicians and was offered the anatomy chair at the University of Padua.

Few accounts refer to his medical writings, most of which were composed after 1660. As most scholarly research on Charleton was carried out during the period in which medicine was viewed as part of a uniform category of 'science', he has been cast as a figure of 'scientific' rather than a 'medical' interest. The 'scientific revolution' narrative is designed around changes in astronomy, physical theory and mathematics. As such, medicine has a problematic relationship with it. Cunningham and Perry note that developments in the 'life sciences' can not be easily fitted into the notion of seventeenth-century 'mathematization and mechanism'.[69] Nevertheless, assumptions founded on the scientific revolution narrative have been assumed to apply to Charleton's medical writings.[70] Only relatively recently has medicine been recognised as necessitating specific historical treatments. With the development of the history of medicine as a discipline, Charleton has been included in reviews of medical thought in the seventeenth century. Yet scholarly research on his medical ideas is lacking. Even those interested in his contribution to medicine have tended to accept that he was a supporter of iatromechanism and experimentalism. Both Robert Frank and Theodore Brown take this approach. Frank sees Charleton in the context of the impact upon seventeenth-century medical thought of Harveian physiology. Brown views the physician within the context of the College of Physicians' adoption of iatromechanism as a response to the Royal Society.

Other areas of enquiry have developed around the history of the body. Within this literature Charleton's mechanical references are often taken at face value, as an indication of the broader shift from a holistic, pre-scientific view toward a strict dualist understanding of the body.[71] Again, the scientific revolution narrative emerges, though in these cases often with a negative slant on 'modern' scientific method. Focus upon the practices of early modern science has characterised Charleton once again as an embodiment of intellectual ferment.

[69] Cunningham & Williams, 'De-centring', p. 413.

[70] This has also been noted by Andrew Wear, 'Medical practice in late seventeenth- and early eighteenth-century England: continuity and union', in A. Wear and R. French (eds), *The Medical Revolution of the Seventeenth Century*, Cambridge, Cambridge University Press, 1989, p. 295. There were of course proponents of the benefits to physic of experimental knowledge. However, this was not the common view.

[71] See for example D. M. Levin and G. F. Solomon, 'The discursive formation of the body in the History of medicine', *Journal of Medicine and Philosophy*, vol. 15, no. 5, 1990, pp. 515-37; D. Leder, 'Medicine and paradigms of embodiment', *Journal of Medicine and Philosophy*, vol. 9, 1984; B. M. Stafford, *Body Criticism: Imaging the Unseen in Enlightenment Art and Medicine*, Massachusetts, MIT Press, 1991.

Despite various changes in historiographic emphasis, Charleton has retained his perch in the historical imagination, as a symbol of transition. Cunningham and Williams argue that many attempts to create new histories of the scientific revolution are effectively re-wordings of the old version, in which a familiar argument is re-made with new kinds of evidence.[72] Instead of using their new historiographical devices to question the nature of revolution and challenge the assumptions contained within existing scholarship, many historians have simply reproduced the story of the scientific revolution. The new historiography deviates only slightly from that which it claims to replace.

It seems that research originally oriented to arguments in the history and philosophy of science has been transferred into the history of medicine, ignoring both disciplinary differences and the diversity of Charleton's views. The writings influential upon his understanding of the body were different from those which directed his ideas about the nature, properties of atoms and similar concerns. This chapter has illustrated that the texts examined within existing historiography have been determined by the 'scientific revolution' narrative within which Charleton has been framed.

The texts I have selected for discussion are, of course, also chosen for their particular qualities. However in this case it is because they offer new insights into how Charleton constructed his identity as a physician. The examination of his medical texts and the detailed consideration of his professional activities reveals different dimensions of Charleton's thought than those offered by various rise-of-empiricism scientific revolution narratives (including those of Shapin and Schaffer). As a physician, he can tell us a complex story about seventeenth-century thought.

A SELECTION OF CHARLETON'S MEDICAL TEXTS

This study does not discuss the author's entire *oeuvre* (although my annotated bibliography includes notes on all published and unpublished works). While I have considered those works upon which his characterisation in existing historiography have been based, I do not wish to reinterpret them.[73] Instead my aim is to examine closely some texts which have received little or no attention in either the historiography of medicine or natural philosophy. Using these works, I explore questions of identity and epistemology which have not yet received

[72] Cunningham & Williams, 'De-centring', p. 409. As Cunningham and Williams note, attempts to revise understandings of the origins of scientific knowledge frequently locate the origins of science in the same period as the "scientific revolution", on the assumption that 'these canonical events, suitably reinterpreted, correspond to the changes they are trying to identify.' Such reinterpretations do not alter our belief in the transition, but rather reshape our understanding of the process involved.

[73] Eric Lewis has argued recently that we should revise our understanding of Charleton's early works, to recognise his eclecticism, see Lewis, 'Early Modern eclecticism'.

serious analysis in relation to the life and career of Walter Charleton. The texts I consider are *Natural History*, *Enquiries into Human Nature*, and *Three Anatomic Lectures*. These three continuous and related works are unique in illustrating their author's approach to certain questions of self-construction and epistemology. Each of the texts falls outside traditional genres, and their atypical natures illustrate aspects of identity-construction beyond those revealed by traditional genres. The similarities between these three works make them a perfect basis for comparison across his changing experiences. Charleton's other texts are not as similar to each other as these are, and therefore comparisons between them are less instructive.

Charleton's medical works numbered nine in total. They were: *Deliramenta Catarrhi* (1650); *Oeconomia Animalis* (1659); *Natural History* (1659); *Exercitationes Pathologicæ* (1661); *Inquisitiones Duæ Anatomico-Physicæ* (1665); *De Scorbuto* (1672); *Enquiries into Human Nature* (1680); *Three Anatomic Lectures* (1683) and *Inquisitiones Medico Physicæ* (1685)—see my annotated bibliography for full details. Some of these works conformed quite exactly to the traditional expectations of medical texts at the time. These expectations were shaped by the medical syllabus, and founded on the assumption that published works were primarily read by those trained in medicine.

There is a large corpus of work from this era which explored medical subjects according to the key questions of cause, symptoms and signs of the disease, the progress of the disease with diagnostic indicators, and finally therapeutics. These were the basic elements of university medical education, as I discuss further in Chapter III. An example of conformity to the genre is *Exercitationes pathologicae, in quibus morborum pene omnium natura, generatione, & causae, ex novis anatomicorum inventis sedulo inquiruntur*.[74] This introduction to the study of pathological 'Physick' examined the nature, generation and causes of most known diseases. *Exercitationes pathologicae* followed the tradition of medical compilation: it provided a nosology, a summary of contemporary arguments, and progressed through the origins, signs and causes of diseases. This volume was composed before Charleton's entry into the Society. *Exercitationes pathologicae* contained a discussion of such issues as the origins of gout and leprosy. It did not, however, incorporate clinical discussions or case descriptions. *Exercitationes* was a compendium of knowledge, and did not present innovative material.

De Scorbuto liber singularis; cui accessit Epiphenomena in Medicatros (1672) followed a specific disease through its causes, signs, symptoms, and progress. This text is discussed more fully in Chapter V. It considered the names given to scurvy, the varieties of the disease, its external and contiguous causes, origins, the diagnostic indications for which physicians must look, and an overview of secondary symptoms. *De Scorbuto* concluded with a selection of cures. In this

[74] London, printed 28 January 1661. These were 'pathological dissertations', in which the nature, generation, and cause of diseases were discussed.

sense its subject matter was circumscribed by an established tradition of how physicians' knowledge should be presented. Considered in the context of contemporary works on the subject of specific diseases, its presentation was formulaic, and the contents indicate clearly that its audience was composed of physicians. The work was perhaps composed in response to Willis' 1667 publication on scurvy in *Pathologia Cerebri & Nervosi Generis Specimina, in quo agitur de morbis convulsivis et de Scorbuto*. Charleton debated the relative significance of various chemical elements, and simply offered disputations of the philosophical principles of Willis' text. He did not draw from patient reports or experimental procedures.

Charleton's *Inquisitiones Duæ Anatomico-Physicæ* (1665), though not precisely medical, nevertheless focused on issues with which a physician might have been concerned. This text consisted of two discourses, the first of which outlined the nature and effects of thunder and lightning. Charleton dismissed vulgar opinions about death being caused by thunder-bolts, and included some additional observations perhaps based on his experiences at the Royal Society. The disquisition was doubtless informed by the talk he gave at the Society on the dissection of a boy killed by lightning. The second discourse was a response to Willis' anatomy of the brain (published early 1664). Again this would have gained its impetus from a presentation at the Royal Society.[75] The publication of *Inquisitiones Duæ* in Latin was perhaps motivated by a desire to present this material, in an impressive and authoritative manner, to roughly the same expert readership who had witnessed the lecture.

Interestingly, those of Charleton's works which examined specific diseases were in Latin. His less clinically-focused works were published in English. The evidence seems to suggest that the works more directly associated with medical practice were differentiated from the broader medical philosophy by the language in which they were composed. Among the medical works, those published in Latin contained a curative emphasis not present in his vernacular publications. Both *Exercitationes* and *De Scorbuto*, because of their therapeutic content, and specific nature, could not be released in the vernacular for fear of seeming to promote the author's clinical prowess. They were written for an audience composed only of physicians. Publication in Latin would have signalled that these works were to be treated as part of the domain of learned discourse, as they were intended only for a circumscribed audience. *Inquisitiones Duæ Anatomico-Physicæ*, by contrast, was probably published in Latin because it was based upon research performed at the Royal Society. It was designed not for a local audience but for Continental readers, interested in the latest findings from England. This latter text was thus a publication intended for a specialised audience. The language of publication is discussed more extensively in Chapter III.

[75] Charleton delivered a lecture at the Society in June 1664, entitled 'Differences betwixt Brains of Men & Brutes'. The lecture discussed precisely those aspects which formed the basis of the second discourse.

Charleton's last published work was *Inquisitiones Medico-Physicæ, De Causis Catameniorum et Uteri Rheumatismo* (1685). This text, like the others which conformed closely to generic expectations, examined the names given to the uterus, both Greek and Latin, and 'their etymology and explanations'. It also explored the genealogy of 'uterine rheumatism'. The author treated the uterus as a 'workshop', in this treatise on the physiology and pathology of menstrual flux, and considered its function. When he traced the causes of Catamenia, Charleton rejected ancient theories regarding cause, but also rejected the more modern theory of the uterine ferment. He invoked Boyle's writings on the blood, and explored George Ent's nutritive juice theory.[76] However, as this text was published only in Latin, its audience was necessarily restricted to readers who were already familiar with other expert writings in the area. Its appeal would have been to those who were medically-trained themselves.

The works considered in this study are less constrained by genre. *Natural History of Nutrition, Life and Voluntary Motion* became one of the century's most popular textbooks of physiology. The Latin edition published almost simultaneously in London was *Oeconomia animalis, novis in medicina hypothesibus supestructa et mechanice explicata*. The *oeconomia animalis* genre had its roots in Dutch medical literature. Charleton modified and adapted the Continental genre, within the specific expectations of an English audience. Evidently he was interested in reconciling ancient and modern authorities and theories. The way in which he supported the legitimacy of new theories through reference to older authority structures reveals something about the status of physicians and their authority base. In 1653, when he composed *Natural History*, Charleton possessed a medical degree, some experience in the King's service and as assistant to Mayerne. Thus his practical experience was limited, and he had no experimental experience to speak of. *Natural History* examined the classical triad of faculties (natural, vital, animal) through their key processes: nutrition, vitality and locomotion. These categories were echoed in other 'animal oeconomies' of the period, but Charleton's eclecticism was distinctive in comparison with the Continental animal oeconomies.

The lectures which constituted the basis of *Enquiries into Human Nature* were delivered in March 1679. They examined the traditional vital functions in anatomical terms (nutrition, life and voluntary motion), with additional lectures on fevers and anatomy of the stomach. The material overlapped substantially with the *Natural History*, which makes a comparison between the two texts highly instructive. Between the publication of *Natural History* and *Enquiries* the author's personal fortunes and experience were transformed. He progressed from private practice and service to the king, to considerable status within the key medical institution of his day. His 1680 lectures were delivered by a man with a secure position at the head of the College of Physicians. Charleton had

[76] Charleton, *Inquisitiones Medico-Physicæ, De Causis Catameniorum et Uteri Rheumatismo*, London, 1685; Leiden, 1686, pp. 72-3.

performed experimental manipulations, and had been part of a thriving community of researchers.

The final of the three medical texts under scrutiny is *Three Anatomic Lectures*, published in 1683. This was a modified translation of a text by Giovanni Alfonso Borelli. Through detailed analysis of the English work, and comparison with the source text, I illustrate how Charleton selected and presented the source material for his audience. By examining his deviations from the Neapolitan's presentation, we gain insight into the author's epistemological orientation, and into what he saw as persuasive presentation of the material to his English audience.

What then is distinctive about the three texts I have chosen for this discussion? *Natural History*, *Enquiries* and *Three Anatomic Lectures* are significant in their divergence from traditional modes of composition. These works were not composed, like the Latin works, according to standard generic constraints, nor were they designed to be read exclusively by a medically-trained audience. All were published in the vernacular. Such publications had to carve a new niche for themselves—in them Charleton presented himself to a 'lay' audience as a professional physician. There seems to be no template for this kind of publication, and therefore he was manufacturing his identity in something of a void. Here he was constructing for a non-expert readership what it was to be a physician. This offers a different perspective from that presented in the Latin works, as it suggests self-construction for an immediate context. Thus the English works raise more questions about audience, and audience expectations, and seem to be far more diverse in their expected readers. They offer an important insight into his authority as a practicing physician in London. These characteristics make them a crucial focus for discussion.

As previously described, Charleton's Latin medical works tended to follow the forms and genres appropriate to traditional education. They encompassed theory, pathology and therapy, and often responded directly to medical works and issues prominent at the time of their composition. The Latin studies often examined traditional questions of physic in ways that did not characterise Charleton's vernacular compilations. The Latin works dealt with traditional medical questions of the origins, causes and signs of each condition, and could therefore be said to fit into an established genre.[77] These medical works, with their more traditional subject matter, reveal less about Charleton's self-construction than do the atypical vernacular works that I examine in detail. Nevertheless they are of course significant indicators of his self-construction, and must be viewed as part of his medical *oeuvre*.

Because the language of vernacular works did not restrict their audience to educated professionals, the author had to devote more effort to establishing for himself an identity in that broader arena. While the Latin medical publications of the later seventeenth century shared a base of learned readers, works in the

[77] None of the Latin works were published anonymously.

vernacular (which were on the increase at this time) were on less assured territory. It is for this reason, I believe, that the physician tended to reveal more explicitly in his vernacular works, the means by which he aimed to increase and maintain his authority. Some of his motivation for these vernacular works might have been Sprat's declaration that Royal Society fellows should use the 'language of artizans' in their publications, to remove the mystery from the medical domain. Perhaps as the Royal Society and College of Physicians vied for greater public currency, Charleton's lectures should be seen as an attempt to promote the College as an institution whose presentations were accessible by the people, as well as by other physicians.

In addition, the three works examined are of particular interest because of the time frame within which they were written. All were composed within a stage of Charleton's career that is infrequently studied. All three works illustrate something of the relationship between his activities and self-presentation. The two later texts are informative partly because they were published after his involvement with the Royal Society experimental community. *Natural History*, by contrast, was published before that involvement. The overlap in subject matter between *Natural History* and *Enquiries*, published on either side of a substantial shift in the author's personal experience, make a comparison irresistible.

It was extremely unusual for anatomical lectures to be published. Yet despite the lack of precedent, two of Charleton's three texts that I examine were published anatomical lectures delivered to professional and training physicians. This immediately makes them notable. The fact that they were printed with the College's blessing is even more noteworthy. These were thus presentations of a distinctive nature. Most of the lectures published were surgical, and the comparison reveals just how uncharacteristic Charleton's lectures were. The two books of anatomic lectures were also genuinely eclectic in content, suggesting that the author, and perhaps also the College, wanted to be seen as knowledgeable across a wide array of contemporary philosophical positions. Thus we witness in these works the influence of the demands of professional medical practice. We can observe in these publications the construction, potentially, of a different kind of public identity from that typically associated with men in Charleton's position. These three medical works have been selected to test some hypotheses about the relationship between practice and self-presentation. I examine whether the author demonstrated adherence to the principles upon which natural philosophical authority was allegedly based. *Natural History*, *Enquiries* and *Three Anatomic Lectures* are spread across crucial decades of his career. The subject matter of the works considered is closely connected, and in some cases overlapping. They offer a unique window onto Charleton's construction of an authoritative identity, not as a natural philosopher, but as a physician.

CHAPTER III

'THE ALEMBIC OF OUR PEN': CHARLETON'S IDENTITY AS A PHYSICIAN

The ensuing discussion examines the profile that Charleton constructed for his readers, and explores contemporary notions about the identity of physicians and virtuosi. The author's presentation of his activities reveals much about his epistemological assumptions, which are central to any understanding of his identity. The manner in which he discovered and presented knowledge of the world around him is indicative of his professional context and the relative meaning he gave to a variety of sources. From these we can understand much about his epistemological values. I examine the identity that Charleton claimed for himself as a physician, and the kind of knowledges that he invoked, in an attempt to untangle the complex relationship between medical and natural philosophical sources of authority. I explore the basis upon which he claimed authority for his pronouncements. Alongside this, it is imperative to consider how he constructed the audience for which his publications were intended. The following analysis of Charleton's representations of his own interests and activities is organised around three main themes: character/temperament; epistemological decorum; publications and language. Within each of these themes I discuss the aspects of virtuoso identity to which I have referred above, and consider their relevance to Charleton's self-construction.

CHARACTER AND AUTHORITY

The relationship between temperament and constitution was central to the self-definition of seventeenth century philosophers. Individual constitution determined the kind of 'genius' a man possessed. As Hooke observed, some constitutions of body 'incline a Man to Contemplation, and Speculation', while others encourage 'Operation, Examination, and making Experiments.'[1] Charleton's writings situated his constitution and temperament in the former category. His audience would have understood the implications of this described temperament for his scholarly professional status. Susceptibility to melancholy was consistent with a social context in which vulnerability signified a refined constitution, and superior intellectual powers. Charleton's *Discourse concerning*

[1] R. Hooke, 'Method of Improving Natural Philosophy', in R. Waller (ed.), *The Posthumous Works of Robert Hooke, M.D., S.R.S., Geom. Prof. Gresh &c.*, New York, Johnson Reprint Co., 1969, p. 9.

the Different Wits of Men outlined the relationship between constitution and character. The finest wits, he observed, were generally contained within 'delicate and tender Constitutions.'[2] Such declarations of susceptibility thus operated as publicly understood signifiers of identity.[3] Hooke's 'Method of Improving Natural Philosophy' outlined the consequences of the individual temperament for knowledge:

> Every man has born within him, or contracted by some way or other, a Constitution of Body and Mind, that does more or less dispose him to this or that kind of Imagination or Phant'sy of things, and every one has some kind of Accident or other, that does more or less dispose him for this Opinion or that Operation of the Mind as well as of the Body.[4]

Charleton presented himself as possessed of an overwhelmingly melancholic constitution. The 'accident' that influenced him was perhaps his lifestyle as a professional physician.[5] Sprat claimed that the successful seeker after knowledge 'should first know himself'.[6] Charleton echoed this, claiming that self-knowledge was the responsibility of the physician: he knows nothing who does not know himself.[7] Before deciding upon his course of life, every man should 'strictly examine his own Genius, and advise with himself concerning the inclination thereof; that so he may at length happily devote himself to that, which he finds most agreeable to the Destination of it'.[8]

Wood described Charleton in 1695 as 'a learned and unhappy man . . . much given to romances'.[9] Though we might take the verdict of the notoriously sour Wood rather lightly, the physician himself referred to 'the often Fermentations and Ebullitions of our Melancholy'.[10] The 'tyranny' of melancholy was caused by 'our native Temperament', and exacerbated by the 'sedentary contemplative condition of life'.[11] Charleton depicted both his individual constitution and the circumstances of his profession as scholarly and sedentary. Thus his professional status was linked firmly to his disposition, and both inclined him toward

[2] W. Charleton, *Discourse Concerning the Different Wits of Men*, London, 1669, pp. 104-5.

[3] Shapin, 'Science and Solitude', p. 195.

[4] Hooke, 'Improving Natural Philosophy', p. 9.

[5] The best solution, according to Hooke, to the problems of inherent disposition and accident, was to discover 'of what Constitution ones self is, and to what one is either naturally or accidentally most inclin'd to believe'. Hooke, 'Improving Natural Philosophy', p. 10. This corresponds with Charleton's Epicurean claim for the importance of self-knowledge in the moral and physical oeconomy.

[6] T. Sprat, *History of the Royal Society of London, for the Improving of Natural Knowledge*, London, 1667, p. 34.

[7] The 'best of mans knowledge' was 'that which teacheth him how to order his Mind, and regulate his Actions'. Charleton, *Epicurus' Morals*, p. 1.

[8] Charleton, *Epicurus' Morals*, p. 52.

[9] Wood, *Athenæ Oxonoienses*, p. 751.

[10] Charleton, 'An advertisement To the Reader', *Darknes*, p. xxii.

[11] Charleton, 'An advertisement To the Reader', *Darknes*, p. xxii.

speculation, contemplation and book learning. In *De Scorbuto*, Charleton cited Seneca's Epistle 8, concerning the character of the enquirer:

> In this place I have stowed myself away, the gates closed tight, so I can advance in many ways. No days end for me in leisure; I appropriate part of the nights for my studies, I do not have leisure for sleeping, but I surrender to it, and having wearied my eyes with late watches and dropping off I keep occupied with my work. Withdrawn not so much from men but also from affairs, and in the first place from my own.[12]

In the *Immortality* dialogues, Lucretius [Evelyn] commented that Athanasius [Charleton] was possessed of 'a Melancholy disposition, and such commonly suffer adverse accidents, to make too deep impressions upon their mind, which is thereupon apt to dejection'.[13] Charleton's deliberate attempt to publicise his own mental vulnerability suggests a strategy to enhance his authority.

Epicurus' moral philosophy (which Charleton promoted in his amalgam from Cicero, Seneca, Epicurus and others) advocated the pursuit of the virtues: Prudence ('or the Dictamen of right Reason'), Temperance, Fortitude and Justice. 'Prudence Private' required the individual to choose 'that course of life, which is most agreeable to the inclination of our Genius'.[14] Charleton asserted a correspondence between his course of life and his constitution. 'Prudence Civill' directed the individual 'to Affect privacy, and yet not to decline publick employments, in the case of the present Necessity of the Comon [sic] Wealth, or the Command of Superiors shall call thereunto'.[15] He followed this template consistently, declaring his personal modesty as constantly overwhelmed by demands of his superiors (in the form of patrons) that he enter the public arena and share his useful knowledge with the broader community.[16] The burden of melancholy marked him with not only the identity of serious scholar but also of a professional physician. This self-description extended to his private correspondence. In 1674, writing to a potential patron and patient, Charleton complained of 'the anxious Labour of the Mind, the perplexity of thoughts, the doubts and fears that usualy [sic] afflict a Learned and Conscientious Physician'.[17]

12 See Charleton's address to the reader, *De Scorbuto*. p. viii.

13 Charleton, *Immortality*, p. 13. Burton commented that 'lack of exercise "dries the brain and extinguisheth the natural heat"; lack of company and distraction encourages mania and wild swings between delight and brooding.' R. Burton, *The Anatomy of Melancholy* (1628), eds F. Dell and P. Jordan-Smith, New York, Tudor Publishing Co., 1927, p. 260. *Immortality's* Lucretius advised Athanasius to abide by the philosophy of Epicurus, and to participate in intellectual diversions in order to 'wear out the Characters your misfortunes and distresses have stampt in your Soul.' Charleton, *Immortality*, pp. 13-14. This was perhaps intended as a promotion of Charleton's translation and compilation of *Epicurus's Morals*, published the previous year. The persona he presented as an author confirmed the theories expounded in this publication.

14 Charleton, *Epicurus' Morals*, see 'The Contents in Scheme'.

15 Charleton, *Epicurus' Morals*, see 'The Contents in Scheme'.

16 However this was a standard device for maintaining proper modesty, in a context in which publishing one's works could be interpreted as immodest self-promotion.

17 Charleton to 'T. M. Baronet', 15 November 1674. This could be Thomas Millington, or Mainwaring. Bodleian Library, MS Smith 13, p. 74.

Shapin has argued that in late seventeenth-century natural philosophy an individual might discover new information in a solitary state (the 'context of discovery'). However to be ratified as knowledge it required a public or multiply-witnessed demonstration—a 'context of justification'.[18] I argue, contrary to Shapin, that for Charleton the context of justification was more closely and more explicitly linked to scholarly textual research than to experimental discoveries. Indeed in various instances in which the context of discovery for Charleton was the (public) experimental community of the Royal Society, the context of justification presented in his work was (private) solitary research. There remained a tradition of identity, available to Charleton, that was self-consciously private and that made itself public through publication rather than through self-description in experiment. Shapin argues that the authority once associated with solitary endeavour gave way, in the seventeenth century, to an emphasis on demonstration before an authoritative audience. However, I believe that Charleton's presentation reflects the continuity of solitary knowledge-production and authorisation. There were, indubitably, natural philosophers who did not follow Shapin's model, but my central interest is in how Charleton negotiated his identity as a physician.

Charleton at times depicted physic as a kind of divine calling, and its insights as restricted to a select few. The anatomy of man, for example, was 'a study so abstruse and difficult, and withal so vast and diffuse; that the last complement of it cannot with reason be expected from the diligence of any one Man, how sagacious and industrious soever.' Therefore the excavation of knowledge about the body was 'divided among those curious Wits, that are by secret instinct disposed to digg in this mine of knowledge'.[19] His reference to a secret instinct asserted the distinctive status and character of physicians, who were uniquely qualified to carry out such investigations.

By contrast with the rhetoric of some natural philosophers, the authority of professional physicians was founded on a strong tradition of ancient learning, and its authority structure echoed that of theology and law. This was an identity marked by the importance of scholarly research and knowledge of ancient medical lore. The prominence of physicians over their less educated medical counterparts in the marketplace (empirics, apothecaries, surgeons) relied upon years of university education, and consequent immersion in classical knowledge. Physicians did not need to impress an audience with their legitimacy: they were an already established part of the social structure. To a community of professionals who had much to gain from association with tradition, an eclectic approach might hold great appeal.

Meditative solitude was Charleton's constant refrain, and it seems he viewed solitude and individual judgement as desirable attributes to emphasise in both

[18] Shapin, 'Science and solitude', p. 207.

[19] Charleton, 'Preface', *Enquiries*, p. xxvi. Charleton believed advances in comparative anatomy had been extensive, in contrast to the mysterious realm of simple anatomy. *Enquiries*, p. xxiv. Charleton's interest in comparative anatomy was attested by his *Onomasticon Zoicon* of 1668.

public and private presentation. We must consider that Charleton was attempting to annex authority deriving from scholarly and solitary research. His own words suggest that he perceived an intimate link between this contemplative life and his status as a practising physician. As we saw in the previous chapter, education for physicians emphasised classical scholarship, and therefore affirmed the status of activities which were solitary in nature. The status of physicians during the seventeenth century assumed a powerful link between learning and authority. The good character of a physician was a vital aspect of his ability to provide learned medical counsel. Only in combination with a serious and learned temperament did knowledge qualify the physician to exercise good judgement. The authority claims and professional status of early modern physicians were linked to the central concepts of 'judgement and advice'.[20] The constancy with which Charleton asserted his melancholy suggests its significance in his bid for an authoritative identity. Given that his self-construction was so distinctly at odds with the profile created by some of the recent historiography for natural philosophy, we might begin to consider that Charleton aspired to a public identity different from the sort of public rhetoric that surrounded the natural philosopher/virtuoso.

TEMPERAMENT AND KNOWLEDGE

In the hierarchy of knowledge accessible to an individual, the specific temperament had to be mastered in order to achieve higher degrees of knowledge. Self-characterisation entailed an epistemological claim. To annex the authority of a speculative philosophical man, Charleton portrayed himself as possessing an impartial and rational, rather than active or empiricist, temperament. The physician's Royal Society activities involved extensive experimentation, and he could have claimed quite legitimately that he was possessed of an active constitution suited to 'examinations'. Yet he presented himself in a manner that emphasised speculation rather than action. This choice suggests that there was equal, if not greater, authority in the identity of a speculative scholar than that of an experimentalist.

This period witnessed talk of the singular 'philosophical spirit', which allowed English citizens to maintain an impartial and balanced perspective. Oldenburg, in correspondence with the chemist Augustin Boutens, opined that English chemical discourse involved 'operations undertaken by men of sense who are free from the vulgar prejudices imposed on the world by some people who undertake to speak dogmatically'.[21] England's naturally philosophic character was frequently constructed by contrast with her Continental neighbours.

[20] Cook, 'Good advice', p. 4.

[21] Oldenburg to Boutens, 11 November 1667, Henry Oldenburg, *The Correspondence Of Henry Oldenburg*, ed. A. Rupert Hall and Marie Boas Hall, Madison, University of Wisconsin Press, 1965, vol. III, p. 590.

Individual temperament reflected national character, and both claimed possession of a constitution uniquely qualified to attain elevated knowledge.[22] Charleton was keen to annex this authority for himself. He depicted his own 'Genius' as,

> so averse to all contests and passionate Altercations, and which alwaies brings me to Philosophicall Discourses only as to Enquiries, not final Determinations, and with perfect indifferency to either side, not caring at all whether my Allegations, or my Opponent's, give the greater light to certainty[23].

The indifferent and impartial temperament of a philosopher made him capable of balancing contradictory opinions and considering all hypotheses. Charleton's praise of fellow English thinkers attributed to them this resistance to judgement and prejudice.[24]

Charleton criticised his Continental counterparts as 'hot and testy', and overly dogmatic. So ardent were these philosophers, 'in defence of their own preconceived opinions, that they account it a piece of disrespect and incivility in any man that seems to doubt.'[25] Thus certain kinds of constitution prevented men from achieving more than basic kinds of knowledge. The lowest form of understanding, and the least desirable, was simple obedience to doctrine. The temperament most suited to the pursuit of knowledge was the cool and rational persona, which was freed from the traps of dogmatism. By implication the elevated intellect was reluctant to exclude alternative explanations. Sprat argued that the knowledge-seeker should be 'well-practis'd in all the modest, humble, friendly Vertues: should be willing to be taught, and to give way to the Judgement of others.'[26] Modesty constituted a vital aspect of intellectual authority.[27]

Charleton, like many of his natural philosophical contemporaries, publicly accepted that it was neither desirable nor necessary for all to be known. The greatest available degree of certainty about most natural knowledge was conditional support of provisional arguments, in this depiction. Through mitigated scepticism it was possible to gain a workable degree of certainty regarding the natural world.[28] This was the description about his nature that the physician's works promoted, and in this way we can identify the values with which he wished to be

[22] Yet while Sprat praised the collaborative nature of English investigation, Charleton did so only in *Immortality*. See Sprat, *History*.

[23] Charleton, *Immortality*, p. 16.

[24] In *Immortality* Isodicastes [Charleton's patron, Henry Pierrepont] was described by Lucretius [Evelyn] as, 'A valiant Assertor of truth, yet far from Tyranny; where he finds an errour, as alwaies reflecting on human frailty, and the obscurity of things in themselves'. *Immortality*, p. 18.

[25] Charleton, *Immortality*, p. 15.

[26] Sprat, *History*, p. 34.

[27] Charleton's modesty in print is contradicted by the observation of his contemporary Wood, that the physician was 'observed by those that knew him, to set a high value upon his own worth and parts'. Wood, *Athenæ Oxonienses*, p. 752.

[28] R. F. Kroll, *The Material Word: Literate Culture in the Restoration and Early Eighteenth Century*, Baltimore, Johns Hopkins University Press, 1991, p. 61.

associated. In these respects Charleton's self-construction echoed that of the 'virtuosi', in that he too presented himself as modest, impartial and an avid searcher after truth. However, he differed in one significant regard—the extent to which experimentalism was one of his declared sources of authority. Here the identity of the physician deviated from the rhetoric of some 'Boylean' natural philosophers.

A HIERARCHY OF KNOWLEDGE

Charleton's epistemology seems to have varied across his works, due in part to his tendency to copy both the methodology and content of contemporary publications. The best we can achieve perhaps is an understanding of the reasons behind his self-descriptions. Divine teachings were to be accepted uncritically, while human knowledge was to be questioned rigorously.[29] Part of the philosopher's proper modesty lay in restricting his questions to the knowable realm: 'All I durst ever aspire unto, is only with pious humility to apply my Reason to such of the Articles in my Creed, as seem to be placed within the Sphere of its comprehension'.[30] Respect for the limits of human knowledge was vital. Charleton promoted the notion, at times, that mankind was unable to access divine knowledge, as a result of original sin. *Ternary of Paradoxes* (1650) outlined the relative status of the different faculties and their possible access to knowledge. Although the *Ternary* expressed Charleton's belief that mankind was destined to darkness, 'without the manuduction of Divine light', yet

> hath the exceeding Mercy of the Fountaine of Light, furnished us with three faithfull Guides, whose conduct if we præcisely follow, we may be reduced to that advantage of knowledge, as will afford us a *Pisgah Sight*, or crepusculous prospect of her reflexive beauties, now, and an assurance of being blessedly engulphed in the Abysse of her Pleasures, in the future.[31]

Even these divine hints would not provide assistance to all, but only to a few dedicated individuals:

> So deeply immersed in Sensuality, are all the wretched Sonnes of Adam, that few can ascend so many degrees toward the height of their primitive Capacity, as with æquall and constant paces to pursue the guidance of these geniall Impressions.[32]

Here he offered an epistemological schema strongly influenced by Helmont. 'Religion, Reason, and Sense' were the three guides toward veracious knowl-

[29] On contemporary hierarchies of knowledge see Lorraine Daston, 'Probabilism' in D. Garber and M. Ayers (eds), *The Cambridge History of Seventeenth Century Philosophy*, Cambridge, Cambridge University Press, 1998.

[30] Charleton, *Immortality*, p. 58. The articles to be demonstrated were 'the Being of God, as Father Almighty, and Maker of Heaven and Earth; and the Immortality of Mans Soul, or Life everlasting.'

[31] Charleton, 'Prolegomena', *Ternary*, p. III.

[32] Charleton, 'Prolegomena', *Ternary*, p. IV.

edge. Religion led 'by the invisible Clue of Faith, on towards the implicit and certaine Apprehension of objects above the reach of the other two.' Below Religion Reason transported man 'through the Circumambages and complex Labyrinth of Discourse'.[33] Sense, 'by a shorter Cut, or blunter method of Preception [sic], directeth to the immediate and actuall Knowledge of the Existence and exteriour or manifest Qualities of Entities'. Sense provided only exterior knowledge, not an understanding of the true natures of things, as the 'Opticks of Flesh and Blood', were 'too dim to endure to gaze upon the naked and entire Glories of Truth'.[34] *Three Anatomic Lectures* declared that, 'Truth is a tree, whose root is in Heaven, and those of which even the wisest of us dim-sighted Mortals here upon earth see nothing but the shadows of its branches'.[35] The 'essence of things' and the 'intimate nature of objects', were hidden from mankind because of the 'Obscurity of Nature and Dimness and imperfection of our Understandings and the Irregularity of our Curiosity'.[36]

While his epistemological approach to questions such as the possibility of certain knowledge varied at some points in his later works, there was a basic continuity in his approach to the hierarchy of human access to understanding. In *Physiologia*, he expressed the argument that we should not 'circumscribe our Intellectuals with the narrow line of our sensible discoveries, but learn there to set on our Reason to hunt, where our sense is at a loss'.[37] Reason was to be applied to all data obtained by sense, for it was Reason that lifted man above his crippled powers of perception. This hierarchical view of knowledge was informed by years of tradition and reiterated by England's social order. The identity of a scholar, which Charleton seems to have annexed reasonably consistently throughout his writings, was reaffirmed in its efficacy by this depiction of the hierarchy of knowledge. Reason, and faith, both of which he continually identified himself with, were important means of access to higher understanding.

EPISTEMOLOGICAL DECORUM

Immortality presented two opposing views on the benefits of adherence to established doctrines. The character Isodicastes [Pierrepont] argued that the ancients had provided 'noble foundations' but 'few compleat Buildings'. He who wished to

[33] Thus language was a manifestation of the reasonable mind and, combined with divine inspiration, a powerful tool for the discovery of truth. Reason, 'from the remote dependencies of Effects upon their particular Causes, ushereth us along by a subalternate Series, or Gradation unto the science of their ultimate Approximation and Individuality.' Charleton, 'Prolegomena', *Ternary*, p. IV.

[34] This inability of sensory information to provide veracious knowledge was the result of 'Originall Sinne', which 'hath so benighted the primitive Clarity of the Intellect, that it deplorably wanders in the infinite seductions of Errour and cannot recover that direct path, which leades to the heaven of Verity, without the manuduction of Divine Light.' Charleton, *Ternary*, pp. III-VI.

[35] Charleton, *Lectures*, p. 104.

[36] Shapiro, *Probability & Certainty*, p. 62. See Charleton, *Physiologia*, p. 5 and *Immortality*, pp. 116-7.

[37] Charleton, *Physiologia*, p. 113.

improve his knowledge 'must advance superstructures of his own'.[38] Lucretius [Evelyn] scorned textual reliance, claiming that 'who so enquires into the operations of Nature, by no other light than that of Books and solitary speculations, shall in the end find his head full of specious Termes, but empty of true and solid Science.'[39] Yet Athanasius [Charleton] defended the 'school-men', claiming that the method of the ancients was sufficient to the demonstration of the fundamentals of religion. Though the writers 'themselves confesse, they were not compleatly Apodicticall', and their polemical reasoning did not 'ascertain equally with Geometrical Demonstrations'; yet their claims 'import either a Physicall or Moral evidence, sufficient to perswade a mind well affected toward truth, and free from the obstruction of prejudice.'[40] This debate over the adequacy of ancient knowledge recognised the need to create modern revisions. Yet it also reminded the reader that reforms should be developed around specific issues, from which moral/religious questions were to be excluded.

Charleton's ideas on such fundamental questions as the possibility of certainty seem to have varied across his *oeuvre*. His self-construction agrees with the virtuoso model in its desire for an impartial identity, but not in terms of a communal context for justification. He made clear that adherence to ancient authority was a feasible method of operation, though he occasionally declared that it could lead to errors. Charleton showed no sign of despising ancient authority because of its age or textual status, thus contradicting the claims of Dear, Shapin and others that these characteristics led ancient texts to be reviled.

ECLECTICISM, SCEPTICISM & PROBABILISM

> *The Virtuosi of our English Universities, it is well known, have of late years, proclaimd open warre against the tyranny of Dogmatizing in any Art or Science.*
>
> Walter Charleton to Margaret Cavendish, May 1669.[41]

Historians such as Peter Dear, Barbara Shapiro and John Henry argue that as the status of ancient authorities declined, so did the certainty of ancient principles, as promulgated by infallible religious authorities.[42] English natural

[38] Otherwise 'he will lie open to the weather of Doubts, and Whirlewinds of various Difficulties, nor will he be ever able to entertain his friends with decency and satisfaction.' Charleton, *Immortality*, p. 53.

[39] Charleton, *Immortality*, p. 5.

[40] Charleton, *Immortality*, p. 5. This echoed John Wilkins' claim that moral certainty attained to those claims that 'every man whose judgement is free from prejudice will consent to'. J. Wilkins, *Of the Principles and Duties of Natural Religion*, 4th ed., London, 1699, p. 8. The ancients' demonstration of the fundamentals of religion could not be improved by more recent geometrical and mathematical demonstration, according to Charleton. However these latter proofs could be sought in other aspects of natural philosophy.

[41] Bodleian Library, MS Smith 13, p. 49.

[42] Dear, 'Totius in verba', *passim*.

philosophers came to reject doctrines, whether ancient or modern, and instead seek not the pronouncement of ultimate truth, but the exploration of plausible alternatives.[43] While certainty was beyond human achievement, philosophical probabilists instead saw a series of gradations of probability, in which moral certainty was the highest degree of knowledge to which man might aspire (knowledge that 'every man whose judgement is free from prejudice will consent to').[44]

Historians studying this aspect of seventeenth-century thought have often drawn heavily upon Sprat's *History of the Royal Society* (assumed to be descriptive of Royal Society methodology).[45] *History* expressed the desire for 'a nonimperious, nondogmatic, noncontentious mode of discussion and presentation.'[46] The fellows should exhibit 'wariness, and coldness of thinking, and rigorous examination'.[47] Society members were to be characterised by their 'indifferent hearing of all conjectures, that may be made from the Tenets of any sect of Philosophy; and by touching every effect that comes before them; upon all the varieties of opinions, that have either been late found out, or reviv'd'.[48]

Shapiro has identified 'a new scientific style hostile to dogmatism, system-making, and assertions of authority.'[49] Royal Society fellows attempted impartial analyses of gathered evidence and calm appraisal of hypotheses. Propositions were to be evaluated in terms of their respective probability.[50] Thus persuasive power was associated with modest self-presentation. Self-doubt was rewarded as evidence of a reliability and virtue in character. Many of Charleton's contemporaries clearly worked toward the construction of such an identity.[51] Joseph Glanvill, in 1676, claimed that 'the Free Philosophers' (Royal Society Fellows) should 'proceed with wariness and circumspection without too much forwardness in establishing Maxims, and positive Doctrines'. Ideally the virtuosi ought only 'propose their Opinions as Hypotheses, that may probably

[43] Henry, 'Scientific Revolution in England', p. 196. See also Shapiro, *Probability and Certainty*, p. 17.

[44] Wilkins, *Principles and Duties*, p. 8.

[45] Brian Vickers claims that the aims expressed in Sprat's *History* must not be read at face value. They were, he posits, merely a plan for the correction of inadequacies within the Royal Society itself. B. Vickers & N. Struever, 'The Royal Society and English prose style', in *Rhetoric and the Pursuit of Truth: Language Change in the Seventeenth and Eighteenth Centuries*, California, William Andrews Clark Memorial Library, 1985.

[46] Sprat, *History*, pp. 33-4. Gaukroger identifies the presence of such a view in a wide range of English natural philosophy in the seventeenth century after Bacon, including a range of orthodox and unorthodox natural philosophies, such as Dee, Fludd, Cudworth, More, Newton and Boyle. Gaukroger, *Francis Bacon*, p. 30.

[47] Sprat, *History*, p. 102.

[48] Sprat, *History*, pp. 104-5.

[49] Shapiro, *Probability and Certainty*, p. 17. The Royal Society agenda 'required great circumspection, modesty and wariness so as to escape the "disguised Lies, deceitful fancies" which resulted from "catching at it too soon".' p. 47.

[50] Shapiro, *Probability & Certainty*, p. 66.

[51] In much secondary literature Robert Boyle is treated as a template for contemporary epistemology and identity.

be true accounts, without peremptorily affirming that they are.'[52] Hooke's 'Method of Improving Natural Philosophy' proclaimed that the natural philosopher should 'proceed with the greatest Degree of Candor and Freedom from Prejudice, not to be byassed by this or that Opinion in making of Deductions'.[53] Boyle maintained that the true philosophers were 'Eclectics' who 'did not confine themselves to the notions and dictates of any one sect, but in a manner include them all, by selecting and picking out of each that which seemed most consonant to truth and reason, and leaving the rest to their particular authors and abettors.'[54] These declarations certainly establish the importance of scepticism, for natural philosophers and physicians alike.

In *Physiologia* Charleton aligned his own investigations with those natural philosophers who 'adore no Authority' (thus who reviled dogmatism). Diogenes Laertius had referred to them as 'the ELECTING, because they cull and select out of all others, what they most approve.'[55] Active exclusion was part of the practice of these philosophers. These men, he claimed,

> pay a reverend esteem, but no implicate Adherence to Antiquity, nor erect any Fabrick of Natural Science upon Foundations of their own laying: but, reading all with the same constant Indifference and æquanimity, select out of each of the other sects, whatever of Method, Principles, Positions, Maxims, Examples, &c. seems in their impartial judgements, most consentaneous to *Verity*; and on the contrary, refute, and, as occasion requires, elenchically refute what will not endure the tests of either right *Reason*, or faithful *Experiment*.[56]

Charleton's published works situated him explicitly within this category, and his writings within the eclectic genre.[57] Charleton's eclecticism had been deeply influenced by both Gassendi and Grotius. He was aware that his declaration of eclecticism 'may yet be censured as superfluous: since not only those Exercises of our Pen, which have formerly dispersed themselves into the hands of the Learned, have already proclaimed as much.'[58] Evidently he expected his readers to recognise the eclecticism manifest in the range and nature of his publications.

With its classical roots in the writings of Diogenes Laertius, eclecticism became increasingly popular in the later seventeenth century, as a result of several factors. Kelley argues that early modern eclecticism was the product of 'the revival of ancient and patristic learning; evangelical religious reform; the

[52] J. Glanvill, 'Of Scepticism and Certainty: In a short Reply To the Learned Mr. Thomas White', *Essays on Several Important Subjects in Philosophy and Religion*, London, 1676.

[53] Hooke, 'Improving Natural Philosophy', p. 20.

[54] R. Boyle, 'The Christian Virtuoso: Appendix to the First Part, and the Second Part' (1744), in T. Birch (ed.), *The Works of the Honourable Robert Boyle*, London, J. & F. Rivington, 1772, vol. VI, p. 700.

[55] Charleton, *Physiologia*, p. 4.

[56] Charleton, *Physiologia*, p. 4.

[57] On Gassendian probabilism, see Daston, 'Probability and evidence', pp. 1117-8.

[58] Charleton, *Physiologia*, p. 4. Potamon the Alexandrian attempted the reconciliation of Aristotelian, Platonist, Epicurean and Stoic philosophies. See Kelley, 'Eclecticism', p. 579.

"liberty of philosophizing," a secular version of the Protestant rejection of dogmatic authority; and the adoption of critical history as the basis for understanding.'[59] The 'liberty of philosophising' gained impetus from Renaissance humanists such as Lorenzo Valla and Erasmus, who had vowed to follow no dictates. Eclecticism, far from being undiscerning, was presented by practitioners such as Charleton as a careful method by which truth could be sifted from opinion and falsehood.

Many historians consider seventeenth-century scepticism to be integrally linked with the replacement of ancient doctrine by newly generated experimental knowledge. Recent works by Shapin and Dear focus upon scepticism insofar as it revealed the need for experimentally derived 'matters of fact'. Empiricist methodology is depicted by these scholars as having circumvented the dead end of doubt. The implication of this argument is that eclecticism was a point through which natural philosophers passed *on their way to* verification through multiple witnessing. In the historiographical search to identify the new rules by which truth could be defined in the late seventeenth century, some important considerations have been neglected. Eclecticism has not been acknowledged as a desirable endpoint. In the view of Shapin and Dear, creation of prestigious communal knowledge was the aim of virtuosi. But in Charleton's case we see an individual deliberately constructing himself as a scholarly, solitary eclectic. I argue for the importance of an eclecticism which took its cue not from rejection of ancient authority, but from the careful reconciliation of ancient and modern knowledge. Indeed as Kelley notes, eclecticism as a method, while it entailed liberation from adherence to doctrine, simultaneously enforced the continuity of ancient knowledge, since it was founded on the notion that truth was the product of collective, rather than individual effort.[60] This seems to mirror the adoption of new philosophies, while simultaneously adhering to the authority of traditional physic, which characterised Walter Charleton's medical oeuvre.

Charleton claimed, in *Immortality*, that he had introduced only ideas that were 'justifiable by right reason, by autoptical or sensible demonstration, and by multiplied experience'.[61] Charleton's eclecticism clearly extended to experimentalism, without necessarily giving it priority. While witnessing of communal activities was part of his epistemological repertoire, this quotation illustrates the extent of his eclecticism, and mixed modes of demonstrative authority it entailed for him. His was an eclecticism of method as well as of sources. This was of course true of many natural philosophers and physicians of his time. As early as 1650 Charleton rejected loyalty to doctrines which 'hold our credulities enslaved to an implicite conformity, by the tyrannous title of Præscription.'[62] Doctrines to which one was firmly committed, he argued, prevented accurate perception. The aim of the natural philosopher was to free himself of such

[59] Kelley, 'Eclecticism', pp. 580-81.
[60] Kelley, 'Eclecticism', p. 583.
[61] Charleton, *Immortality*, p. 52.
[62] Charleton, 'To the reader', *Deliramenta*, p. vii.

constraints, and exercise his mind without the limits of prejudice. One should reject such 'Philosophicall and Religious opinions, as we have been accustomed from the minority of our own Understandings'. They were an obstacle to true knowledge, for once trapped within a prejudice, even though we may be 'greatly deluded, yet cannot the arm of the strongest reason bend us from our accustomed judgement.'[63] The removal of expectations and 'opinions', declared Charleton in 1650, 'shall be the constant businesse not onely of my studies, but also of my earnest prayer.'[64] Stephen Gaukroger has revealed the importance of eclecticism in Bacon's thought, and has argued persuasively for the continued significance of eclecticism in seventeenth-century natural philosophy.[65] It seems Charleton echoed Bacon's rejection of scholastic adversarial disputation in favour of 'Moderation or the Middle Way'.[66]

Charleton portrayed his friend Glisson as one who reviled doctrines, determined to deal only with probabilities: 'his modesty is so great, as that he expressly professes his own want of full satisfaction concerning the truth of sundry particulars therein contained'. Glisson, in other words, refused to accept that he had the final answer to his questions. His theories were presented 'as positions, not of apodictical evidence, but great probability, and worthy to be embraced, only till time shall have brought more credible ones to light.'[67] To claim certainty would be to assert possession of something to which man was not permitted access. The role of the sceptic was not to create theories, but to destroy established doctrines by illustrating the possibility of alternatives.[68] This approach was invoked by Charleton, who regularly posed solutions without heeding whether they contradicted others recently posited.[69] His aim was thus presented as the proper expression of theories not the generation of atheoretical matters of fact.

CHARLETON'S ECLECTICISM

Charleton's statements of eclecticism echoed those of his natural philosophical contemporaries, and in this regard he conforms with the 'virtuoso' model put

[63] Charleton, 'To the reader', *Deliramenta*, p. ix.

[64] Charleton, 'To the reader', *Deliramenta*, p. x.

[65] Gaukroger, *Francis Bacon, passim.*

[66] F. Bacon, *The Works of Francis Bacon*, ed. J. Spedding, R. Leslie Ellis and D. Denon Heath, 7 vols, London, Longmans, 1857-61, vol. VI, p. 754. Cited Gaukroger, *Francis Bacon*, p. 11.

[67] Charleton, *Immortality*, p. 40. This is in fact illustrated in Glisson's lectures on the brain, from which Charleton copied much of his own lecture on voluntary motion in the 1680 *Enquiries.*

[68] H. Van Leeuwen, *The Problem of Certainty in English Thought, 1630-1690*, The Hague, Martinus Nijhoff, 1970, p. 95.

[69] In 1661, Boyle's spokesman Carneades in *The Sceptical Chymist* stated that, 'In case that some of his arguments shall not be thought of the most cogent sort, that may be, he hopes it will be considered, but that it ought not to be expected, that they should be so. As the author's aim was 'but to propose doubts and scruples, he does enough, if he shews, that his adversaries arguments are not strongly concluding, though his own be not so either . . .' R. Boyle, 'The Sceptical Chymist' (1661), *Works*, vol. I, pp. 460-61.

forward by Shapin and others. My argument is not that these historians have wrongly portrayed the nature of eclecticism, but that they have underestimated its contemporary status. According to Shapin's formulation for 'virtuoso' identity, the aim of men like Charleton was to generate fixed, atheoretical 'matters of fact', within a collaborative experimental context. Yet Charleton clearly promoted the importance of multiple competing theories between which he did not attempt to choose. He identified with an existing tradition of eclecticism, prevalent in late seventeenth-century natural philosophy. Some argue that seventeenth-century scepticism was connected with the creation of an epistemology centred around the generation of 'matters of fact'. However Charleton seems not to have had any strong concern with the generation of experimental matters of fact. An eclectic approach to knowledge constituted a source of authority parallel to, and for Charleton more desirable than, experimental prowess. Eclecticism as an epistemological approach was in his case far more developed than is often recognised. He seemed peculiarly capable of negotiating knowledge without reference to experimentally generated certainty. Thus Charleton's determined and thoroughgoing eclecticism places him outside the Shapinian 'virtuoso' model, and it seems plausible to suggest that it reveals major problems with that model's explanatory accuracy overall.

This physician lacked the basic motivation which Shapin claims drove natural philosophers in their search for authority through experiment. An authoritative identity for a physician was not reliant upon experimental practice—he had an established source of authority in the scholastic traditions of learned physic. While experiment was indubitably one of his epistemological tools, it did not have primacy in his proposals, and his aim does not appear to have been the creation of fixed empirical truths. Charleton presented himself as achieving insight in solitude, and there is no evidence to suggest that he saw community as a central epistemological source. He could have drawn reference to his extensive experimentation in the community of Royal Society fellows as a 'context of justification' for some of his findings. But it seems that for Charleton experiment was only one of *many* possible sources of knowledge, and not necessarily the primary one. That experiment did not replace the importance of scholarship and theology is a truism, but observation of the ways in which the physician negotiated various sources of authority is instructive.

INNOVATION & AUTHORSHIP

Shapin and Schaffer's profile for natural philosophers has virtuoso natural philosophers insistent upon the promotion of new, rather than classical, knowledge. Yet Charleton's work positively avoided any claims to innovation, and made no such assertions about himself. Instead he drew from a tradition more akin to that of Renaissance humanist eclecticism. The eclectic tendencies of the Renaissance humanist tradition led to the combination of multiple authorities, obscuring the distinctions between specific texts.[70] Charleton was continually censured for his reliance upon the writings of others.[71] However the continuity of this textual reliance throughout his oeuvre, and the fact that he still (after prior criticism) identified himself as engaged in the creation of compilations, in turn suggest that the method held desirable status as a signifier of authority. He must have been criticised for the habit in relation to *Ternary*, for the preface to *Deliramenta* contained an enthusiastic defence of the former work.[72] It was, he claimed, simply not in his character to generate new knowledge: 'every Brain is not constellated for new Discoveries; nor can every Age boast the production of a Copernicus, Gilbert, Galileo, Mersennus, Cartesius, or a Harvy'.[73] His own contribution, he protested, was in the realm of compilation. Once again, he positioned himself within a scholarly tradition, in an overt glorification of ancient authority.

His introduction to his next publication, *Darknes*, declared it to be fully dependent upon others' writings. Charleton described it as a 'gratefull commemoration of those venerable Authors, from whose replenish't Treasuries, the richest Contributions toward this Hospitall for the cure of the miserable Infidels, were derived.'[74] He declared himself 'adliged' to the ancients, 'as well by the bond of Gratitude, as Honour'. The first was for 'the due tribute of Commemoration, and an open profession of our beholdingnesse to them', the second, 'in order to the prævention of being reputed Plagiary.'[75] His rebuttal of the charge was explicit, and his depiction of his own scholarly role defensive. It was common to excuse the inadequacies of ancient authors on the basis of the constraints of their pagan era. Charleton adopted such a defence of Epicurus in *Epicurus' Morals*.[76] But he went further, promoting the consonance of ancient theories with the principles of natural theology. He continually defended the ancients' latent understanding of divinity: when Galen praised the structure of the body, he claimed, he was in fact lauding *divine* construction. Charleton was quick to defend the direct relevance of ancient knowledge on a theological basis.

[70] P. O. Kristeller, *Renaissance Thought II, Papers on Humanism and the Arts*, New York, Harper & Row, 1965, p. 37.

[71] Wood noted that the physician 'hath written many books (but [a] great part of them are collected from other authors)'. Wood, *Athenæ Oxonienses*, p. 751.

[72] Charleton claimed that the censure he had received was undeserved: 'there are many Stars, and those of the greatest magnitude, now shining with full splendor, in the sphere of Learned Authors, whose beams were in great part derived from others, whom the revolution of Time had made lesse vertical, and declined towards the West of Oblivion.' After all, 'Who can blast the verdant Laurels on Virgils front, by saying, though truly, that He converted more then 1000 of Homers verses to his own use?' 'To the Reader', *Darknes*, p. xix.

[73] 'Providence introducing such, as Time doth new Stars, single and seldom.' Charleton, 'To the Reader', *Darknes*, p. xxi This was contrary to Bacon's claim that all men were capable of the discovery of truth, through use of appropriate methods.

[74] Charleton, 'To the Reader', *Darknes*, p. ii.

[75] Charleton, 'To the Reader', *Darknes*, p. xv.

[76] Charleton, *Epicurus' Morals* was preceded by an introductory 'Apologie for Epicurus', written to 'a Person of Honour'.

Charleton declared openly his closeness to scholastic methods, declaring that *Darknes* 'præsents, in Epitome, and entire, what many moneths reading of the Schoolmen would have exhibited at large, and in fractures.'[77] He saw his own exercise as comparable to that of 'those Bookmen, whose vigilant nostrills are acquainted with the odour of the Lamp, need not bee informed, how many anxious hours have been spent on the sedulous Extraction of this Vial of precious Truths, from the choycest Flowers of so many excellent Gardens'.[78] *Darknes* however was no simple reiteration of others' arguments, but the result of a selective process, a distillation of essential elements from a greater mass. 'It hath', he claimed, 'been no small part of our care, to Refuse, as well as Elect' material for inclusion. Here he invoked hermetic imagery, the distillation of the essence from consulted texts, as they passed through 'the Alembic of our Pen'.[79] Far from merely echoing of the works of others, he claimed to have called upon inspired judgement, resolving the mysteries and deeper truths of the texts. Thus he separated himself from association with the 'schoolmen', while simultaneously ratifying their technique through his own use. *Physiologia* evidently also received criticism. Athanasius, the autobiographical character in *Immortality*, complained that he had reaped no reward for his labours in composition, 'but most severe, inhumane, uncharitable, unjust Censures', including the accusation of 'usurping other men's Notions, Maxims and Experiments for my own, without so much as naming the Authors'.[80] Charleton however showed no sign of a shift in style as a consequence of the criticisms he received, and his later works illustrate exactly the same tendencies toward compilation.

Immortality claimed that the author had not introduced 'any Alterations . . . but what carry their utility with them'. Thus 'every intelligent man may easily perceive, that it hath been the Reformation, that drew on the Change; not the desire of Change, which pretendeth the reformation.'[81] Echoing the Baconian discourse of public utility and purposive human endeavour in natural philosophy, this declaration also defended the author's impartiality. He was not interested in discovery for its own sake. This defence implies the existence of contemporary criticism of gratuitous innovation. Charleton sidestepped such accusations by claiming that he was led by the material itself, and that he was not interested in generating a radical break for the sake of it.

By contrast with the Shapin model in which innovation was central to natural philosophers' claims about themselves, Charleton again explicitly defined himself in a role far from that of an innovator. His denial of any originality is important when considering first, the extent to which he fitted such a model; and second, the reasons why he might have constructed himself thus: the benefits he gained from being seen *not* to introduce innovations. His self-portrayal as sensitive,

[77] Charleton, 'To the Reader', *Darknes*, p. xxi.
[78] Charleton, 'To the Reader', *Darknes*, p. xviii.
[79] Charleton, 'To the Reader', *Darknes*, p. xix.
[80] Charleton, *Immortality*, p. 10.
[81] Charleton, *Immortality*, p. 52.

contemplative and capable of profound understanding defined him as perfectly equipped to comprehend precisely original and divinely ordained principles: the mysteries of the temple of Aesculapius. He did not believe that medical knowledge itself was divine, but rather that it reflected the underlying presence of divine law. The ability to interpret revelations in medical knowledge was in the character of the physician. This reaffirms the centrality of his character definition, for in the individual character resided the potential to access great/divine insight.

An understanding of how Charleton constructed himself within his writings is vital to a reading of his medical works. We can observe a strategy by which he ratified his abilities as a physician. In some ways Charleton's eclecticism accords with the avoidance of doctrines associated with Royal Society virtuosi. Yet he did not replace eclecticism with degrees of certainty derived from experimental 'matters of fact', as some scholars allege. Indeed he showed no desire to generate such atheoretical truths, but rather was content to discuss multiple theories, often giving conflicting theories equal weight. In this, his self-description differs considerably from that suggested by Shapin *et al.* Unlike the epistemological precariousness ascribed to the newly fledged sort of natural philosopher, the physician had much to gain from the synthesis of traditional earning with more recent innovation. In this sense an appreciation of his professional context is vital to our understanding of Charleton's thought.

PUBLICATION AND LANGUAGE

To comprehend the content of Charleton's works we must examine the audience for which they were intended. The physician's texts reveal something of this audience, and his attitude toward publication. Here we need also to consider the importance of his institutional affiliations. The College of Physicians did not traditionally encourage original research or publication.[82] The status of College Fellows was not reliant upon literary activity, and indeed many prominent members produced no publications. There was a danger that physicians publishing specialised medical material might give away some of the knowledge specific only to physicians, and which constituted the basis of their professional authority. The demarcation of the status of physicians from those of apothecaries, surgeons and quacks was an important aspect of the seventeenth-century medical marketplace. This was perhaps one reason why specific and detailed medical works, which outlined the traditional areas of physicians' specialised knowledge, were often published in Latin. Frank claims that devotion of energy to original research and literary activity was virtually inverse to College status, and that publications by College members emanated from the pens of those in the lower

[82] Cook argues that the College of Physicians disapproved of the attempts to gain publicity through publication. See Cook, *Trials*, p. 119. Dew concurs that it was uncommon for physicians to publish, as it was 'ungentlemanly to have one's name in print'. Dew, 'Politics of the Body', p. 11.

levels of the hierarchy.[83] Certainly many prominent physicians did not publish at all on medicine during their lifetimes (eg. Charles Scarburgh), while others published very little (eg. George Ent). However, contrary to Frank's argument there are others whose prolific publications throughout their medical careers (for example, Thomas Willis and Richard Lower) suggest that this was not a substantial deterrent. Perhaps one of the main motivations to publish was to attract patrons. Charleton's solitary self-presentation, in the dedicatory epistles to patrons preceding his works, portrayed him as spending all his hours devoted to study and learning, between the distractions of his clinical work. The patron was always handsomely praised in his dedications, and lauded for having supported the advancement of understanding through the kind encouragement of Charleton's lonely task. This depiction of a dreary life would have been an appeal for further support—a wealthy patron could lay claim to a keen interest in ideas, through his support of solitary industry in another.

Thomas Belayse, the second Viscount Fauconberg (1627-1700), was one of Charleton's patrons. Though a financial, rather than an intellectual aid to his research, the author's dedication to *Natural History* portrays Fauconberg as a 'Grand Exemplar' of truth and reason, whose approval constituted verification of fact: 'whatever position hath once received the stamp of your Assent and Approbation' will be accepted as 'Authentique and Current', by even 'the most scrupulous'.[84] Yet the dedications always proclaimed the author's reluctance to publish, a widespread device used to avoid seeming too ready to promote oneself. Charleton's dedication to an unnamed patron, in the anonymous *Ephesian Matron* (1668), declared that he was powerless to obey, 'even the strongest inclinations of my own Genius, when your commands have once told me, I must apply my self to complaisance with Yours; I have strugled [sic] hard to master my natural aversion to Writing'.[85] From an unusually prolific author this seems wonderfully ironic. It was common for the author to describe his work as the product of his patron's command. *A Brief Discourse Concerning the Different Wits of Men* was described on its title page as having been 'Written at the Request of a Gentleman, Eminent in Virtue, Learning, Fortune.'[86] The dedication always emphasised the physician's obedience to the wishes of his patron. In his prefacing letter, Charleton pleaded the inadequacy of his work under the restrictive deadline he had been given.[87]

Professional physicians were financially dependent upon clients and patrons. Hence it was in their interests to imitate and flatter genteel interests and

[83] Frank, *Virtuosi*, p. 65.

[84] Charleton, *Natural History*, p. ii. Fauconberg was a prominent supporter of Cromwell, but converted to royalism at the Restoration, and rose rapidly to privileged status.

[85] W. Charleton, 'A Letter Concerning the Ephesian Matron: To a Person of Honour' in *The Ephesian and Cimmerian Matrons: Two Notable Examples of the Power of Love and Wit*, London, 1668. He claimed to have read the story in Greek, Latin, German and French.

[86] This note was included on the title page of the 1669 edition, which still did not list Charleton as the author, although it did add 'And now Published with Consent of the Author.'

[87] 'You are obliged to look upon it as a Specimen rather of my Obedience, than of my Learning'.

preoccupations. The likely contenders for the role of patron were wealthy educated men, and it was entirely logical that one way for physicians to win their favour was to be seen to embody the classical scholarly tradition in which the gentry had been educated. Physicians thus imitated the gentry in order to attract patronage. Publication was one method by which a man in Charleton's position could attract attention and potential patronage, but this was a delicate game. The published author had to appear reluctant to have his words in print, and therefore often pretended to have been forced into publication by high-ranking patrons. This protestation was to be found in the 'Dedication' that preceded the body of the text. It was always couched in the most flattering possible terms so that the patron looked like a true lover of knowledge, and the physician merely his or her humble servant. Dedications were also a method of publicly thanking a generous patron, and therefore paying back in praise something of the debt the physician owed in financial terms.

Charleton's prefaces often lamented unfair censure, and as a young physician he depicted his audience as unreasonably critical. I have found few sources in which his views were directly refuted, though Charleton presented himself as needing to defend himself continually against opponents. The English writer Henry Layton took issue with Charleton's *Immortality*,[88] providing a summary of the theological implications of the physician's case and disputing it at several points. He concluded that Charleton's reasoning had not been successful in demonstrating the soul's immortality. *Oeconomia Animalis* was criticised by Antony Deusing, who disagreed with Charleton's claims that the workings of the mind were mysterious.[89] The controversy was described by Peter Alvares.[90] It is interesting, given Charleton's eclecticism, that Alvares praised the Englishman's clear thought against Deusing's credulous pursuit of every existing theory. Alvares was careful to point out that ancient, as well as contemporary, authorities supported Charleton's argument.

In speculating about the physician's audience as revealed through his own words, we must recognise the strategies by which he may have presented to his audience. Perhaps his argumentative presentation was simply a manifestation of the scholastic method, in which arguments were presented, possible protestations suggested and then answered, by the author. Or his tendency to make his work sound deeply controversial might have been intended to boost sales by

[88] Henry Layton, 'Observations upon Dr. Charltons treatise intituled, The immortality of the humane soul', in *Observations upon Mr. Wadsworth's book of the souls immortality*, London, [s.n.], 1670, pp. 201-15.

[89] Antony Deusing, *Oeconomia Corporis animalis: ac speciatium de ortu humanæ dissertatio . . . Opposita conceptibus . . . D. Gualt. Charletonis*, Groningæ, 1661.

[90] Peter Alvares, 'Epistola Petri Alvares De Charletonis, & Deusingii Controversia' appeared in Benedict Blottesandaeus, *Deusingius Heautontimorumenos, sive epistolæ selectæ eruditorum, quæ immaturis Antonii Deusingii*, Hamburgi, Typis N. Molybdii, 1661, pp. 1-8. Deusing seems to have been convinced that the workings of the mind could be revealed, and cited a range of theories to demonstrate this.

implying that any person up to date with contemporary public debate ought to be familiar with his writings. Charleton's preface to *Deliramenta* referred to the 'malevolent, severe and uncharitable' censures he received for his 'precedent exercise'. The autobiographical character Athanasius in *Immortality* lamented that since the publication of a recent work (*Physiologia*), he had been 'embroil'd in as many troubles and distractions, as malice, persecution, and sharp adversity could accumulate upon me'.[91] As his living came from medical practice we can assume that the medical community, if not his patients, were displeased with his publication.[92] Athanasius remarked bitterly that he had been 'wholly depriv'd of all the chief endearments of life',[93] and that 'for almost these two last years, I have been continually toss'd up and down by a Tempest of Calamity'.[94] This 'storm' of misfortune raised by 'the cruelty and rage of my Enemies', may have been his exclusion from the College of Physicians.

Charleton's claims regarding the detractions of his enemies allowed him to assert the selflessness of his own enterprise. By repeatedly claiming that he had gained no benefit from publication, he reaffirmed his pure motivations, and therefore promoted his intentions as non-materialistic. We might consider these complaints of censure as exaggerated. In 1657 he declared, defensively, that he had written his earlier works (*Spiritus*, *Ternary*, *Deliramenta*, *Darknes*) 'out of pure devotion to knowledge; and commendable ambition to be serviceable to the Commonwealth of Learning in proportion to my talent'.[95] He did not suggest the importance of his publications for the practical improvement of health. Charleton's contribution seems to have been pitched at the world of learned scholarship. Their utilitarian benefit was more removed: advancement of the body of learning rather than improvement of individual cases. To suggest the practical benefit of the medical principles he outlined might be regarded as clinical self-promotion.

He sulked that, as a result of unfair criticism, he would thereafter be 'employing all my Collections, Observations, and Speculations Philosophicall, only to the furnishing the little Cabinet of my own brain.'[96] His concerns were explicitly non-medical, but showed a preoccupation with elevated intellectual and scholarly status. Charleton tended not to portray himself as a participant in a knowledge-producing community, but rather as a solitary contributor, largely in a

[91] Charleton, *Immortality*, p. 11.

[92] Perhaps the reason for their displeasure was the fact that, as a physician, he had published inappropriately by straying into other fields?

[93] Charleton, *Immortality*, p. 11.

[94] 'My Anchors are lost, my Vessell leaks, the Winds hurry it from the land, and I hourly expect to sink down-right.' Charleton, *Immortality*, p. 12. The metaphor of fortune as an ocean on which the individual was tossed was similarly invoked by William Harvey, in correspondence with George Ent. Gweneth Whitteridge includes the quotation in 'William Harvey: A Royalist and No Parliamentarian', in C. Webster (ed.), *The Intellectual Revolution of the Seventeenth Century*, London, Routledge & Kegan Paul, 1974.

[95] Charleton, *Immortality*, p. 10.

[96] Charleton, *Immortality*, p. 10.

scholarly context. He claimed, 'that *Sapere domi*, to endeavour the acquisition of Science in private, ought to be the principall scope of a Wise man'. He would thereafter work only for his own intellectual benefit.[97] This posture was perhaps intended to reiterate Charleton's high scholarly intentions in publication, defining his endeavour as the selfless attempt to increase knowledge, and reducing the likelihood that his work might be seen as the self-promoting device of a practitioner.[98] What he stressed instead was a sombre, bookish, melancholy identity produced by university education and deep contemplation. His texts promoted Charleton's possession of characteristics desirable in a professional physician—such as a scholarly, serious temperament, which was itself a promotion of his authority. Yet he would avoid the inadmissible step of promoting his actual services as a practising physician. Charleton's depiction of the dangers and detractions consequent upon publication raises the question of why he chose to publish so prolifically, across such diverse subjects. The frequency of his publications suggests a sizeable audience of well-educated individuals interested in theories about the body's internal operations. This audience was probably not medically trained, but was certainly familiar with other works in the field. His declared motivation was a desire for the improvement of himself and his society. The status war between physicians, apothecaries and surgeons would have affected the identity of writers such as Charleton in ways very different from the determinants of the identity of natural philosophers. Recognition of Charleton's medical status is crucial to any interpretation of his life and work.

ANONYMITY AS IDENTITY

One resort for the unfairly censured author was anonymity. An author might publish anonymously to allow his work to be assessed without prior expectations about the author and his status. Charleton reported exactly this reasoning for publications that appeared without his name. His anonymous works included: *Immortality of the Human Soul* (1657 edition); *Ephesian Matron* (1668 edition); *Two Discourses* (1669); *Natural History of the Passions* (1674); *Plato, His Apology of Scocrates* (1675); *The Harmony of Natural and Divine Laws* (1682) and *Life of Marcellus* (1684). These works provide some clues as to his perception of the dangers of authorship. Some were subsequently printed under his name (*Immortality, Ephesian Matron, Two Discourses* [twice]). Charleton's

[97] However, as *Immortality* was composed a mere two years after the publication of *Physiologia*, and was followed by twenty-two subsequent publications, one can assume that Charleton's faith was restored.

[98] This reflected the moral principles expressed in *Epicurus' Morals*. Those who 'have by force of Reason repressed their Natural Ambition and vain Affectation of Popularity; or having learned, by their own and costly Experience, the certain troubles, and uncertain duration of Grandure, have withdrawn themselves from the storm, that frequently threaten men of Publick Charges . . . good reason is there, that these should esteem the quiet of a Private condition'. Charleton, *Epicurus' Morals*, p. 61.

prefaces to these works depicted his audience as excessively critical, and not to be trusted to treat his work civilly. The preface to *The Harmony of Natural and Divine Positive Laws* depicted the author's anonymity as a defence against immoderate detraction, for 'when we remain ignorant upon whom to fix the blame of our Frustration, commonly that ignorance turns to our advantage, by mitigating our Resentments, and keeping our displeasure from transgressing the limits of Humanity and Moderation.' If the author showed appropriate modesty (that is, if he was 'duly conscious to himself of Human Frailty and diffident of his own Learning and judgement', he ought to be commended. This depiction of Charleton's motivation for publication reiterated his fundamental modesty.[99]

It seems anonymity could be a defensive tactic; an anonymous text might gain acceptance possibly denied through association with the author's title. If favourably received, the author could reveal himself. *Immortality* was first published when Charleton's reputation might have been precarious due to his *Physiologia*. He may have wished for the later book to be taken on its own merits, rather than immediately critiqued by association with his authorship. The book was later published under his name, when his reputation had improved. None of the physician's medical works were published anonymously. This might mean that his deviations (into moral philosophy, translation, etc) were deemed inappropriate (by the College) to his professional role. The anonymous works were largely moral tracts, associated with Epicurean philosophy. The latter's atheism was widely criticised in late seventeenth-century English society, and Charleton had been censured for his earlier association with mortalism in *Physiologia*. The anonymity of these publications suggests that Charleton thought they might have detracted from his authority as a physician. *Natural History of the Passions* (1674), *Plato, His Apology of Socrates* (1675) and his translation of Plutarch's *Life of Marcellus* (1684)[100] were perhaps published anonymously because they were non-medical works, published in his first decade with the College—at this time he may have wished to be associated exclusively with the College's enterprise.

Charleton's complaints about audience criticism, and constant lamentations about the difficulty of compiling his works, combined with his protestations that publication was forced upon him by his patrons, were all part of a larger strategy through which he asserted a desirable professional character. By portraying his motivation as completely averse to public acclaim, and by asserting the critical reception of his works as the result of sheer envy, he constructed an identity both selfless and relevant. By promoting his modesty and arguing that his interest was only in learning, he was able to annex the serious identity that would

[99] In the preface to the anonymous *Harmony of Natural and Divine Positive Laws*, the publisher (Walter Kettilby) depicted the author writing 'to no other end, but to confirm his Faith by inquiring into the Reasonableness and Purity of it, and to augment his Piety toward God.' His posture was 'worthy a Philosopher and a Christian' as, surely, was his publication.

[100] Charleton's translation of 'The Life of Marcellus', was published anonymously in *Plutarch's Lives*, ed. J. Dryden, London, 1684, vol. 2, pp. 401-68.

have benefited him as a professional physician whose goal was better understanding and not personal acclaim. Thus rather ironically, public declarations of one's modesty could operate as a means of securing greater credibility.[101]

It is clear that, despite his protestations, Charleton very much wanted to be in print. It was a crucial way of attracting patrons, and of paying tribute to existing patrons. As previously suggested, the expenses of publication may have been one reason for his decaying fortunes—late in life he continued to publish despite the fact that his patrons had died off.

LATIN & THE VERNACULAR

The language of Charleton's publications also needs closer discussion, for the language in which a text was written reveals something of its audience and aims. Robert Frank claims that until the late seventeenth century, anatomical works were generally published in the vernacular. The lack of anatomical innovation in England resulted in a need for compendia of Continental innovation, designed to brief English readers on the discoveries of their European neighbours.[102] This appears to have been the kind of program to which Charleton contributed. When the English had something to communicate to the wider scientific community, it was expressed in Latin.[103] Many early English scholarly medical publications were in Latin. Glisson, Ent and Wharton published exclusively in the learned language,[104] as did the prominent physician Thomas Willis. These were innovative works, not simply compendia of existing knowledge. Charleton's studies of specific diseases, in Latin, were read by continental audiences (as illustrated in Oldenburg's correspondence regarding *De Scorbuto*). These works conformed to the traditional genre of medical writings, which covered specific clinical issues.

The use of learned language indicated that the audience of the work was educated and specialised. Unlike some colleagues, Charleton was interested in vernacular publication. The fact that the majority of his published works were in English has been taken as evidence of his intention to write 'for the benefit of a wider and possibly non-medical audience'.[105] However, it seems that the language in which Charleton wrote may have been more dependent upon subject matter than simply upon his audience. Of his twenty-six publications, only nine were in Latin. His Latin oeuvre included all of the works on specific diseases, including his study of scurvy, his tract on uterine rheumatism, and his early

[101] See also Shapin, *Social History of Truth*.
[102] R. G. Frank, 'Viewing the Body: Reframing man and disease in Commonwealth and Restoration England', in W. G. Marshall (ed.) *The Restoration Mind*, Newark, University of Delaware Press, 1997.
[103] Frank, 'Viewing the Body', p. 76.
[104] The English translation of Glisson's *De Rachitide* was unauthorised.
[105] Webster, *Great Instauration*, p. 272.

work on pathological causes. Expressed in the learned language, they entered a different domain, of learned discourse. Note the contrast with Thomas Willis, who included extensive clinical discussion, but published only in Latin. Only Charleton's works on less explicitly clinical concerns were released in English. His studies of disease also examined traditional questions (regarding causes and signs of disease) in ways that did not characterise his English compilations.[106] It seems the works more directly associated with medical practice were differentiated from the broader medical philosophy by the language in which they were composed.

Charleton commenced his literary career with a translation of Helmont (*Ternary* and *Deliramenta*, 1650), followed by the translation of Gassendi (*Physiologia*, 1654). He later published vernacular versions from the Greek writings of Diogenes Laertius, as *Epicurus's Morals* (1656, 1670). He also retold the story of the *Ephesian Matron* from Petronius (1655), and Plutarch's *Life of Marcellus* (1684, 1688). Finally, his *Three Anatomic Lectures* offered an overview of Borelli's *De Motu Animalium*. These works gave the common reader access to 'forreigne' tracts.[107] Making public inaccessible writings seems to have been one of Charleton's objectives. He claimed that one of his motives was the 'discharge of that duty, which, as a Scholer, I owe unto the Publique, in the Explanation, Improvement, and Communication of obscure Truths'.[108] Thus he presented the task of translation as one of his responsibilities as a disseminator of otherwise concealed knowledges. I have suggested in an earlier chapter that translations may have allowed Charleton to investigate some of the new theories in circulation, without claiming to adopt the ideas. He depicted himself following in the 'footsteps of those Worthies, who have infinitely both enriched and ennobled our Language, by admitting and naturalizing thousands of forraigne Words, providently brought home from the Greek, Roman, Italian, and French Oratories.'[109]

Sprat's *History of the Royal Society* advocated clarity of expression and use of the vernacular: 'the language of Artizans, Countreymen, and Merchants, before that, of Wits, or Scholars.'[110] Perhaps as a consequence of Sprat's directive, Charleton was often defensive about his use of Latin. In the early seventeenth century translators often faced criticism for making learning too common, or for infringing upon learned institutions (such as the College of Physicians).[111] Charleton was quick to note that he had been forced by the complexity of the subject matter to compose some of his texts in Latin. The nature

[106] This fuels Frank's theory that publications in the vernacular were often intended for the a common audience to catch up on contemporary developments, while Latin works were designed to convey new research to a Continental readership.

[107] That Charleton's works in both English and Latin were available overseas is demonstrated in existing correspondence (between Sluse and Oldenburg and between Pratt and Hobbes).

[108] Charleton, 'To the reader', *Deliramenta*, p. ii.

[109] Charleton, 'To the reader', *Deliramenta*, p. iii.

[110] Sprat, *History*, p. 113.

[111] C. Hill, 'Science in Seventeenth-Century London', *The Listener*, vol. 67, 1962, p. 943.

of the subject was such that even if it *could* be expressed in the vernacular, *Deliramenta Catarrhi* could only have been understood by 'the leading part of learning, and chiefly those, who have more then looked into the mysterious Temples of Aesculapius'. Consequently 'no English, how plain soever, could have driven their ample meaning into those skulls which are so thick, as not to admit it in Latine.'[112] Charleton here claimed a special status for medical knowledge as integral to medical training: a kind of knowledge which was not be accessible to lay folk, but which required years of training to give it meaning. Contrary to Sprat's programme, this also contradicted his own statement that English language was as powerful and versatile as Latin.

Various contemporaries complained that the vernacular was incapable of providing counterparts to some Latin expressions. In 1644 Digby claimed to use Latin technical terms because of the 'scarcity of our language'.[113] Charleton's 1680 anatomy lectures at the College of Physicians occasionally resorted to Latin because 'the nature and quality of the Subjects treated of . . . cannot be fully expressed in our yet imperfect Language'.[114] Where English terminology was not sophisticated enough to convey technical medical concepts, he invoked Latinisms. The use of Latin terms was thus necessary in translation, and one could not expect to get by without familiarity with Latin.[115]

Charleton claimed that he had received considerable criticism for the translations of *Ternary* and *Deliramenta* from Helmont. *Deliramenta*'s preface referred to the 'malevolent, severe and uncharitable' censures he received for his 'precedent exercise', and the prefaces to both texts defended the translation from Latin to the vernacular. Charleton acknowledged 'the Popular opinion', that translations were 'but slender Adumbrations, or pale Counterfeits' of the original texts.[116] But he refuted the implication that the fault lay with the English language. The latter was surely capable of expressing ideas thought to be explicable only through Latin.[117] From 'the venerable Majesty of our Mother Tongue', he announced, 'may be spun as fine and fit a garment, for the most spruce

[112] Charleton, 'To the reader', *Deliramenta*, p. v.

[113] Digby, *Two Treatises*. Vickers notes the late seventeenth century conviction that modern languages such as English were 'unstable', while Latin was 'unchanging and universal'. Vickers, 'The Royal Society and English Prose Style', pp. 27-28. The necessity of Latin within English writing was reinforced by theologian Clement Barksdale in his verses prefacing Charleton's *Darknes*: 'He, that will our Learnedst Writers scan, Must be both Latinist and Grecian.'

[114] Charleton, 'Preface', *Enquiries*, p. xxx. His language, he warned, would be 'harsh and ungrateful . . . to ears unaccustomed to any but their Mother tongue'.

[115] This declaration echoed Thomas Browne: 'I confess the quality of the Subject will sometimes carry us into expressions beyond meer English apprehensions.' Browne implied that his work was not addressed 'unto the people . . . but unto the knowing and leading part of Learning.' It would thus be understood only by those already knowledgeable. T. Browne, 'Pseudoxia Epidemica, or Enquiries into very many received tenents and commonly presumed truthes' (1646), *The Works of Sir Thomas Browne*, ed. Charles Sayle, London, G. Richards, 1907, vol. 1, pp. 117-8.

[116] Charleton, 'Epistle Dedicatory', *Ternary*, p. ix.

[117] Vernacular publications would 'stagger that Partiall Axiome of some Schoolemen, that the Latin is the most symphoniacall and Concordant Language of the Rationall Soule.' Charleton, *Ternary*, p. x.

Conceptions of the Minde to appeare in publick in, as out of any other in the World'.[118] Thus he defended the addition of Latin technical words in his vernacular publications, and when he translated *from* Latin he affirmed the adequacy of the English language. The apparent contradiction between these views is explicable in terms of the different purposes they served. It seems the physician had no fixed view on the necessity of either language, but alternated as it suited his purpose.

Charleton popularised Continental ideas in his vernacular publications, which compiled others' work in a form accessible to an educated lay audience. He claimed the translation of significant works to a broader readership as one of his scholarly responsibilities, in keeping with Royal Society ideals of clarity and accessibility (as expressed by Sprat). This self-description promoted his natural philosophical prowess. Yet he was not ultimately committed to vernacular communication, as he published later in Latin to suit a trained medical audience. The language of his publications, as Frank argues, may have been linked to the nature of his topics. The Latin works, while essentially traditional in their treatment, discussed specifically medical questions. The vernacular works, on the other hand, were often compendia of recent hypotheses; overviews to introduce readers to the field. It should be noted that he only followed this directive to write in 'the language of countreymen' in relation to certain subjects. Charleton's specialised medical writings were not part of any popularising oeuvre. This would have left him open to criticism for appearing to advertise his services, and for making specialised physicians' knowledge public. This was part of the protection and demarcation of physicians' professional status from the incursions of other groups. Latin was also part of the distinctive arena of knowledge which underlay the identity of physicians. In this sense the reasons for publishing in Latin were antithetical to the linguistic democracy encouraged by Sprat.

It seems that subject matter was more of a determinant of the language of publication than was audience, and that medical subject matter may have been published in Latin where it related to topics on which Charleton might be perceived to be advertising his services were his discussions published in the vernacular. It would be simplistic to say that he wrote either for a learned or for a lay audience. He wrote for both, and his audience can best be deduced from specific works, rather than from his *oeuvre* as a whole. Charleton does appear to have presented himself within the parameters specified by Sprat, in some of his English compendia. Thus vernacular works linked him with natural philosophy, more explicitly than with medicine.

This chapter has examined how Charleton situated himself within contemporary intellectual milieux, and explored the relationship between the identity he constructed and the kinds of knowledge that he invoked. The identity that Charleton constructed through his writings is suggestive of the kind of

[118] Charleton, 'Epistle Dedicatory', *Ternary*, p. ix.

authority he hoped to annex for himself and his texts. I began this discussion with the claim that Charleton has been mis-represented in existing historiography, which has focused almost exclusively on the substantive content of Charleton's works, and not on how he portrayed his own intentions. Shapin, Dear and others argue that the newness of natural philosophy as a discipline, and the distinctive theological and political context of the time necessitated an innovative gambit for authority. The community of natural philosophers had to generate a profile of themselves which legitimated their claims to knowledge, and which subjugated existing forms of authority to their greater knowledge-making powers. Yet such a construct obscures the requirements of a professional physician such as Walter Charleton, whose income derived from convincing his patients that he was well-versed in classical knowledge. In contrast with Robert Frank's claims that authority for physicians was associated with experimental activities, Charleton constructed an identity reliant upon meditation upon textual sources—far from that of an active, experimental investigator.

Many scholars, including Shapin and Dear, have argued that seventeenth-century natural philosophy shifted from a scholarly textual to an active experimental basis for knowledge. Although we know that Charleton was an active participant in experimental life, he did not emphasise it as an aspect of his public identity. Indeed I argue that to do so would have jeopardised his professional standing. As a physician, he stood to gain little from experimental status, but a considerable amount from sound philosophical credentials. Experiment would have been associated by many with the manual practices of mere surgeons, which offered very good reasons *not* to be identified with manual practices and anatomical manipulations, as they were carried out in the Royal Society. This may aid our understanding of his scholarly attitudes, and his determined eclecticism.

For centuries the authority of physicians rested upon their classical learning. While in Shapin's model the identity of 'virtuoso' natural philosopher relied upon the construction of a novel form of authority, physicians had no reason to construct new kinds of authority. Charleton's resilient adherence to traditional scholarly/speculative identity demonstrates that despite his involvement in the very practices that the scientific revolution had been thought by some to consist of, he nevertheless saw sufficient benefit in presenting himself to his *audience* as traditionalist. While the presence of tremendous intellectual ferment in this era is unquestionable, innovation for its own sake was evidently not imperative to his survival as a physician.

Recent historiography has also argued that the seventeenth century witnessed a shift from private to public sites for the legitimation of knowledge. Yet Charleton constructed a melancholic scholarly, passive and contemplative identity, within which he asserted the private nature of his theories, and the individual nature of their demonstrations. I argue that this reflected his desire to retain the authority of classical knowledge, upon which his profession was founded. A speculative and scholarly characterisation implied desirable intellectual features, supporting his claims to be an impartial judge of hypotheses. Presumably this passive and meditative identity afforded him a status more desirable than that of the active experimentalist. Thus Charleton situated himself within an intellectual milieu often ignored in studies of this period. Emphasis upon experimentalism has obscured the contemporary significance of speculative eclecticism.

Shapin's approach to the question of authority is challenged by Charleton's relentless eclecticism. Teasing out how an identity was negotiated is more revealing than assuming that it conformed to a model with fixed principles.[119] To comprehend such a truly eclectic figure, we need a more elastic understanding of the negotiation of authority and identity than has been provided by recent historiography. I suggest that Charleton's profound eclecticism resulted partly from his desire to balance the authority of a physician with that of a natural philosopher. While many around him advocated the rejection of classical learning, Charleton's works continually attempted the reconciliation of old knowledges with new. In this sense we can see him as a professional physician circumscribed by the traditional expectation of his patients, yet fascinated by the atmosphere of innovation and experiment.

Charleton's case suggests the possibility of an authoritative identity for a physician outside the experimentalist agenda. His eclecticism, far from being a stepping stone on his path toward rejection of established authority, was a means by which he was able to combine modern and ancient knowledges, and to retain the traditional authority of a physician in a world of innovation. Historians have failed to comprehend how professional physicians might differ in their practices and therefore in their identity from natural philosophers. Further research is required on the construction of the identity and authority of the physician-philosopher, and the need of men like Charleton to distance themselves from manual/experimental practices.

When dealing with the life and works of a physician whose livelihood derived from attracting and treating patients, we see the 'ferment of ideas' not only as they were circulated within the learned community, but also the relationship between the holder of ideas and the demands of the audience upon which he depended. Patients played a determining role in Charleton's self-presentation, and their expectations in turn were shaped by centuries of tradition. Its is useful to observe the self-construction of a keen/innovative follower of diverse aspects of the New Philosophy, through his attempts to maintain an identity acceptable to patients who would have had no interest in (or who might have held a strong desire to avoid) experimental activity. William Harvey complained to Aubrey that, after his treatise on the circulation of blood, he 'fell mightily in his practice' and 'twas beleeved by the vulgar that he was crack-brained'.[120]

[119] Shapin's *Social History of Truth* portrays behaviour within the natural philosophical community as determined by commonly understood rules and their 'counter-maxims'. This over-simplifies a situation which was not rigid in the first place. Shapin's study neglects the presence of important exchanges outside his set framework of experimentalism. Charleton seems to be evidence of the failings of a 'model' approach to the complexities of history.

[120] Aubrey, *Brief Lives*, vol. 1, p. 300.

Because physicians competed with a range of other medical approaches, they could not afford to mystify their clients with new theories and treatments.[121] A certain mode of conduct was appropriate to physicians, and Charleton was answerable to the expectations attendant upon his professional role.

This discussion has revealed several outcomes for our understanding of Walter Charleton and his identity as presented in his published works. He presented his role as that of revealing knowledge to a wider audience, and this is certainly borne out by his publication of compendia of foreign ideas. These were presented in a manner consistent with virtuoso ideals, both in vernacular publication and the aim of clarification. However, having said this, all of Charleton's Latin works were medical (or at least taxonomic, as in the case of *Onomasticon Zoicon*), rather than philosophical or theological in emphasis. It seems these medical texts were written exclusively for a learned audience. So while Charleton was a populariser in many respects, he maintained the important language and audience distinction between medical and non-medical texts. Specialised medical knowledge was published only in Latin, and in these cases was presented according to traditional principles of order and hierarchy.

Charleton published no Latin works, indeed no medical works, anonymously. His anonymity was presented as a shield against detractors and so, like much of his work, aided his claim to be interested in knowledge for its own sake. This reinforced his status as a free agent, impartial and not bound by doctrines or authorities. In the anonymous works he explored theological issues, or offered versions of classical tales. These were not authoritative texts, but rather works that expressed his prolific range of interests. Although he published widely, we can see his use of different modes in the variety of genres into which he strayed. Though conversant in several languages, Charleton promoted the adequacy of the vernacular when he stood to benefit, in eyes of his audience, from doing so. However, he was capable of an expedient defence of whichever language he was using at the time.

This chapter has considered a range of Charleton's writings, private and public, to establish his divergence from the model. In subsequent chapters I focus on three works to explore in more detail the public construction of his identity.

[121] Wear, 'Continuity and Union', p. 302.

CHAPTER IV

'THE ANIMAL OECONOMY': *NATURAL HISTORY* (1659) IN THE CONTEXT OF ENGLISH PHYSIOLOGY

The previous chapter examined the integral relationship between self-presentation and epistemology, and set the framework for this and subsequent chapters, which consider questions of self-presentation and epistemology in relation to specific medical works by Charleton. I have already outlined some issues relating to generic self-presentation, including how the appropriate character for physicians was shaped within their education, textual traditions and modes of practice. I have identified the contemporary trends toward eclecticism that influenced Charleton's medical writings, and suggested that his publicly presented identity as a physician might be influenced by questions of etiquette in relation to authorship and audience. This chapter analyses the first of three specific medical works, in terms of the epistemological and historiographical issues highlighted in the introductory chapters.

Natural History of Nutrition, Life and Voluntary Motion was published by Henry Herringman in 1659. It was released in Latin almost simultaneously. Both editions were dedicated to Thomas Viscount Fauconberg and Sir George Ent. The numerous editions of *Natural History* illustrate that it was popular, in both languages. The book shows a departure from the matter theory and moral philosophy on which he had previously published, and thus also signals the point at which historians of his work seem to lose interest. My discussion aims to illustrate the links between the determinants of identity and the kinds of epistemological and methodological devices upon which the physician drew. I examine influences upon Charleton's work, and outline the context of physiological thought in seventeenth-century discourse in order to situate the text on the 'animal oeconomy' (*oeconomia animalis*, the name given to the living system) in its milieu, and to highlight his negotiation of contemporary theories. The *oeconomia animalis* as a genre of physiological investigation had its roots in Dutch medical literature. Charleton's text modified and adapted the Continental genre, within the specific expectations of an English audience, and according to the English approach to living systems.

The chapter explores Charleton's integration of traditional knowledge of the animal oeconomy with more recent anatomical discoveries. Clearly he was interested in combining ancient authority with contemporary developments, and the manner in which he supported the legitimacy of new theories through reference to older authority structures reveals much about the status of physicians and their authority base. I analyse how he negotiated existing and innovative theories through language, and the extent to which he presented his statements as authoritative. In addition, it is important to look at whether Charleton's epistemology and methodology presented a coherent philosophical program, the plausibility of which he wished to prove. I discuss the epistemological status of varieties of demonstration (empiricism, reason, analogy, textual authority). My discussion does not explore all of Charleton's physiological discussions, but focuses on those that I believe offer the greatest insight into his thought and self-presentation.

In 1653, when he composed this work Charleton possessed a medical degree, some experience in the King's service and as assistant to Mayerne. It was therefore prior to his practical involvement in medical and anatomical experiment. This first original medical text makes an important point of comparison with Charleton's later medical lectures, because between it and those later texts his practical experience was transformed.

THE CONTEXT OF *NATURAL HISTORY*: MATERIALISM & PHYSIOLOGY

In order to understand Charleton's approach to physiological issues, it is useful to consider the historical trajectory of physiological thought in the context of broader intellectual trends. According to Luyendijk-Elshout, Charleton initiated the title '*Oeconomia Animalis*' for physiological studies.[1] 'Animal oeconomies' were published predominantly in Latin, rather than English, but even Latin editions were rarely printed on Charleton's side of the channel.[2] His work was thus part of a Continental, rather than a local, tradition.[3] The genre of animal oeconomies was largely shaped by the Dutch Cartesian school, particularly such writers as Leiden physician Cornelius van Hogehlande, and Theodoor Craanen.[4] The emergence of the *oeconomia animalis* genre has been

[1] A. M. Luyendijk-Elshout, 'Oeconomia animalis, pores and particles', in T. H. Lunsingh Scheurleer & G. H. M. Posthumus Meyjes (eds), *Leiden University in the Seventeenth Century: An Exchange of Ideas*, Leiden, Brill, 1975. However, Folter and others argue for C. Hogehlande (1590-1651) as the initiator with *Cogitationes*, Leiden, 1646. See R. J. de Folter, 'A newly discovered *Oeconomia Animalis*, by Pieter Muis of Rotterdam (1645-1721)', *Janus*, vol. 65, 1978, p. 184.

[2] With the exception of William Cockburn's *Oeconomia corporis animalis*, London, 1695.

[3] See the various Dutch examples, including: Benjamin van Broeckhuysen, *Oeconomia corporis animalis sive cogitationes succinctae de Mente*, Nijmegen, 1672, Amsterdam & Leiden, 1683; Theodoor Craanen, *Oeconomia Animalis ad circulationem sanguinis breviter delineata*, Gouda, 1685, Amsterdam, 1703; J. Bohn, *Circulus Anatomico-Physiologicus, seu Oeconomia Corporis Animalis*, Leipzig, 1686 and C. Bontekoe, *Metaphysica, et liber singularis de motu, nec non ejusdem Oeconomia Animalis*, Leiden, 1688, 1692.

[4] Hogehlande attempted to synthesise Cartesian ideas into an account of the animal oeconomy, calling for the explanation of natural motions and operations by purely mechanical explanation. Hogehlande, *Cogitationes*, pp. 135-7. The work influenced Willis. See R. Sloan, 'Descartes, the sceptics, and the rejection of vitalism in seventeenth-century physiology', *Studies in the History & Philosophy of Science*, vol. 8, no. 1, 1977, pp. 21-2. By the mid- to late century Dutch scholarly and

attributed to the rise of the mechanist world view, as it often entailed the application of particulate or corpuscular theories to the body's fluids. Folter notes the common characteristics of the genre as including: a concern with physiology; an origin in mechanistic philosophy, with influence from both iatrochemical and mechanical ideas; centrality of particles, ducts, pores and motion; prevalence during the later seventeenth century; and publication in the form of short treatises.[5] Charleton's text coincided with several of these criteria. Folter cites Charleton's animal oeconomy as 'an attempt to explain physiological processes mechanically', though he attributes this character to all animal oeconomies as a generic classification.[6] However, Charleton's approach was not as dogmatic as that of the Dutch 'oeconomists'.

The early seventeenth century saw a growth of interest in matter theories. Paracelsus' revisions of peripatetic elements, and the revival of ancient atomism contributed to this.[7] While Aristotelian physics posited the presence within the body of innate 'real qualities' and 'substantial forms',[8] newer theories suggested quite different characterisations of matter and its properties. Charleton has often been thought to have participated in this revolution. Brown describes *Natural History* as 'an epitome of current English thinking on the "animal oeconomy" '.[9] Physiological researches in late seventeenth-century England revolved around the traditional questions of reproduction, locomotion, nutrition, the origin and nature of bodily heat, respiration, the motion of the heart, and the nature and composition of the blood. New technologies revealed the agency of numerous minute structures in bodily processes, in addition to the elements of bodily structure identified by Galenic physiology. The late seventeenth century therefore witnessed a shift in emphasis, from occult and fluid properties,

---

experimental research on the animal oeconomy had incorporated the new strands of Cartesian mechanism, in the rational and experimental research at the University of Leiden. Lindeboom claims that it was at Leiden University, under the influence of Franciscus Sylvius (1614-1672), that physiological experiments were first used to test hypotheses. See Gerrit Lindeboom, 'Dog and Frog: Physiological experiments at Leiden during the seventeenth century', in Lunsingh Scheurleer & Posthumus Meyjes (eds), *Leiden University*. Among those who experimented upon the typical physiological processes of circulation, respiration, digestion, and reproduction, was Johannes Walaeus, whose work was read by Charleton. Another substantial innovator was Reinier de Graaf (1641-1673), whose experiments on the pancreas and on reproduction were influential throughout Europe. Jan Swammerdam (1637-1680) performed experiments on respiration and muscle contraction. Lindeboom, 'Dog and frog', pp. 282-8. Henricius Regius' *Fundamenta physices* (1646) was among the first to declare open support for Cartesianism.

5 Folter, 'A newly discovered *Oeconomia Animalis*', p. 188.

6 Folter, 'A newly discovered *Oeconomia Animalis*', p. 184.

7 In addition, the discovery of Lucretius' *De rerum natura* in 1417, and increased attention to Laertius and Epicurus, combined with new interest in Plato's geometrical matter theory. M. Boas, 'The establishment of the mechanical philosophy', *Osiris*, vol. 10, 1952, p. 423.

8 These characteristics, attached to matter or substance, resulted from sense perception which was thought to 'penetrate to the ultimate reality of matter'. Secondary qualities, Boas observes, were believed 'to be real, innate and intrinsic in bodies.' Boas, 'Mechanical philosophy', pp. 414-7.

9 Brown, *Animal Oeconomy*, p. 68.

to the systems of fibres through which fluids moved, and even the nature of the fibres themselves.[10] Charleton's text illustrates the inclusion of fibrous agency in physiological function.

The civil war brought the end of various kinds of restriction and censorship on printed material in England, and the 1650s saw an influx of alternatives to traditional academic philosophy.[11] These included translations of Paracelsus and Helmont (aided significantly by Charleton himself), and English editions of the materialist philosophies of Descartes and Gassendi. Cartesian physics influenced various English natural philosophers, including Joseph Glanvill, Henry Power and John Mayow.[12] However, English physiology was founded on Harveian methodological principles, and empirical researches tended to encourage retention of the idea of active faculties within matter, and led away from stricter Cartesian mechanism. The product of this English combination of Continental theories and empiricism was a kind of hybrid, 'a mechanical philosophy in which it was accepted that the fundamental particles of matter could have special chemical properties'.[13]

Interest in the animal oeconomy gathered toward the end of the sixteenth century, and the mid-seventeenth saw the emergence of an explicit focus upon the study of life-preserving processes. English approaches were distinguished from Continental texts by several characteristics. French natural philosophers did not engage in an experimental program with the enthusiasm of their English counterparts.[14] Many French commentators on medical topics took their inspiration from rational, rather than empirical, medicine. In France Cartesian mechanism gained influence during the later seventeenth century, offering new solutions to tradition questions of physiological operation.[15] Henry argues that although Continental developments and innovations should not be underestimated, specific conditions within England resulted in the generation of a singular approach to understandings of the animal oeconomy.[16] In addition to this

[10] Fournier gives the examples of Willis and Power in England. Fibre-oriented researches included those of Grew, Malpighi, Ruysch, Leeuwenhoek, Baglivi and others. See Marian Fournier, *The Fabric of Life: Microscopy in the seventeenth century*, Baltimore, Johns Hopkins University Press, 1996, p. 129.

[11] T. Brown, 'Physiology and the mechanical philosophy in mid-seventeenth century England', *Bulletin of the History of Medicine*, vol. 51, 1977, p. 36.

[12] See Joseph Glanvill, *The Vanity of Dogmatizing*, 1661; Henry Power, *Experimental Philosophy*, 1664; and John Mayow, *Tractatus Quinque Medico-Physici*, 1674.

[13] Henry, 'Scientific Revolution', p. 186.

[14] Henry, 'Scientific Revolution', p. 182.

[15] Boas, 'Mechanical philosophy', p. 452. Descartes' ideas were reworked by Hogehlande, Regius and others. Descartes' inspiration was largely philosophic but his account of the animal oeconomy devoid of the Galenic faculties proved persuasive to many who were more practically oriented. Cartesian physiology emphasised the need to search out the mechanical origins of specific operations within the body. Brown, *Animal Oeconomy*, p. 85.

[16] Henry attributes this partly to the peculiar characteristics of English theology. Henry, 'Scientific Revolution'.

distinctiveness, I argue that the unique situation of professional physicians was one factor which led them to generate an eclectic, rather than dogmatic approach to animal oeconomy.

Substantial anatomical contributions were published mid-century. Many of these, both local and Continental in origin, influenced Charleton, such as Johann Vesling's *Syntagma anatomicorum*[17] and the works of Jean Pecquet, Jean Riolan and Thomas Bartholin. Researches increased in the 1660s under the aegis of burgeoning scientific institutions.[18] In England this included the work of Thomas Wharton[19], Francis Glisson, Christopher Wren, Charles Scarburgh[20] and Thomas Willis.

Physiological understanding in the late seventeenth century supposedly responded to natural philosophy's increasing interest in the materialist characterisation of functions. Brown claims that seventeenth-century intellectuals saw the need for a revision of previously held notions of the animal oeconomy. Secondly, he identifies the prevailing belief 'shaped by the mechanical philosophy, that repairs ought to be made by borrowing from the physical sciences.' Mechanical explanation through the figure, motion and size of the minute constituents of matter could be applied to physiological actions previously attributed to perceptive qualities of matter. Such materialist explanation removed sentience from the animal oeconomy. By the early eighteenth century, 'the main task of the physiologist', claims Brown, seemed to be the application of the latest techniques of physics to the most manageable problems of the animal oeconomy.'[21] The category of 'physiologists' is questionable. Research in what we now know as physiology was conducted by a range of interested non-specialists.

Removal of the mystification surrounding the faculties of the body was an aim expressed in natural philosophical literature.[22] 'Attraction', many now felt, should be explained without resort to the inherent tendencies of the matter attracted, nor to a sympathetic relationship between attractant and attracted. The validity of

[17] This volume went through multiple reprints, and was translated into most European languages. In England it was published as *The Anatomy of the Body of Man*, London, 1653.

[18] Researchers included Marcello Malpighi, Nicolaus Steno and Reinier de Graaf.

[19] Helped by Harvey's colleagues Glisson and Ent, Wharton pursued extensive investigations of glands, published in 1656 as *Adenographia sive Glandularum totius corporis descriptio*. This was dedicated to the College, and to the individuals within it who had been so important: Baldwin Hamey, Francis Glisson, Ralph Bathurst and George Ent. Wharton and Charleton were noted for their exposition of the lacteals, by the Continental commentator Hieronymus Barbatus in 1667, whose *Sanguine et eius Sero* Charleton edited.

[20] With the help of Charleton and Wren, Scarburgh was responsible for some of the work on human musculature that informed the final lecture of Charleton's 1680 *Enquiries* at the College. Scarburgh performed splenectomies, and 'both he and Charleton investigated the application of statics to muscular motion.' Frank, *Oxford Physiologists*, p. 24.

[21] Brown, *Animal Oeconomy*, p. 1.

[22] Boyle devoted much effort to the generation of alternatives to peripatetic explanation through occult qualities, as had Descartes. The works of both men were increasingly prominent in natural philosophy throughout the later seventeenth century (their influence is clear in Charleton's 1680 lectures).

materialist physiology was debated around certain key questions: reproduction, nutrition, muscular motion (whether communicated from the soul or initiated in the muscular fibres), cardiac motion and respiration.[23] The central question was the origin of sentience, and the role of a 'prime mover', or God.[24]

Charleton's *Physiologia* applied Gassendian atomist principles to the explanation of sense perception and the natural qualities of matter (such as occult qualities, electricity and magnetism), and rejected many things that he appeared to accept in *Natural History*. Because of his important role in the dissemination of Gassendian physics in England, Charleton is often seen to have been comprehensively atomistic in outlook. However, this view can be refuted by consideration of his determined avoidance of monolithic explanation, throughout *Natural History*. Although seventeenth-century physiological texts often questioned Aristotelian physics and classical pathology, orthodoxy was not simply displaced by a new system. Alternative explanations were drawn from a range of chemical, mechanical, atomistic and vitalistic theories.[25] The creation of an alternative physiological system—a coherent and innovative animal oeconomy—is often assumed to have been the aim of seventeenth-century physiology and natural philosophy (and specifically depicted as the aim of a mechanist body model). However, while the old Aristotelian and Galenic dictates are often thought to have been replaced by new physical theories of natural philosophy, this was not the case. Despite great changes in the profile and practice of anatomy during the sixteenth century, and despite humanist attacks on scholastic physics, the traditional Galenic view of the interior body, and its tripartite hierarchy of physiological functions, remained fairly constant.[26]

Many physiological processes required distinctive kinds of explanation: heat, generation, vivification, nutrition, excretion, the specific allocation of fluids to parts, perception, the action of the will, and a multitude of other complex issues. Actions of the body were traditionally believed to be goal-directed, performed by a hierarchy of bodily parts in accordance with their specific needs: Galenic physiology ascribed to the organic parts of the body the ability to attract, hold and assimilate nutriment, and to repel that which could not be integrated.[27] The rise of materialist physiology led to the questioning of sentience, occult powers, secret virtues, and real qualities appropriate to specific actions.

[23] French, *History of the Heart*, p. 71.

[24] Fournier, *Fabric of Life*, p. 92.

[25] Frank, *Oxford Physiologists*, p. 90.

[26] These located (vegetative) venous functions in the liver; (vital) arterial functions in the heart; and (animal) sensory and motor operations in the brain. See R. Porter, *The Greatest Benefit To Mankind: A Medical History of Humanity from Antiquity to the Present*, London, HarperCollins, 1997, pp. 184-5 for a synopsis. Galenic theory at this stage was combined with additions from both Arabic commentaries and humanist scholars in the fifteenth century, as well as later anatomical researches. Frank, *Oxford Physiologists*, pp. 2-3.

[27] O. Temkin, *Galenism: Rise and Decline of a Medical Philosophy*, Ithaca, Cornell University Press, 1973, p. 89. The parts of the body possessed the power to attract whatever they required, and to repel substances which might do them harm.

Rather than attaching himself to a system of explanation for the animal oeconomy, Charleton explored a range of hypotheses from various sources. He did not depict himself as a proponent of a revolutionary agenda, but was an interested reader of new approaches, and an active participant in the new experimentalism. His medical agenda was not the comprehensive overthrow of traditional knowledge, but the reconciliation of old and new hypotheses. *Natural History* revised traditional medical understandings and integrated some atomistic emphases, such as the importance of particle motion and figure. Charleton's depiction of interior motion in terms of non-sentient process did not reflect a determination to overthrow the entire peripatetic framework. Attacks on Galenic medical knowledge reflected neither the presence of coherent alternative systems, nor comprehensive rejection of traditional understanding. Medical writers often aimed to reconcile innovative knowledge with the existing physiological system, rather than attempting to generate a new one.[28] Charleton did not advocate a complete revision and reconstruction of physiology.

DIRECT INFLUENCES ON CHARLETON'S PHYSIOLOGY

Amongst the English writers who explored new ways of understanding human physiology was physician George Ent (1604 -1689), trained at Cambridge and Padua.[29] The small number of his publications belie Ent's importance in the medical world of Charleton's day. He was a close friend to Harvey.[30] Yet his *Apologia pro circuitione sanguinis* (1641), which defended circulation theory, also examined the physiological principles of Harvey's opponent Descartes.[31] Ent's versions of physiology took some Cartesian hints toward replacing traditional teleology, vital principles and sentient matter with implicitly mechanistic physiological ideas.[32] Like many physicians of this period (such as Glisson,

[28] For example the textbooks generated for students of physic: Caspar Bauhin, Jean Fernel and Jean Riolan.

[29] Ent performed a series of roles at the College of Physicians, and in the Royal Society. His first appointment, as Gulstonian lecturer, was in 1642. Following this he was censor for twenty-two years, and registrar from 1655 to 1670. Ent was president of the College of Physicians between 1670 and 1675, and subsequently 1682 and 1684. He was knighted in 1665 for his contributions to medicine.

[30] A major instigator in the field of comparative anatomy, he worked with both Harvey and Severino. Webster, *Great Instauration*, p. 319. His letter of dedication, printed as a preface to Harvey's *De generatione animalium*, 1651, implies that Ent had a role in encouraging the great physiologist to publish the work. Ent became the primary link between Harvey and the Royal College throughout the period of the Republic. Ent's only other medical publication was an analysis of Thurston's text on respiration (with which he had some research involvement): G. Ent, *Animadversiones in Malachiæ Thrustoni*, 1679 and 1685.

[31] Brown, 'Physiology and the mechanical philosophy', p. 27.

[32] However, these aspects of Ent's work went without comment in the 1640s, as the impact of French mechanical philosophy had not yet been felt in English physiological thought. Brown, 'Physiology and the mechanical philosophy', p. 30.

Willis and Wharton), Ent was interested in Helmontian/Paracelsian chemical medicine. This eclecticism of approach was similar to Charleton's.

The latter's intellectual debt to Harvey was enormous, and *Natural History* was heavily dependent upon ideas and demonstrations drawn from both *De motu cordis* and *De generatione animalium*.[33] The inclusion of the circulation of the blood necessitated new approaches to some of the traditional processes (such as the action of the liver). Harvey influenced Charleton directly through their association at Oxford, and the latter was also interested in the views of several followers of the elder physician. As mentioned, Ent was a great personal influence on Charleton.[34] Charleton was to dedicate to him not only the English and Latin versions of his animal oeconomy (1659/1669), but also *Exercitationes physico-anatomicæ* (1659) and *De scorbuto* (1672). The latter dedication referred to a substantial friendship between the two men. Ent was significant to Charleton as an active colleague, co-experimentalist and a guarantor of academic credibility. He was at various times invoked by Charleton to witness and authorise his assertions. The young physician's other major mentor was Francis Glisson (1597-1677), prominent in the College of Physicians, founding fellow of the Royal Society, and friend and colleague of Harvey.[35] Glisson's *De Rachitide* (1650) was the first pathological work to apply the circulatory model,[36] and his subsequent *Anatomia hepatis* (1654) incorporated circulation into hepatic physiology.[37] Glisson leaned

[33] Though often dismissed as a relatively minor addition to medical knowledge, the latter was the major source for Charleton's ideas on nutrition and generation. See W. Pagel, 'Harvey and Glisson on irritability, with a note on Van Helmont', *Bulletin of the History of Medicine*, vol. 41, 1967, pp. 497-514. Charleton was tremendously impressed with Harvey's work, referring to him as 'Hippocrates the second', p. 28. In 1657 the younger physician commented that he was 'so well satisfied of the Verity of this *Harveian Circulation*' that he intended to undertake the justification of Hippocrates' aphorisms on the nature and sanation of diseases 'by reasons and considerations deduced merely from this one Fountain, the Hypothesis of the Circulation of the blood'. *Immortality*, p. 35.

[34] The British Library copy of Glisson's *Tractatus* is addressed by the author to Charleton.

[35] On Glisson see Pagel, 'Harvey and Glisson on irritability'; J. Boss, 'Helmont, Glisson and the doctrine of the common reservoir in the seventeenth century revolution in physiology', *British Journal for the History of Science*, vol. 16, no. 3, 1983, pp. 261-72; J. Henry, 'Medicine and Pneumatology; Henry More, Richard Baxter, and Francis Glisson's *Treatise on the Energetic Nature of Substance*', *Medical History*, vol. 31, no. 1, 1987, pp. 15-40; O. Temkin, 'The Classical roots of Glisson's doctrine of irritation', *Bulletin of the History of Medicine*, vol. 38, 1964, pp. 297-328; J. Boss, '*Doctrina de circulatione sanguinis haud immutat antiquam medendi methodum*: an unpublished manuscript (1662) by Francis Glisson (1597-1677) on the implications of Harvey's physiology', *Physis*, vol. 20, no. 1-4, 1978, pp. 309-35; G. Giglioni, 'Anatomist Atheist? The "Hylozoistic" Foundations of Francis Glisson's Anatomical Research', in O. P. Grell and A. Cunningham (eds), *Religio medici: medicine and religion in seventeenth century England*, Brookfield, Scolar Press, 1996.

[36] Francis Glisson, *De Rachitide sive morbo puerili qui vulgo The Rickets didictur*, 1650.

[37] The research toward this volume was performed by Glisson with Ent and Wharton. Glisson's description of the liver in this work has been one of the sources of his lasting fame. Webster, *Great Instuaration*, p. 317. See also G. Giglioni, 'Anatomist Atheist?', p. 115. Glisson had delivered the Gulstonian lectures to the Royal College of Physicians on the function of the liver, in 1641. These formed the basis of the 1654 *Anatomia*.

towards the hermetic monism of Van Helmont[38] and, like Ent, followed a neo-scholastic philosophy substantially different from the mechanistic tendencies of Hooke and Boyle.[39] Charleton's reliance upon the theories of Glisson has hardly been explored.[40]

THE EPISTEMOLOGY OF *NATURAL HISTORY*

There were a number of traditions upon which physicians drew, and also undeniably new emphases (arising from the new scientific communities) which had an impact on medical knowledge. Charleton negotiated a variety of epistemological options. He invoked a range of methods, which I will explore in a series of case studies throughout this chapter. Charleton was reliant upon authoritative texts, both ancient and modern. His physiological concerns were clearly shaped by a long medical tradition, and he was reverential toward ancient authors. His work is characterised by compilation and reconciliation of disparate elements of physiological theory from a range of textual and non-textual sources. He drew much knowledge from his extensive reading. It is difficult to ascertain how much apparently first-hand information had in fact been drawn from the texts of others. He often copied experimental outcomes from the writings of others (such as Harvey, Ent and Glisson). For example, his discussion of generation focused on the development of chickens,[41] most of which information came directly from Harvey's works. The manner in which he did so supports the idea that the primary authority came not from the experimental instance, but from the authority of the individual from whose text they were taken (see my discussion later in this chapter).

Charleton attributed some status to empirical demonstration, but it was by no means the primary form of verification. His early physiological treatise includes some first hand anatomical knowledge. Although no substantiating records place him at the scene of Oxford experimental activities in the 1650s, it seems likely that he was privy to some experimental performances. *Natural History* referred to vivisections performed on dogs, rabbits, snakes, eels and frogs,[42] as well as comparative anatomical observations on cows, horses, boars and other larger animals.[43] The author detailed several anatomical demonstrations

[38] Brown claims that the main contribution of *Anatomia* was its illustration that careful, empirical studies pointed to actions and properties unique to living things and irreducible to physical explanation. Brown, 'Physiology and the mechanical philosophy', p. 43.

[39] Brown, *Animal Oeconomy*, p. 119.

[40] His later *Enquiries* made detailed reference to the monistic physiology expressed in Glisson's *Tractatus de natura substantiæ energetica*, 1672. Pagel traces the direct lineage of monistic matter theory from Helmont to Glisson Pagel, 'Harvey and Glisson on irritability'. This lineage should be extended to include Charleton.

[41] See Charleton, *Natural History*, pp. 40, 60.

[42] See respectively *Natural History*, pp. 67, 82, 115, 89. Charleton also referred variously to 'dissections of living creatures' (though he was not always specific about which species), p. 141.

[43] Charleton, *Natural History*, pp. 24, 160.

concerning the actions of the vessels of the body under anatomical interventions, such as ligatures.[44] Dissections of the heart were also mentioned,[45] along with sundry other observed details.[46] Frank indicates that Oxford was the main site for this kind of investigation in the 1650s.[47] Equally powerful was analogical demonstration, applied to many aspects of physiology which were not anatomically demonstrable. I have highlighted this method in the discussion that follows. Experimental examples were sometimes invoked for their analogic resonances, as I shall illustrate with various anatomical examples through this chapter.

Charleton rarely referred to specific clinical observations. *Natural History* only once included the confirmation of a proposed hypothesis 'by an observation of our owne'.[48] A patient's suffering from a 'pricked' nerve was included to affirm Glisson's theory of distribution of nourishment.[49] This diversity of approaches suggests that Charleton was an eclectic not simply of ideas, but also of method. I will discuss these varieties of epistemology as they refer to particular demonstrations throughout *Natural History*. The following discussion explores the ways in which Charleton demonstrated his claims about the natural world. I examine the relative status of empirical evidence and reasoned analysis. In doing so I hope to illuminate the status that Charleton himself attributed to various epistemological devices. Ultimately I believe this indicates the ways in which a physician could negotiate an authoritative professional identity in late seventeenth-century English society.

THE CONTENT OF *NATURAL HISTORY*

The topics of Charleton's *oeconomia animalis* (and those of the Dutch school) were slightly modified versions of classical physiology. *Natural History* examined the classical triad of faculties (natural, vital, animal) through their key processes: nutrition, vitality and locomotion.[50] It divided the natural faculty into nutrition, generation, digestion and growth. Vital faculties were examined through the subdivisions of the journey of the chyle, the heart's action, depuration of the blood, circulation and respiration. The animal faculties, generation and action of the animal spirits, were considered through reference to the issue

44 Charleton, *Natural History*, pp. 74, 83.

45 Charleton, *Natural History*, p. 80 [mis-numbered as p. 78], again on p. 88.

46 '. . . we have the testimony of our sense; that the rudiments of . . . Tumors, are like Eggs included in a membranous filme', Charleton, *Natural History*, p. 159.

47 Frank, *Oxford Physiologists*, *passim*.

48 Charleton, *Natural History*, p. 157.

49 This theory was also ratified by the anatomical observations of others. Though the presence of nutritious juice as asserted by Glisson was not anatomically established, 'we have been assured by judicious and credible persons, that they have seen no small quantity of the Nutritive juice, exstilling out of the nervous Chord of the Thigh in a man . . .' Charleton, *Natural History*, pp. 168-9.

50 Luyendijk-Elshout, 'Oeconomia animalis', p. 299.

of locomotion.[51] These were the primary areas for discussion in coming to an understanding of the operation and laws of the human body,[52] and as categories were echoed in other 'animal oeconomies' of the period. Charleton's eclecticism was distinctive in comparison with the Continental animal oeconomies.[53]

Natural History traced the progress of matter through the body. Its ten sections covered the essential life processes and the operation of the body. The sections were as follows: Nutrition, Chylification,[54]the Journey of the Chyle, the Generation of Blood,[55] the Uses of the Blood, the Causes of its Motion and the Depuration of the Blood.[56] He then tackled Respiration, Lymph ducts, the Distribution of Nourishment and the question of Voluntary Motion. Galenic physiology focused on processes such as nutrition, respiration and locomotion,[57] performed by the faculties located within body's three principal cavities: bowel, thorax and head.[58] Each was the site of fundamental transformations: the generation of the vital and animal spirits, and purification. The substantial transformations in the animal oeconomy, based on shifts in the balance of the four primary qualities, were determined by active powers of attraction, retention, assimilation and expulsion, resident in every part of the body.[59] The processes portrayed by these faculties were presented within *Natural History*, and its structure was governed by the logic of their sequential operation. Although Charleton's physiology incorporated more recent alternative hypotheses, the structure of his text echoed the priorities of Galenic pathology, adopting a processual model in the examination of internal function.[60] Below are a series

[51] Luyendijk-Elshout, 'Oeconomia animalis', *passim*.

[52] Hobbes declared that 'Man's nature is the sum of his natural faculties and powers, as the faculties of nutrition, motion, generation, sense, reason &c.' Thomas Hobbes, *Elements of Law, natural and politic*, ed. J. C. A. Gaskin, New York, Oxford University Press, 1994, Part I, 'Human Nature', p. 21.

[53] Broeckhuysen's *Oeconomia Animalis* (1672) relied heavily on Cartesian physiology, and thus excluded locomotion. Similarly Theodoor Craanen's *Oeconomia Animalis* adhered strongly to Cartesian corpuscularism.

[54] Chylification was the earliest stage of digestion.

[55] Sanguification involved the generation of blood from the chyle.

[56] Depuration was the removal of excrement from the blood.

[57] J. Bylebyl, 'Nutrition, quantification and circulation', *Bulletin of the History of Medicine*, vol. 51, 1977, p. 369.

[58] Anatomical texts of the period were also structured around these cavities, as discussed in chapter seven. However, the sequence of anatomical examinations was partially governed by the speed of putrefaction of the organs of the cadaver upon which they were demonstrated.

[59] Bylebyl, 'Nutrition', p. 371.

[60] The physiology of antiquity implied an inevitable sequence. Blood was generated (sanguification) and the natural spirits extracted for distribution through the veins. These were created out of chyle that passed into the liver from the gut. The mixing of blood with pneuma, and generation of vital spirits was performed in the heart, from whence they were distributed through the arteries. The brain hosted the creation of animal spirits (out of ennobled vital spirits), which were sent out through the nerves. Porter, *Greatest Benefit*, p. 77. Charleton's text therefore followed a structure apparently demanded by the subject itself.

of individual case studies, which explore aspects of Charleton's methodology and self-presentation in relation to physiological questions.

CASE STUDIES IN EPISTEMOLOGY

Demonstration—The case of the chyle

Natural History's discussion of chylification invoked both empirical and rational sources of authority. The progress of chyle was examined in some detail, but the majority of this discussion was concerned with the role of chyle, rather than blood, as the origin of milk in lactating women. This discussion raised methodological questions about varieties of demonstration. In part this was the result of iatrochemical debates over the nature and origins of physiological fluids, but it also implied some issues of structural anatomy. The topic highlights Charleton's negotiation of material and immaterial agency in the animal oeconomy. His preferred theory posited the presence of chyle-carrying *venæ lacteæ* in the breasts and womb. Although he believed it to be 'highly probable' that they existed, no anatomist had 'hitherto been so happy in his searches, as to discover . . . [their] secret wayes or passages'. However, he concluded that verification through anatomical evidence was secondary to the verification of reason. Although such vessels 'have long concealed themselves from the eye of the body, yet are they obvious to the eye of the Mind: and the acuteness of our Reason may herein supply the dullness of our sense.' The fallible perception of the senses was supplemented by the incisive penetration of reason, and the probability of the solution discovered through 'sundry and weighty Arguments'.[61] This form of rational deduction was far from the indifferent experimentation that allegedly defined natural philosophical enquiry.

Throughout his career Charleton maintained the importance of knowledge derived from the insights of the mind in addition to that gleaned through observation. The 'Engyscope of the Mind' made real knowledge about internal operations available. This epistemological principle, founded on classical notions of the hierarchy of the human faculties, underlay a long tradition of medicine, and continued to determine much of Charleton's understanding of the internal body. The 'weighty' arguments to which he referred included the claim that the transformation of blood into milk was contrary to the 'apodictical doctrine of the circulation of the blood', as reinforced by the authority of Vesalius and the 'witness' of our own eyes.[62] Although the inadequacies of observation were the very reason for the argument, Charleton supported the authority of Vesalius as a witness. The chyle was similarly carried to the womb, Charleton claimed, and substantiated this by extensive reference to the authority of great men, namely

[61] Charleton, *Natural History*, p. 19.
[62] Charleton, *Natural History*, p. 20.

Hippocrates and Harvey.[63] This accords with the theory of Peter Dear, Simon Schaffer and others that the testimony of trusted individuals constituted an accepted component of seventeenth-century demonstration.[64] Experiments were not, in these instances, definitive. Where Charleton referred to experiments it was often the text, in which the experimental account appeared, that constituted the source of authority. This is particularly noticeable in his use of Harvey's writings.

Explanatory doctrines—The role of 'sympathy'

The final reasons concerned the 'resemblance in manifest qualities' between chyle and milk.[65] Charleton noted 'the many resemblances', which suggested that chyle was the origin of milk.[66] There was a 'great Sympathy or consent betwixt the womb and paps'. Acknowledging that the notion of sympathy had been rejected by many contemporaries as an occult action, Charleton reinvented sympathy as a material action. Sympathy, in his formulation, was not possible without direct physical connection, in this instance vessels linking the breasts and womb.[67] Though such anatomical characteristics had not been discovered through observation, Charleton believed it was 'highly consentaneous to truth that there are such vessels'.[68] He accepted the presence of a sympathy founded on a physical connection between two organs, despite the absence of any physical proof. This kind of reasoning was not exclusive to Charleton.[69] The example illuminates an eclectic blend of empiricism and doctrine. The influence of atomism led him to shift the basis of sympathy to physical links, rather than occult or active powers. The method by which he demonstrated his claims was entirely theoretical. In conclusion he observed 'what need we thus anticipate, by conjecture, when we dayly expect the discovery of the wayes through which they passe, by Anatomists, who now a dayes excite themselves in strict enquiry after them?'[70] This seems to imply that anatomical proof would replace conjecture. However, this statement concluded a discussion in which the absence of anatomical evidence in no way impeded the development of probable explanation. While anatomy was one form of demonstration, it did not provide the logic on which Charleton's assertion was founded. There is no support here for Shapinian expectations regarding the empirical context for the creation of atheoretical matters of fact. Charleton simply assumed that anatomy would eventually demonstrate the point that he had

[63] Charleton, *Natural History*, pp. 26-29.
[64] See Simon Schaffer, 'Godly men', p. 59.
[65] These included their common fattiness, tendency to corrupt, and their colour in addition to their 'reciprocall convertibility', Charleton, *Natural History*, pp. 24-5.
[66] Charleton, *Natural History*, p. 25.
[67] Sympathy between parts of the body was either through communion (the presence of vessels connecting the two), or through society (their common office).
[68] Charleton, *Natural History*, p. 30.
[69] As French illustrates, Harvey reasoned in just such a manner. French, *History of the Heart*, p. 67.
[70] Charleton, *Natural History*, p. 32.

proven through reason. Anatomy was a *post hoc* form of explanation: irrelevant to the real understanding of the relationships between bodily parts. This example illustrates the continuity of the idea of the penetrating 'eye of the mind', in accordance with the hierarchy of knowledge described in *Ternary*. It also illustrates a significant lack of epistemological continuity, and a materialist negotiation of occult processes and selective anatomical evidence.

A question of certainty—The case of sanguification

Sanguification was a key physiological transformation. As the process by which chyle became blood, it invoked questions of vitality and the origin of the life which characterised the blood within the body. Various organs had been posited as the site of sanguification: Galen cited the liver, Aristotle the heart and 'some Anatomists' had named the veins. Charleton rejected each, and instead followed Harvey's argument that sanguification arose from the blood itself (providing examples from Harvey's embryology).[71] He invoked as explanation the 'similary' principle—one of the doctrines upon which traditional scholastic physic was founded. It was, he claimed, 'canonical' that 'all Naturall agents endeavour . . . to assimilate to their own nature, the thing upon which they act'. The absolute certainty with which this principle was known generalised to the conclusions drawn from it: 'it seems of equal certainty, that the activity of the vital Blood, is most properly consigned to the work of Sanguification.'[72] Here a classical medical doctrine was treated as unquestionable by Charleton, and his conviction regarding the physiological principle was readily projected onto associated conclusions. Charleton's demonstrations that the body was nourished by vital spirits were founded largely on such principles as, for example, 'Betwixt the thing nourished, and its nutriment, there ought to be a certain Analogy, or Similitude',[73] and that nutrition could progress only from less to more perfect states, and never the converse.[74] These laws, founded on ancient principles, were invoked to prove the more modern of the two hypotheses. This was not an overthrow of ancient authority, but a reworking of familiar principles to new theories, and a reconciliation of traditional methods and innovative perspectives.

The blood's role as the agent of its own transformation reflects prevailing interest in the processes and nature of the blood in the wake of Harvey's researches. Circulation, often cast as the origin of seventeenth-century physiological mechanism, had its origins in Aristotelian philosophy as an agent of

[71] Charleton, *Natural History*, pp. 34-7. Among his proofs he cited experimental vivisections.

[72] Charleton, *Natural History*, p. 39.

[73] Charleton, *Natural History*, pp. 58-9, 61. In this doctrine he implied the scholastic distinction between spermatic and sanguineous parts of the body. The delineation was also applied by Glisson, after George Ent, who had posited in a meeting at the College of Physicians that certain parts of the body were not fed by the blood. A. Cunningham, 'Historical context of Wharton's work on the glands', in *Thomas Wharton's Adenographia*, Oxford, Clarendon Press, 1996, p. xlvi.

[74] Charleton, *Natural History*, p. 55.

preservation and regeneration. In Charleton's time, the circulation model was often applied to the perfecting transformation of the spirits.[75] This spirituous analogy of circulation (posited in Regius' *Fundamenta physices*) was also important to Charleton's animal oeconomy. Circulation effected the transformation from blood into spirits: the original vital liquid was translated into blood,[76] which underwent 'rarefaction', separating its 'spirituose' constituents.[77] Through each circulation, the spirits grew 'more and more subtile and agile; and so must at length be brought to the requisite height of volatility.' Following this they diffused to the extremities, which they 'warm and vivify' before dispersing themselves into air.[78] Charleton's depiction of the transformation of spirits followed Glisson's division of the state of spirits into three categories: fixation, fusion, and volatility.[79] This hierarchy, applied by Charleton, Willis and various others, constituted the progress of the spirit (from chyle to dispersal), and described the conditions through which matter passed in the perfecting transformation.[80] In *Natural History*, the circulation of the blood was significant as a chemical process of blood-perfection, rather than a mechanical motion necessary for the generation of heat. In this way apparently mechanical physiological processes could incorporate, or indeed express, more traditional medical doctrine.

The presence of faculties—The case of the blood's motion

The question of the attraction of the blood to the heart was examined in some seventeenth-century investigations. This physiological issue rekindled conflict between Aristotelian and Galenic interpretations.[81] Following the contemporary opposition to 'faculties' and 'powers' in explanation, Charleton confidently claimed that 'in Nature there is no such thing, as the motion of a body by attraction'. This, he believed, had been 'by solid and irrefutable arguments proved' by Ent, and by himself in 'our discourse of Occult Qualities' (his discussion in the

[75] Clarke notes that, given the similarities between the cardiovascular and nervous systems, it seemed highly probable that they would operate according to the same processes. E. Clarke, 'The Neural Circulation: The use of analogy in medicine', *Medical History*, vol. 22, 1978, p. 298. Also, W. Pagel, 'William Harvey and the purpose of circulation', *Isis*, vol. 42, 1951.

[76] Charleton, *Natural History*, pp. 41 and 44.

[77] Charleton, *Natural History*, p. 93, also pp. 64-65. Willis described exactly this process in relation to the blood-carrying vessels in the head, which he thought acted as 'distillatory organs, which by circulating . . . and as it were subliming the blood, separate its purer and more active particles from the rest', Willis, *Practice*, p. 72.

[78] Charleton, *Natural History*, p. 63. The origin of this transition was not the heart but the *chorion*, or bowel, p. 45.

[79] Clericuzio, 'The internal laboratory: The chemical reinterpretation of medical spirits in England (1650-1680)', in A. Clericuzio and P. Rattansi (eds), *Alchemy and Chemistry in the 16th and 17th Centuries*, Boston, Kluwer, 1994, p. 60.

[80] Clericuzio, 'Laboratory', p. 63. This formed the basis of the idea of the body as an alembic in which chemical transformation took place, a metaphor which underlay much of Willisian physiology.

[81] See French, *History of the Heart*, on the causes-of-motion debate.

Physiologia).[82] The heart, he argued, was a 'great and weighty machine', which operated largely according to mechanical principles. In his proofs regarding the uses of the blood, Charleton had cautioned against the tendency to 'ascribe the actions of things to their Qualities or Faculties; thereby indicating the Formal Reason or Manner; by which the substance operateth'. On the contrary, he claimed, 'it is the very substance it self, to which those Qualities are inherent, that really performeth the action.'[83] This was a call for the recognition that agency inhered in matter, not in its formal 'qualities'. The question of the origin of agency in the body recurred throughout his discussion of the animal oeconomy. These statements have been invoked by some to suggest Charleton's participation in a linear trend toward the rejection of immaterial explanation in favour of mechanical and observable phenomena. However various of his other statements imply that he saw no conflict between a broad rejection of faculties and a more specific acceptance of their actions.

Despite his rejection of the ancient explanation by attraction, Charleton argued that the heart was 'endowed with a certain Motive-virtue inhærent and essentiall, called the Pulsifick Faculty, which is conjoyned, as a concomitant cause, with the blood it self, in giving it a due motion.'[84] The pulsifick faculty regulated incoming and outgoing blood. His explanation of the motion of the blood thus invoked exactly the 'powers' against which he had warned. Although the fabric of the heart was important, it constituted only a remote cause. This suggests that the fabric of the heart and its fibres was insufficient to explain all the necessary aspects of its functions. The faculty was still necessary to certain elements, such as the regulation of functions.[85] This was consistent with his methodological assertion that the causes of the heart's motion were to be discovered by a combination of 'ocular Inspection, and Reason'.[86] Charleton acknowledged that the assertion of the need for a faculty might not be the final resolution of the issue, but affirmed its plausibility as the most likely explanation thus far:

> Notwithstanding these reasons alleaged, we dare not set up our rest in this doctrine of the Ancients, concerning a Pulsifick Faculty implanted in the heart: only we have recited it, as the most probable Conjecture of all others, touching this abstruse Argument, the proxime Cause of the Motion of the blood. Nor shall we adhere to it longer, than untill we shall be so happy to meet with a more satisfactory solution of that admirable Phaenomenon. In the mean time, Modesty commands us to declare that we find this knot to be too hard and intricate for the teeth of our weak understanding.[87]

[82] Charleton, *Natural History*, p. 84. Ent's view was expressed in G. Ent, *Apologia pro circuitione sanguinis,* London, 1641.

[83] Charleton, *Natural History*, pp. 58-9.

[84] Charleton, *Natural History*, pp. 84-5.

[85] His reasons for the need of a faculty to explain this regulation were several, and included observations of the heart's motion after being cut; and that the heart, as the hierarchical leader of the oeconomy, must have a regulating role. Charleton, *Natural History*, p. 85.

[86] Charleton, *Natural History*, p. 87.

[87] Charleton, *Natural History*, p. 86. Charleton cited other authorities who had also chosen to retreat from resolution of this difficult issue: Fruscatorius, Galen, and Scaliger.

Though he was vigilant about over-reliance upon any single hypothesis, it was appropriate to illustrate an 'impartial' acceptance of the most probable hypothesis when it was known. His acceptance of the ancients was at times self-conscious, as in this instance. However, in many other cases his adherence to traditional doctrines seemed uncritical.

Digestion involved the attraction of active and useful substances and excretion of inert or harmful matter.[88] Some kind of perception was necessary to distinguish appropriate objects of retention or refusal. For Charleton this was achieved through the action of an 'Acid humor' which was 'endowed with an incisive, penetrating faculty'.[89] Evidently the faculty was not superseded as a form of explanation. Nor was it rejected explicitly because of its implication of adherence to ancient doctrines. The physician reaffirmed his eclecticism and modesty in his retreat from any definite conclusion. Faculties constituted a useful explanation of processes traditionally understood as goal-directed. Charleton invoked corpuscular configurations to replace faculties in some, but not all, instances. He did not reject occult operations *per se*, but reconfigured the animal oeconomy to exclude them from particular physiological functions. He did not provide a coherent model of materialist physiology in 1659, but dealt individually with the range of hypotheses on specific operations. In many cases he simply adopted the theory put forward by one of this contemporaries, and presented it beside another, conflicting theory without apparent recognition of a conflict.

A faculty of attraction—The case of respiration

Respiration was a key research area at Oxford in the 1650s and later in the Royal Society.[90] Charleton raised and rejected a number of existing hypotheses, including Galenic theory, which postulated that the lungs attracted air.[91] The dilation of the chest and lungs in respiration, he announced, was 'not from any Motive Faculty congenial to the Lungs'.[92] Nor, he claimed, did it arise from 'the impulse of the blood out of the Heart into the Lungs'.[93] He rejected the Cartesian argument that air was forced into the lungs by thoracic expansion.[94] Robert Boyle had also claimed that the lungs were moved by the motion of the diaphragm and thorax. Boyle had postulated the existence not of a faculty, but the 'spring of the air' as its cause. While Boyle was not acknowledged in the English version of

[88] L. King, *The Growth of Medical Thought*, Chicago, University of Chicago Press, 1963, pp. 127-8.

[89] Charleton, *Natural History*, p. 12.

[90] See Frank, *Oxford Physiologists*, *passim*. Of particular importance were the researches of Robert Boyle, Robert Hooke and John Mayow.

[91] The idea of 'the Attraction of Aer into the Lungs . . . is a meer dream'. Charleton, *Natural History*, p. 131.

[92] Charleton, *Natural History*, p. 133. Aristotle had asserted the similarity between the lungs and a bellows.

[93] Charleton, *Natural History*, p. 134.

[94] Frank, *Oxford Physiologists*, p. 143. See Charleton, *Natural History*, p. 135.

Charleton's animal oeconomy, Charleton did refer to Johannes Vesling's *Syntagma Anatomicorum*.[95]

The lungs, therefore, as they were 'not moved either by themselves, or by the Heart, or by the Thorax', could only be moved 'by some other part in the Breast, in which as in the first original, the motion of Inspiration doth begin'. This part, Charleton concluded, was the diaphragm, which was itself 'moved by an Ingenite Faculty'.[96] Charleton used the ancient analogy of a bellows regarding the action of the lungs. Having dismissed the belief that the air moved itself, or was attracted into the lungs, inspiration was explained by the 'similitude of the flux of air into a pair of Bellows'. This analogic explanation was highly appropriate, 'there being no other difference betwixt the repletion of the Chest, and the repletion of a pair of Bellows, with aer, but only this; that the Bellowes are opened by an externall force, and the Chest dilated by an internal.'[97] In rejecting Aristotelian interpretation of the relationship between lungs and bellows, Charleton did not reject the metaphor, but the attribution of motive power, or the manner of 'impulsion'.[98] His redefinitions did not reconfigure the metaphoric description, but rather the underlying explanation of its operation (i.e. the nature of causal attribution). He rejected the notion of a dilating faculty in the lungs, only to relocate the faculty in the diaphragm. This exemplifies the integration of old metaphors into a new context. A significant shift in explanation had occurred beneath the continuity of terminology. In this interesting example a metaphor was invoked to maintain continuity between traditional physiology and the theory that the author supported.

A corpuscular explanation—The role of the liver

As a key organ in the Galenic physiology, the liver's function was hotly debated in seventeenth-century physiological discourse. Galen stated that the liver manufactured blood from chyle. However, even before Harveian circulation theory shifted understanding of hepatic function, the adequacy of this explanation had been questioned.[99] Francis Glisson's 1654 *Anatomia hepatis* asserted that the

[95] According to Boyle, the lungs filled because they were dilated, rather than dilating because they were filled. Frank, *Oxford Physiologists*, p. 143. Boyle's observations on the spring of the air were made in 1659 and published around 1660. Johannes Vesling's *Syntagma* was published in English as *The Anatomy of the Body of Man*, London, 1653.

[96] Charleton, *Natural History*, p. 135.

[97] Charleton, *Natural History*, p. 132.

[98] Though Charleton agreed that 'the Lungs are filled with aer, and emptied again, or elevated and depressed alternately, as Bellowes are, yet it is doubtfull, whether (as the hand which moves the bellowes, by opening and shutting them, is the cause both of the influx and efflux of the aer in them) there be not some other part of the Chest, besides the Lungs, which being first dilated and contracted, is the cause, why the Lungs are opened and shut? or more plainly, whether the expansion of the Lungs be from an ingenite Faculty?' Charleton, *Natural History*, pp. 133-4.

[99] French, *History of the Heart*, p. 65.

liver's role was 'depuration', the removal of bile matter from the blood.[100] Charleton followed this theory of the liver's role. *Natural History*'s explanation of the functions of the liver did not invoke faculties.[101] Charleton listed and rejected physiological explanations that invoked sentience or occult qualities to explain hepatic action. It was impossible that 'each particular Excrement is so directed, by any Intelligent Faculty' with a power to distinguish,[102] nor that the 'similitude of substance' or 'neerness of analogy' would draw them together.[103] His final refutation of 'Attraction Similary' was that 'in Nature there is no Motion by Attraction, but all from Impulsion.'[104] The idea of an intelligence from the brain responsible for the action of the minute parts, was rejected because such a faculty was self-aware, and 'no mans soul is conscious of any such act, as the distinction of Excrements'.[105]

The denial of this explanation for specific action necessitated an alternative. Charleton's discussion of the liver deviated from the explanations through chemical transformation which had characterised much of his physiology. Instead his explanation echoed the Boylean corpuscular hypothesis which centred on the magnitude and configuration of particles. The operation of the liver was based on sieve-like separation of parts of the blood as excrement.[106] Purification of blood in the liver resulted from:

> the Correspondence of Magnitude and Figure betwixt the minute particles of this or that peculiar excrementious humor to be separated from the blood, on one side; and the small passages leading into, and insensible pores in this or that part, peculiarly constituted for the separation thereof, on the other.[107]

While this process of purification appears explicitly mechanical, it was aided by 'the help of that particular Fermentation, which each humor doth suffer either neer unto or in the place of its separation'. Fermentation specific to each humor was invoked to explain the process of transformation.[108] Thus there was a chemical element to the process. Charleton here incorporated a corpuscular explanation, though such a method appeared nowhere else in *Natural History*. It seems

[100] N. Mani, 'Glisson's hepatology and Wepfer's work on apoplexy', in L. G. Stevenson (ed.), *Celebration of Medical History*, Baltimore, Johns Hopkins University Press, 1982, p. 39. See also Cunningham, 'Historical context', p. xxxix.

[101] Charleton, *Natural History*, p. 111.

[102] Charleton, *Natural History*, p. 98.

[103] Charleton, *Natural History*, p. 98. This referred to the Galenic idea that similarity between substances drew them together.

[104] Charleton, *Natural History*, p. 99.

[105] Charleton, *Natural History*, p. 99.

[106] Charleton, *Natural History*, p. 110 onwards.

[107] Charleton, *Natural History*, p. 100.

[108] Thomas Willis was probably the most notable proponent of fermentation theory. Willis' works on fermentation was released in 1658 (after the alleged 1653 composition of *Natural History*). However, Charleton probably knew of Willis' experimentation and principles at Oxford in the earlier 1650s. Charleton, *Natural History*, p. 100.

he followed particular philosophies or explanations on particular questions, which suggests that he did not think in terms of 'framework' explanations. Here corpuscularism was combined with the more familiar determinants of physiology, the humours.

The role of analogy in explanation—The case of lymphatics

Charleton's discussion of the lymph ducts was short, and covered the discovery, varieties, situation, and the nature and purpose of the liquor they contained.[109] Although only briefly introduced in the exercitation on lymph ducts, the purpose of lymphatic fluid and its circulation within the body were thoroughly explored in his account of 'the distribution of nourishment through the Nerves'. Here he set out to demonstrate Glisson's theory that 'the Nerves are vessels carrying the Nutritive juice to the parts'.[110] His lengthy discussion of reasons in support of the Glissonian hypothesis[111] included the depiction of the nutritive juice through analogy with egg-white. The movement of nutritive juice through the nerves, was illustrated by analogy with the manner in which 'the White of Eggs is brought into the womb of the Hen, by the Nervs'.[112] However as it had never been found itself the very nature of the nutritive juice itself had been demonstrated by analogy with egg-whites. It was described as a 'roscid Humor, not much unlike the white of an Egg'.[113] This depiction of it was designed partly in order to distinguish the *succus nutritius* completely from blood. The use of the egg analogy to demonstrate the motion of the *succus nutritius*, the nature of which was founded upon analogy with the white of an egg, illustrates a circularity of method. Here the means of demonstration became the proof of the accuracy of the claim demonstrated.

There were several problems with the Glissonian notion of movement of nutritive juice through the nerves. One was the lack of evident passageways within the nerves. Another was that dissection and vivisection did not support the results predicted by the hypothesis. A third complication was that no such liquor had ever been found in the nerves.[114] Though Glisson's theory could clearly be refuted by anatomical investigation, Charleton maintained that 'these Difficulties are not weighty enough to counterbalance the Reasons formerly

[109] The English version of Charleton's physiology credited innovation to Thomas Bartholin, *Vasa Lymphatica*, 1653, and also George Joyliffe. Joyliffe never published on lymphatics, though he lectured to the College of Physicians in 1650. Charleton also credited Glisson's *Anatomia hepatis*, and Olof Rudbeck. Charleton's own mention of the research performed at the College of Physicians in the 1650s suggests the importance of Wharton and Glisson, with additional work from George Ent, John Bathurst, and Baldwin Hamey. Cunningham, 'Historical context', p. xlviii.

[110] Charleton, *Natural History*, p. 182.

[111] Charleton, *Natural History*, see pp. 157-69.

[112] Charleton, *Natural History*, p. 162.

[113] Charleton, *Natural History*, p. 158.

[114] Charleton, *Natural History*, p. 163.

alleaged; forasmuch as they may be easily solved.'[115] The lack of 'manifest hollowness' he believed, did not preclude the possibility of distillation through micro-fibres. As proof he invoked the analogy of capillary action through an 'Indian cane' (to which 'our sense is witnesse') and insensible nutrition in the leaves of plants (of which 'our Reason assureth').[116] These reasons considered, Charleton concluded that 'we may lawfully conclude; that it is not sufficiently evinced, that the Nerves are impenetrable by the succus Nutritius, only because they have no manifest cavity.'[117] This exemplifies the partial application of anatomical knowledge. Invoked to demonstrate contentious points, turned against opponents in other instances, it was not a final determinant of certainty. Here the inadequacy of anatomical knowledge was highlighted, to affirm a preferred hypothesis. Yet competing theories were sometimes attacked for their *lack* of anatomical demonstration.[118] This illustrates the relatively low epistemological status attributed to available anatomical proofs. Anatomical evidence was not inherently conclusive. It could be dismissed by an argument founded on analogy, where Charleton considered the latter 'more weighty'.

'A war of opinions'—The debate over spirits

> So great a War of Opinions among the Princes of Phylosophy, is a strong argument, that the thing about which they contend, is not yet sufficiently understood.[119]

Considerable discussion surrounded the question of the soul's relationship to physical matter in the late seventeenth-century. Cartesian and Hobbesian materialism, the revival of Epicurean atomism and its emergence in new physical theories, had raised the spectre of the soul's mortality.[120] The implications of materialism were unfolded in controversial discussions in natural philosophy and popular interest. Within the context of disputes over the agency of matter, the existence of spirits came under scrutiny. Spirits were traditionally viewed as the instruments of the soul within the animal oeconomy.[121] Among the three faculties (animal, vital and natural), Galenic physiology stated the existence of

[115] Charleton, *Natural History*, p. 164.

[116] Another attack on the validity of knowledge obtained through vivisection implied that the violent struggles of a dying animal could cause failure of the predicted effect. *Succus nutritius* could not be found in dead bodies because the straining of the parts in the throes of death forced the juice out of the nerves into the parts. Charleton, *Natural History*, p. 166-9.

[117] Charleton, *Natural History*, p. 168.

[118] See the later refutation of Willis' spirit hypothesis. Charleton, *Three Anatomic Lectures,* 1683.

[119] Charleton, *Enquiries*, p. 505.

[120] The 1640s and 1650s witnessed a rush of literature on the mortalist controversy. Thomas Browne's *Religio Medici* set the stage for ongoing disputes in English theology and natural philosophy in the latter part of the century. See G. Williamson, 'Milton and the Mortalist heresy', *Studies in Philology*, vol. 32, 1935.

[121] Frank, *Oxford Physiologists*, p. 7.

two varieties of spirits that regulated the animal oeconomy.[122] Animal spirits, instruments of the rational soul, relayed messages from the seat of the animal faculty in the brain. Vital spirits, instruments of the vital faculty in the heart, communicated nutrition and vitality through the circulatory system. The ancient doctrine, that animal spirits from the brain were transmitted to the parts, was followed by many seventeenth-century theorists (including Regius and Descartes, who saw the flow of animal spirits as a hydraulic system).[123] Antiquity had taught that the soul acted on the muscles through the transmission of the animal spirits. These were the 'immediate instrument of the Soul in voluntary Motion'.[124] Galenic understanding posited the nerves as passageways through which the animal spirits passed, on their way from the brain (where they had been prepared and purified out of blood) to the extremities.[125] The animal spirits transmitted sense impressions from the external world to the brain, then acted as the brain's agents for the control of the muscles.[126] The muscles moved when 'filled and distended with a greater gale of spirits, issued out of the store-house of the Brain'.[127]

The emergence within late seventeenth-century English natural philosophy of the idea that matter itself was possessed of neither sentience nor secret powers challenged this teleological understanding of the body. The period witnessed a 'metamorphosis of meaning in the notion of spirit.'[128] Clericuzio argues that Glisson, Charleton and Willis all believed matter to possess an 'internal principle of organization, life and sensibility', in the form of spirit. This spirit, he argues, was understood by all three authors in particulate terms, through its spe-

[122] The natural faculty, in the liver required no intermediary, as venous blood was the common material of nutrition. Frank, *Oxford Physiologists*, p. 7. Charleton refuted this idea of blood as nutrition, in favour of the *succus nutritius*.

[123] R. Descartes, *De homine*, Leiden, 1662, p. 21.

[124] Aristotelian and Galenic texts depicted *pneuma* as a product of the intake of air by the body. It was rarefied and vaporous matter, able to perform sensory and nervous functions for which ordinary matter was too sluggish. For an account of their evolution, see J. J. Bono, *'The Word of God' and the 'Languages of Man', Interpreting Nature and Texts in Early Modern Science and Medicine*, Madison, University of Wisconsin Press, 1995, pp. 99-100.

[125] Frank, *Oxford Physiologists*, p. 4. For the history of the idea of the hollow nerve, see Clarke, 'Neural circulation', p. 301. The principle of neural circulation, as an echo of Harveian blood circulation, originated with Henrik Regius. H. Regius, *Fundamenta physices*, 1646. On this topic, see E. Clarke, 'The doctrine of the hollow nerve in the seventeenth and eighteenth centuries', in L. G. Stevenson & R. P. Multhauf (eds), *Medicine, Science and Culture*, Baltimore, Johns Hopkins University Press, 1968, p. 129. Regius had drawn his conclusions from the bubbles of 'spiritous fluid' observed on the exterior of a snail, viewed from below while it moved along a glass surface. The experiment was reported by Charleton and echoed by Power, *Experimental Philosophy*, p. 38.

[126] Frank, *Oxford Physiologists*, p. 4.

[127] Charleton, *Natural History*, p. 188.

[128] Clericuzio, 'Laboratory', p. 51. Increased interest in the chemical composition of blood also generalised to a concern with the exact definition of the chemical nature of the spirits. See for example the writings of William Croone, *De motus musculorum*, Amsterdam, 1667.

cific chemical properties.[129] In many instances Charleton used the term 'spirit', but invoked processes performed by matter without sentience. Familiar terminology could be used to persuade readers of innovative explanations. These substitutions of terminology allowed border crossing between different theories and approaches. Indeed Charleton seems not to have perceived linguistic boundaries as dividing competing theories about the nature of the animal oeconomy.[130]

Some alternatives to ancient doctrine explained muscular action in physical terms, rejecting entirely the hypothesis of animal spirits, in any form, as 'both improbable and unnecessary'.[131] A second major explanation of muscular motion emerged, which stated that the nerves were under perpetual tension. Communication from the brain consisted of vibrations transmitted along the nerves to and from the brain. Francis Glisson noted in *Anatomia hepatis* that 'All irritation indicates the existence of perception', which in turn indicated the presence of nerves.[132] His 1672 *Tractatus* developed the theory of natural perception in much greater detail. Glisson stated that all matter was possessed of a sentience, or 'natural perception', independent of consciousness.[133] This 'irritability' theory, which posited the energetic nature of matter, was profoundly influential upon *Natural History*.[134] However it was to Harvey, not Glisson, that Charleton attributed the theory of fibre irritation in his discussion.[135] Charleton drew from Harvey, from the Helmontian corpus, and probably from Glisson's *Anatomia hepatis*, which he had read by the time he composed his *Natural History*. He thus combined the theory with matter theories from sources as ancient as Galen and as recent as Paracelsus, Helmont and Harvey.[136]

[129] Clericuzio, 'Laboratory', p. 59.

[130] In many instances Charleton criticised other theorists' use of spirits, where he himself invoked a largely identical explanation which merely replaced spirits with another agent.

[131] Charleton, *Enquiries*, p. 495.

[132] F. Glisson, *Anatomia hepatis*, London, 1654, ch. 44, p. 397. See also Henry, 'Medicine and Pneumatology', p. 18.

[133] Giglioni, 'Anatomist Atheist?', p. 121. All matter, whether organic and inorganic, was defined by an inherent capacity for motion.

[134] Charleton, *Natural History*, p. 119.

[135] Harvey's *De generatione animalium* (1651) had made a range of references, some contradictory, to the irritability of tissue. See W. Harvey, *Disputations touching the generation of animals*, trans. G. Whitteridge, Oxford, Blackwell, 1981, pp. 296-7, and 297-8. Indeed, Charleton paraphrased Harvey when he claimed: 'For whatever is wholly destitute of sense, is wholly uncapable also of being irritated to performe any action or motion, in order to its safety. Nor can we, indeed, otherwise discern what is Animate and sentient, from what is Inanimate and void of sense; but only by some Motion excited in it, by something molesting and irritating it: which motion doth continually both follow and argue sense.' Charleton, *Natural History*, p. 121. See Harvey: 'We have no other sign by which we can distinguish an animate and sentient creature from one that is dead and senseless except by its movement which is provoked by some offensive object and which follows immediately upon the experience of this sensation and so proves the existence of sensation.' Harvey, *Disputations*, pp. 297-8. Harvey's ideas about irritability are discussed by Henry, 'Medicine and Pneumatology', p. 19.

[136] Frank, *Oxford Physiologists*, p. 31. Though many ideas expressed in *Natural History* were Paracelsian or Helmontian, Charleton rarely made reference to Helmont, and not once to Paracelsus. He was undoubtedly familiar with the Helmontian corpus, but preferred to invoke Glisson.

The question of agency was endlessly debated, and Charleton illustrated some of the ways in which the issues were approached. The interchange of language between faculties/spirits and corpuscular explanation highlights the lack of any discrete theory-change. Charleton distinguished between the two prevailing explanations of muscular action, of sentient spirits and of fibrous agency.[137] Voluntary or animal motion was the result of a 'forcible and copious influx of Animal spirits, at the command of the soul'.[138] The second kind of muscular motion was caused by natural perception or irritable fibres. In this case 'the Fibers of the muscle spontaneously recontract themselves, after they have been extended, or restore themselves to their native tenour'. This kind of motion, known as restitution, was 'common to all tensile bodies'.[139] It could potentially occur in the muscles without the knowledge of the rational soul.[140] The following discussion examines Charleton's exploration of the properties and processes of each kind of muscular motion, and considers the kinds of evidence he invoked. He seems to have perceived the two theories as competing, but not irreconcilable. In his medical writings, specific physiological hypotheses were negotiated without reference to the theological disputes often invoked by historians as characteristic of this period.

Nayler has illuminated the close relationship between Charleton's discussion on voluntary motion in this work, and the dissertation by the German Jacob Müller, *De natura motus animalis et voluntarii*.[141] Müller's dissertation was published in 1617, and a further tract, 'De usu musculorum' was printed in 1628. It was this later text that Charleton appears to have consulted. Müller argued for the agency of spirits in muscular motion.[142] Mysteriously Charleton cited Müller only in the Latin version of his animal oeconomy, and not at all in the English edition.[143] His only English reference to Müller was in *Immortality*, which declared the College of Physicians had received 'a hint' from Jacob Müller, on the 'nature of Animal and Voluntary Motion.' This had contributed to an understanding of 'the Geometry observed by the Creator in the Fabrick of the Mircocosme', and had also led to the 'verification of the Anatomical

[137] This distinction was also later used by Alfonso Borelli, *De motu animalium*, vol. 1, p. 14.

[138] Charleton, *Natural History*, pp. 202-3.

[139] Charleton, *Natural History*, p. 203. The ideas of restitution and natural perception were expressed by Glisson (upon whose *Tractatus de natura substantiæ energetica,* Charleton's discussion was based) and by Harvey.

[140] However, in 1659 Charleton argued that the muscles possessed equal but contrasting restitutive impulse. One impulse could triumph over the other only through the assistance of the soul.

[141] See Nayler, 'Insoluble Problem', p. 100.

[142] Nayler suggests that Charleton's application of Vitruvian principles led him to alter Müller's emphasis somewhat. Charleton, *Oeconomia Animalis*, pp. 281-2; *Natural History*, p. 203. Charleton also drew from Fabricius' *De musculi utilitatibus,* with which Müller does not seem to have been familiar. Nayler, 'Insoluble Problem', pp. 110-14.

[143] Charleton, *Oeconomia Animalis*, pp. 284-5. His reference to contemporaries indicates the existence of opposition to the gale of spirits theory [Croune and Scarburgh]. See Nayler, 'Insoluble Problem', pp. 100-102.

assertions by demonstrations Mathematicall'.[144] Yet Charleton did draw reference in his *Natural History* to various English contemporaries (Glisson, Harvey and others), and it is possible that his avoidance of acknowledgement of Müller was part of a broader strategy to promote English innovation—particularly that of Glisson. It appears that the central questions of muscle physiology for Charleton were different from those that motivated Müller.[145] The German was less explicit about his reasons for the exploration of the topic, and his aim was not overtly philosophical.[146] Charleton made no claim of originality for his material on muscle physiology. Subsequent scholars have attributed to him the mechanical theory of motion. Nayler highlights the adoption of flawed accounts by later scholars, perpetuating a misconception regarding Charleton's claim to originality.[147]

Competing theories demonstrated—Spirits and voluntary motion

All motion, in Charleton's view, was instigated by the soul, in response to perception of 'good or evill' around her.[148] His discussion of voluntary motion posited the mediation of animal spirits between the immaterial soul and corporeal instruments of motion: 'we (with all the Ancients) conceive, that the Animal Spirits sent from the Brain, by the Nerves, into the Muscles, are the Immediate instrument of the Soul, whereby she doth impress an actuall motion upon the Muscles.'[149] Yet he noted that the nature of the *Instrumentum Medium* through which the Soul caused muscular movement, was contentious, and the necessity of spirits was something which 'many, especially of late yeers, have seemed very much to doubt'.[150] This reflected the recent physiological controversies surrounding the work of Descartes, Harvey, Scarburgh and others.

Charleton cited the identity between cause and effect as a proof of the necessity of spirituous agency in neural communication. Every instrument, he claimed, 'ought to be accommodate, as well to the nature of the Agent which is

[144] As a consequence College fellows had progressed 'towards the explication of the reasons and manner of the Motions of the muscles, by the principles of Mechanicks'. Charleton, *Immortality*, p. 37.

[145] Nayler, 'Insoluble Problem', p. 102

[146] Müller, paragraph 30: Nayler, 'Insoluble Problem', p. 108.

[147] She views Webster's failing as particularly serious, largely because he was aware of Müller's work. Nayler, 'Insoluble Problem', p. 123. See Webster, 'Solomons House', p. 400, fn. 20.

[148] Charleton, *Natural History*, p. 183.

[149] Charleton, *Natural History*, p. 183. This could be demonstrated anatomically; severing a nerve effectively prevented motion: 'because the intercourse betwixt the brain and that particular Muscle is wholly destroyed'. p. 199. On seventeenth century muscle physiology, see J. A. Bennett, 'A note on theories of respiration and muscular action in England c.1660', *Medical History*, vol. 20, 1976. For a historical overview, T. Kardel, 'Stensen's myology in historical perspective', in *Steno on Muscles*, Philadelphia, American Philosophical Society, 1994.

[150] Charleton, *Natural History*, p. 183.

to use it, as to the effect to be produced by the use of it'.[151] As voluntary motion was performed on a rapid impulse from the Soul, therefore:

> betwixt the incorporeal Agent, the soul, and those corporeal instruments, the Muscles, there must be some Intermediate instrument, such as is capable of being so transmitted from the Brain, into the Muscles, with the greatest velocity imaginable, and of setting them instantly a-work according to the determination of the soul.[152]

The proof of the presence and necessity of the spirits was their physical fluidity: 'no part of an Animal can be thought capable of such easie and expedite Mobility but the spirits, which flow through the body in less than a twinckling of an eye'.[153]

The rapid movement of spirits within the body necessitated a system of pathways through which they might pass. The animal spirits were transmitted from the brain through the nerves, and neural anatomy was consequently understood in terms of this requirement. The containment and direction of the spirits by the nervous system dictated particular requirements in structure: 'Lest the spirits might flow into this flesh, indeterminately or at randome and scatteringly; there ought to be such peculiar vessells or Conduits, which . . . may both carry the spirits thither, and preserve them from straying or dispersing by the way'.[154] The anatomical form of the muscles was concluded from the hydraulic theory of the spirituous oeconomy.[155] Charleton concluded, with Galen, that the muscles must be 'soft', 'spongy', and 'distinguished with multitudes of Fibers', in order to 'most easily and readily admit the Gale of spirits flowing into its substance, and be by them filled or distended.'[156] He also decided that the components of musculature were coated with a membrane, to preserve 'the spirits immitted into the body of the muscle moved, from passing quite through, or dispersing themselves, which they are apt to do.'[157] Thus anatomical configuration was speculated according to the logical necessity of spirituous access. Anatomical understanding was guided by a concern to discover the physical evidence of a hydraulic system for the conveyance of spirits.

Charleton clearly recognised the opposition between the two theories of muscular motion that he expounded. However, he made no attempt to conclude his discussion with an attempt to claim certainty for one theory over the other. Again his eclectic method allowed him to explore both theories, without declaring his allegiance to either.

[151] Charleton, *Natural History*, p. 187.

[152] Charleton, *Natural History*, p. 187. This contrasts with the spiritualist model of Thomas Willis, who attributed sentience and self-determination to the spirits.

[153] Charleton, *Natural History*, p. 187. Charleton's language here followed closely that of Croune, *De motus musculorum*, p. 33, sect. 35.

[154] Charleton, *Natural History*, pp. 189-190.

[155] On this relationship between Galenic physiology and anatomical understanding, see Frank, *Oxford Physiologists*, p. 7.

[156] Charleton, *Natural History*, p. 189.

[157] Charleton, *Natural History*, p. 192.

THE 'COHERENCE' OF *NATURAL HISTORY*

Charleton's declarations were often more exploratory than prescriptive. Consequently it would be unwise to assume a transition in his understanding. He presented himself in accordance with the identity of an impartial observer of contemporary debate. The English physician put forward contemporary ideas in a relatively uncritical fashion, without attempting to assimilate them within a coherent explanatory framework. Apparent 'contradictions' become less puzzling when we sacrifice the quest for a 'consistent' philosophy within Charleton's *oeuvre*. If there is a unifying theme to his physiology I suggest that it is the re-examination of bodily and internal motion according to largely, but not exclusively, non-sentient principles. Charleton does seem to have generally aimed at the removal of the notion of attraction from physiological explanation.

Natural History offered explanations which removed goal-orientation from the organs themselves, as a means to illustrate the importance of figure, fit and motion (the corpuscular values). But as a thoroughgoing eclectic, averse to system-building, Charleton did not attempt a synthesis along these lines. I argue that his status as a physician also led him to avoid such revisions. Although *Natural History* removed most specific goal-orientation from the organs, Charleton was prepared to accept that perception inhered in fibres of the body itself, a conclusion he had not accepted in his simple matter theory, the *Physiologia*. Motion was not necessarily the most important element of the animal oeconomy, it was simply the most consistently challenged element. Charleton was keen to illustrate that the proximate cause of motion was material, not immaterial, and that the immediate surrounding circumstances were more comprehensively explicable through matter and motion. He appears at times to promote the removal of sentience, yet the case studies examined in this chapter illustrate that he did not work toward any coherent synthesis (which one might think of as the point of a systematic removal of sentience).

Nayler comments on the apparent self-contradictions within *Natural History*: for example Charleton's ambivalence regarding the existence of animal spirits.[158] Similarly, he tended to refute explanation by 'faculty', generally invoking Ent in support of the rebuttal. Yet the diaphragm was extended through the action of the 'ingenite Faculty'.[159] Nayler also notes Charleton's refusal to treat as mutually exclusive the notion of natural or involuntary motion of the organs themselves and the dependence of voluntary motion upon a flow of animal spirits.[160] The physician was 'attempting to balance opinions' rather than to provide a

[158] He referred to 'supposed' animal spirits in relation to Glisson's argument on nutritive juice. But he was willing to incorporate them where they proved useful in explanation. Charleton, *Natural History*, p. 167. *Oeconomia Animalis*, p. 236. Nayler, 'Insoluble Problem', p. 128. The Latin version of the physiology did not qualify the existence of the spirits as 'supposed'. This is noted by Nayler.

[159] Charleton, *Natural History*, p. 137. Nayler, 'Insoluble Problem', p. 131.

[160] Nayler, 'Insoluble Problem', p. 130.

consistent explanation of neural and muscular operation.[161] Though he was influenced by contemporary ideas about the removal of sentience from the body, he seems not to have wanted to sacrifice the explanatory power that they offered.

Charleton's *Natural History* illustrates his thoroughgoing eclecticism, and his determination to avoid the use of a single doctrine in explanation. As we have seen in the previous chapter, the seventeenth-century English natural philosophical context encouraged avoidance of constraining hypotheses, and here his method was consistent with that of a natural philosopher. However, I have argued that his reasons for eclecticism lay in the importance (for professional reasons) of retaining traditional authority. His writings illustrate that significant shifts in explanation could be masked by a continuity of terminology. Innovations could be made acceptable through the use of familiar terminology. Likewise, changing terminology and metaphor could mask the continuity of explanation.

If the animal oeconomy genre was the product of the application of Cartesian mechanism to human physiology, then Charleton generated a distinctively English version. The fact that his version was less mechanistic than those of his Dutch contemporaries, may reflect the lesser influence of Cartesianism in England. Charleton's animal oeconomy was not firmly attached to any doctrine, in contrast to the mechanical emphasis of contemporaries on the Continent. Again, this may have been influenced by the singular preoccupations of English epistemology. Thus he was more of a rationalist than the 'empiricist natural philosopher' model suggests, but less so than Continental counterparts.

This chapter has examined Charleton's published views concerning questions thought by historians to have signified a major shift toward mechanism. These include corpuscularism, (liver), mechanism (circulation), the faculties (motive, versus ingenite), the hierarchy of knowledge (chyle), the status of analogy (lymphatics) and the role of the spirits (voluntary motion). Charleton dabbled in each of these explanatory theories within a text that was published in the vernacular for a non-learned audience. By doing so, I believe he was able to explore innovative theories that interested him, without becoming associated with any of them specifically. *Natural History* shows the fluidity of his epistemology, and I hope illustrates my earlier points about the ways in which his modes of demonstration differed from those presented by many natural philosophers writing during the same period. I have argued that the reasons for such deviations were related to Charleton's professional status as a physician. The subsequent chapter looks at the experimental activities in which Charleton participated during his years at the Royal Society, and further develops my discussion of the complex relationship between identity and epistemology.

[161] Nayler, 'Insoluble Problem', p. 131.

CHAPTER V

'THE REPUBLICK OF LETTERS': CHARLETON'S IDENTITY IN THE ROYAL SOCIETY & COLLEGE OF PHYSICIANS

Natural History, Charleton's first original medical text, makes a useful comparison with his later published medical lectures. As we've seen, the work illustrates the author's reliance largely upon textual sources. However *Natural History* was composed before the author had empirical experience of his own, and before he was part of an experimental community. The previous chapter examined Charleton's presentation of arguments, and his self-construction under these circumstances, through a consideration of some key examples from the text. This present chapter shifts focus from specific texts in order to illustrate the nature and range of Charleton's activities in the Royal Society and College of Physicians. I analyse how such institutional involvement affected his presentation.

Historians who have considered his involvement in the Royal Society have depicted Charleton as an active promoter of the experimental philosophy. His involvement in experimental investigations has led some historians to depict him as a predominantly empirical natural philosopher. Barbara Shapiro describes him as 'the most explicit defender of the empirical sciences against . . . logic and mathematics.'[1] Theodore Brown portrays Charleton as 'first guiding the early Royal Society to real experimental physiology'.[2] His epistemological views are seen by these scholars as revealing a classic 'virtuoso' profile, in the sense that he placed priority upon experiment.

But are such characterisations accurate? And if they are, how helpful are they in increasing our understanding of his published works? I suggest that to portray Charleton as an avid promoter of experimental method is to miss more important and relevant dimensions of his self-construction. If the experimental activities of physicians were constitutive of their professional identities, then Charleton's activities in the Royal Society might identify him as an empirical natural philosopher. However, by examining his self-construction in published texts, we observe that his keenness for experimentation did not mean that he gave priority to the resulting empirical knowledge. Those who position practices as the centrepiece of identity neglect the important tension that emerges here between practices and presentation. If we perceive textual self-representation as a constituent of Charleton's public identity, we are left with another question: what was at stake

[1] Shapiro, *Probability and Certainty*, pp. 28, 34-5, 67.
[2] Brown, *Animal Oeconomy*, p. 68.

for him in the transfer of his epistemological authority away from active experimentation? Some possible answers to these questions are posed and explored in this and subsequent chapters. Attention to experimental practices has obscured the ways in which Charleton's identity was affected by factors external to the community of virtuosi. His identity was forged within the context of multiple epistemological influences, specific to the practice of a professional physician.

The most substantial evidence we have of Charleton's activities is provided by the Royal Society records, and this is the source consulted by historians charting the extent of his experimentalism. The present chapter traces the physician's activities as described in those records. However, in order to understand how he viewed these activities in relation to his professional status, we must look at how he presented them. This chapter highlights the relationship between his published writings and his practices, so that we can examine the relationship between a community's knowledge-making practices and self-presentation. Subsequent chapters compare the style and epistemology of his medical writings after his involvement with the Royal Society and the College of Physicians, and consider whether participation in experimental societies significantly changed the content and style of Charleton's writings toward explicitly empiricist values.

Shapin, Dear and others claim that certain writers promoted their experimental experience at all costs, even fabricating laboratory scenarios in order to legitimate their knowledge claims.[3] However, these historians work on a limited sample of experimental natural philosophers (specifically Boyle and Newton) to generate their model, which they generalise to encompass Charleton. But despite being an 'experimentalist' in practice, this physician did not represent experimental practice as central to his identity. I argue that Shapin's formulation of identity and authority does not represent the values upon which the physician acted (leaving aside the larger question as to whether it is defensible in other cases). Medical discourses were circumscribed by considerations different from those of natural philosophy. The divergences between Charleton's practices and his constructed identity illuminate oft-neglected tensions in late seventeenth-century self-representation. The following discussion explores the differences between the physician and natural philosopher, and considers some of the features, epistemological, theoretical, social and economic, which distinguished them. I analyse the patterns of authority invoked by Charleton. I do not claim that all physicians constructed their identity in this manner. However this particular physician exemplifies an authoritative self-construction for late-seventeenth-century English physicians which deviates from the models generally invoked by historians (both those who emphasise action and those who emphasise literary self-construction). I believe such divergences must lead us to challenge the idea that physicians can be viewed as fitting the Shapinian model. My analysis highlights the frequently unacknowledged gap

[3] As Dear claims, 'the specificity and consequent verisimilitude of the presentation lent the described experience an authority functionally equivalent to, though different from, that deriving from an authoritative ancient text'. Dear, 'Totius in verba', pp. 154-5.

between practices and textual presentation, and reveals the need for considerable further research. In sum, it raises serious questions regarding the accuracy of the model proposed by Shapin and others.

PHYSICIANS AS EXPERIMENTAL NATURAL PHILOSOPHERS: A PERVASIVE MYTH?

One reason for the assumption that seventeenth-century physicians subscribed to Royal Society empiricist epistemology is that, until the inception of that institution, records of actual medical practice were minimal. The records kept by the College of Physicians were purely legal and regulatory. As a professional body concerned with regulation, the College had no need to discuss and debate epistemological issues, nor to generate a policy on the epistemological primacy of experimental observation.

The medical profession maintained a relatively secretive attitude regarding methods, ideas and investigations.[4] There exists no alternative body of literature on the activities of learned physicians, to provide a different model of their identity and activities from that of the Royal Society records. But while membership of the Royal Society was important, it was only a part of the professional life of a physician. As far as the physicians themselves were concerned, the Society provided them with no money, and constituted perhaps a relatively small part of their professional world. For many historians of science, the Royal Society assumes a dominant position, not because of its significance for physicians like Charleton, but because of the long shadow it casts through the subsequent history of scientific investigation. The history of science, interested in the practices and technologies by which scientific knowledge is generated, has frequently neglected the literary technology involved in this process. Recent recognition of the importance of textual strategies has not encompassed the situation of professional physicians.[5]

THE IDENTITY OF THE COLLEGE OF PHYSICIANS

An important element of the public image of the College of Physicians was the level of education of its fellows. Emphasis upon education was reinforced by the

[4] A.R. Hall, 'Medicine and the Royal Society', in A. G. Debus (ed.), *Medicine in Seventeenth Century England*, London, University of California Press, 1974, p. 427.

[5] Roger French takes up the issue of the relationship between natural philosophy and specifically medical requirements for demonstration and persuasion. R. French, *William Harvey's Natural Philosophy*, New York, Cambridge University Press, 1994. His analysis is more nuanced than that of Robert Frank, who seems largely concerned to emphasise the importance of a Harveian research agenda. See Frank, *Oxford Physiologists*. Frank's detailed study of collaboration within a 'scientific' community overstates the importance of a characteristically 'Harveian' research agenda in Royal Society activities. This emphasis leads him to neglect other important aspects and influences in contemporary research.

College's entry requirement of a medical degree from an English university. College fellows frequently attacked ill-educated opponents or applicants, persons possessed of 'no learning', or 'of a mean and mechanical education'.[6] The institution thus deliberately contrasted its own priorities with the practices of unlearned practitioners in the medical marketplace, and with the practices of surgeons. The deliberate academic construction of the College is exemplified by Charles Goodall, whose 1684 account of the College was contemptuous of that 'sort of men not of academical but mechanical education', whose aim was to make the faculty of physic 'a prey to ostlers, cobblers and tinkers'.[7] Charleton's writings illustrate a negotiation of both traditional and innovative strategies in promotion of medical knowledge.

Although important, hands-on experience was not part of the College's public self-presentation. Public identity stressed physicians' philosophical orientation in contrast with the purely empirical knowledge of the surgeons. As Charleton exemplifies, physicians might deliberately cast themselves as professional men whose knowledge was primarily derived from the study of texts. His relatively infrequent recourse to experimental demonstration suggests that the latter was not the most persuasive method by which to secure an authoritative identity. Physicians emphasised their university training, as a consequence of their need to distinguish themselves from the range of unlicensed practitioners, quacks, barber surgeons and apothecaries. As such the distinctions between their identity as learned scholars and those of quacks or advertising empirics, were crucial. Royal Society experimentalism could be identified with the category of an 'empiric'. This would have stripped Charleton of the status associated with a learned physician. My earlier chapters have explored the ways in which their education separated physicians from other medical practitioners and surgeons.

In the 1660s the College was hammered by repeated institutional challenges from the Society of Chemical Physicians, competition with the Apothecaries, internal strife, the disasters of the Great Fire and Plague and the emergence of a new institution: the Royal Society. The institutional clash at this point seems to have been between the College of Physicians and chemical physicians and apothecaries. These groups had emerged as direct competitors in the medical marketplace, during the later seventeenth-century, with the removal of censorship and increased popular interest in hermeticism and Paracelsus. These alternatives constituted the major challenge faced by traditional Galenic therapy during Charleton's lifetime.

The College of Physicians has been variously portrayed as an active experimental body which contributed to scientific revolution enthusiasm for know-

[6] *Annals*, vol. 5, 54a. Cook, *Decline*, p. 203.

[7] Goodall, *The College of Physicians Vindicated*, pp. 1-2, 22. As Cook observes, these self-authorisations indicate how the College wished to present itself, and are no neutral guide to reality. Cook, *Decline*, p. 28. On this issue see also Henry, 'Doctors and Healers', p. 192; and Wear, 'Continuity and union', p. 319.

ledge, and as an intransigent organisation struggling to survive in the new intellectual ferment. College Fellows have been cast as both 'progressives' and 'retardants' of progress.[8] Those who see the College as primarily intransigent note, as Hall observes, that the institution did not specifically encourage research and education in anatomy, physiology and pharmacology.[9] The concerns of the College historically centred on the defence of tradition, and of Royal authority. Its licensing authority was sanctioned by over a hundred years of existence in London. The scholastic educational prerequisites for College entry are seen to reflect its emphasis upon textual, rather than empirical, knowledge.[10]

Webster has revised this interpretation of the College's apparent intransigence, invoking Charleton's *Immortality* to demonstrate a strong experimental life there.[11] The work portrays the College as a dynamic organisation in which the pursuit of new knowledge was a primary objective. This makes an interesting comparison with his own later lectures, which certainly invoked empirically-derived knowledge, but indicate a rather ambiguous perspective on its centrality in his own identity and epistemology.

As we have limited textual information about their practices, this debate can continue endlessly. I argue that our most important (and indeed our only) means of understanding the importance of what the College did, is to consider what men like Charleton *said* about it. Webster has used Charleton's description to understand the College's activities, but I believe that he has not sufficiently interrogated the physician's meaning. Thus I make a distinction between their activities (which are poorly recorded), and their professional identities as constructed in their published works (and that which they encouraged their members to adopt). For the purposes of my discussion the actual extent of experimental innovation at the College is less significant than the extent to which members emphasised it as part of their identities.

Cook argues that the College was tolerant of the fluctuations of religious sympathy among its members, and was able to 'bend with the prevailing winds'.[12] However divisions within the College were caused by debates concerning 'the question of what kind of medical learning would be used as the foundation for regulating medicine in London'.[13] This debate slightly sidesteps my

[8] See Webster, 'Solomon's House'.

[9] Hall, 'Medicine and the Royal Society', p. 425.

[10] Those who argue for the profound intransigence of physicians include C. Hill, *Intellectual origins of the English revolution*, Oxford, Clarendon Press, 1965, pp. 28-9, 74-84; R. F. Jones, *Ancients and Moderns*, St Louis, Washington University Press, 1961, p. 213. Charles Gillespie redresses the neglect of dynamism among physicians, in his 'Physick and Philosophy: A study of the influence of the College of Physicians of London upon the foundation of the Royal Society', *Journal of Modern History*, vol. 19, 1947.

[11] Webster, 'Solomon's House', *passim*.

[12] H. Cook, 'Institutional Structures and personal belief in the London College of Physicians', in O. Grell and A. Cunningham (eds), *Religio Medici: Medicine and Religion in Seventeenth-Century England*, Brookfield, Scolar Press, 1996, p. 103.

[13] Cook, 'Institutional Structures', p. 110.

question of the College physicians' presentation of themselves. I am less interested in exploring the identity of such a complex institution, which I believe was far from uniform in its constituents' opinions.

Little serious scholarship has examined the relationship between the Royal Society and College of Physicians. Brown claims that the College adopted mechanism to annex the authority associated with the Society, and sees Charleton within this context.[14] However Cook argues more persuasively that the College response was not imitation but a sense of the potential epistemological conflict with the new organisation. The emergence of the Royal Society created tensions at the College. Some physicians were concerned that the new science undercut the authority of learned physic. It placed empiricism, associated in medical circles with lowly medical practitioners, above the authority of texts.[15] Not only was experimentalism irrelevant to the status of learned physic, but it implied the primacy of methods associated with a lower category of medical personnel. The mere 'empiricks' were a group against whom the physicians contrasted their own actions and authority. A medical program founded on the methods espoused by the Royal Society would have verified the claim of empirics that 'experimentalism was better than the education of learned physicians'.[16] In this sense there was potentially a profound difference in the methods of persuasion espoused by the two institutions (even where the reality of their methods was more similar than their polemical self-descriptions). Charleton's activities in the College of Physicians have received very little attention. While the span of his contributions at the Royal Society was only eight years, his active involvement in the College of Physicians endured from 1676 until his death in 1707, a period of just over thirty years.

THE ROYAL SOCIETY AND THE COLLEGE OF PHYSICIANS

The founding of the Royal Society offered many physicians the chance to develop intellectual and theoretical fascinations. In contrast to the College of Physicians, the Royal Society held no jurisdiction or administrative functions. It had an active experimental life, and was not limited to regimented demonstrations. Porter observes that medical practitioners made up the largest and most active professional group within the early Society.[17] Yet the Royal Society had

[14] Brown, *Animal Oeconomy*.

[15] Cook, *Decline*, p. 133.

[16] Cook, *Decline*, p. 181.

[17] R. Porter, 'The early Royal Society and the spread of medical knowledge', in A. Wear and R. French (eds), *The Medical Revolution of the Seventeenth Century*, Cambridge, Cambridge University Press, 1989, p. 272. Of the forty individuals invited to join the Royal Society at its inception, fourteen were physicians. Fellows of the College were to be 'admitted as supernumeraries, upon condition of submitting to the laws of the society, both as to payment on their admission and the weekly allowance, and the particular works or tasks, that should be allotted to them.' Birch, *History*, vol. 1, p. 5.

no specific concern with medical issues or interests.[18] The high proportion of physicians among the Fellows did not lead to experimental research which altered the identity and activity of physicians.[19] Rather, the Society's physicians directed their interests away from medicine, to fulfil the expectations of an investigative intellectual community which had no particular professional commitment toward physic.[20] Unprecedented opportunities were offered for physicians to experiment outside their field of training, while at the College of Physicians, experimental activities had reportedly declined in the 1660s.[21]

Considerable debate surrounds the extent of medical experimentation at the Society. Gillespie argues that physicians, individually or on committees, contributed to the majority of the Society's research projects in its first year.[22] Stearns concludes that medical experiments followed close behind 'physical' researches, as some of the most frequently performed at the Society,[23] while Porter finds minimal evidence of active medical experimentation.[24] The extent of medical investigations can be endlessly revised according to divergent definitions of what constituted 'medical' research. Regardless, the fact that physicians performed experiments does not mean that those experiments were central to the generation and legitimation of medical knowledge. The use of experimental research is as relevant as the amount performed. Of vital importance is the question of how physicians presented themselves in relation to experimental activities. Physicians' involvement in experiment does not imply that they defined themselves by this experimentation. This contrasts with the definition of the

---

Among the many physicians from the College of Physicians who joined the Royal Society were Charleton, George Ent, Christopher Merrett, Thomas Coxe, Francis Glisson, Nehemiah Grew, Daniel Whistler, Timothy Clarke, Nathaniel Henshaw, William Petty, Jonathan Goddard, Martin Lister, Jasper Needham, Charles Scarburgh, Thomas Willis, Lord Dorchester and others. Hall, 'Medicine and the Royal Society', p. 424. See also Gillespie, 'Physick and Philosophy', p. 216.

18 Hall, 'Medicine and the Royal Society', p. 421. Early activities at the Society suggest diverse interests, with no specific agenda, medical or otherwise. The Fellows reflected diverse interests and training, comprised as they were from the nobility, gentry, practical professions (engineering, ship-building etc) and other areas of expertise.

19 Clark, *Royal College of Physicians*, Oxford, Clarendon Press, 1964, p. 309. The Society's focus was natural knowledge, which excluded theology and law. Physicians thus constituted the group with the highest academic standing within the Society. By contrast Frank believes that this breadth of research ultimately redefined the role and status of the physician-philosopher in English society. Frank, 'Virtuosi', pp. 99-100.

20 According to Brown, the Society's approach to experimental research in general owed more to the natural magic tradition than to the physiological methods of Harvey. Brown, *Animal Oeconomy*, p. 68. See Gillespie, 'Physick and Philosophy', on the non-medical interests of physicians.

21 Frank, 'Virtuosi', p. 99. See also Porter, 'Early Royal Society', p. 273.

22 Gillespie, 'Physick and Philosophy', p. 217.

23 In conjunction with 'experiments in natural history and chemistry' performed by physicians in the Society, 'medical interests accounted for 43 percent of all experiments; 44 percent of all papers read at the Royal society were on medicine and natural history'. R. P. Stearns, 'The Relations between Science and Society in the Later Seventeenth Century', in *The Restoration of the Stuarts, Blessing or Disaster?*, Washington, Folger Shakespeare Library, 1960, pp. 72-3.

24 Porter, 'Early Royal Society', p. 288.

natural philosophical community by their experimental interests, as Shapin, Dear *et al.* argue.

CHARLETON'S ROYAL SOCIETY INVESTIGATIONS

Natural history

Charleton was proposed as candidate for the Royal Society in January 1661, and was admitted in May. From the beginning he was 'much in evidence at the meetings'.[25] His early contributions confirm his involvement in a range of natural history investigations, and demonstrate no identifiable medical research agenda. His early investigations covered such diverse subjects as the nature and speed of sound, the effects of poisons, and the freezing of salt water.[26]

Physiology & anatomy at the Royal Society

Brown argues that when experimental physiology eventually became part of the Society agenda, Charleton was a principal instigator, along with Timothy Clarke and William Croune.[27] Clarke's lead in anatomical experiments gained the attention of Croune and Charleton, neither of whom had previously contributed anything more substantial than 'the odd curiosity and the occasional experiment'.[28] However, Charleton's role in anatomical investigations has been overestimated by Brown. His animal trials were directed chiefly toward interests such as taxonomy, splenectomy (removal of the spleen, generally from a living dog) and grafting. He participated in some injection trials in late 1664. Overall his entry onto the anatomical stage was more gradual than Brown implies. The latter argues that Charleton, Croune and Clarke experimented on respiration.

[25] Rolleston, 'Walter Charleton', p. 407. This is noted in the diary entries of other members, and Royal Society records.

[26] For example, he was involved with observations on the nature and speed of sound, (on which prominent physician Charles Scarburgh was also working). 20 August 1662; Birch, *History*, vol. 1, p. 105. After experiments with Jonathan Goddard and William Croune, in September Charleton presented an essay 'concerning the velocity of sounds', entitled 'Apparatus Phonocampticus, or what enquiries are principally to be made by such, who would attain to the certain knowledge of the nature of echos' 17 September 1661. Charleton incorporated an anatomical argument on the significance for the human neuroanatomy of the effects of sound in his 'Certain Differences betwixt the brains of men and brutes', in 1664. Months after joining, Charleton presented a paper on poisons (19 June), and proposed experiments of freezing salt water (25 September).

[27] Brown, *Animal Oeconomy*, p. 69. Clarke performed various comparative anatomies, working on dogs, rabbits and frogs. Croune's fish dissection at the Society in January 1662 led Charleton to experiment with dissections also (presenting a pike's jaw in October 1662). Croune (Croone), who had already worked with Charleton on sound experiments, was to become an anatomist of some note.

[28] Charleton's natural philosophical contributions continued throughout his involvement with the Society, and seem to have been a substantial interest rather than one immediately displaced by experimental research.

He claims that Charleton's experiments on fish gills in March 1665 (and probably therefore his study of a diving bird) should be understood as evidence of the prevailing interest in respiration physiology.[29] If Charleton was involved in the study of respiration, which preoccupied many Fellows, these activities were not noted in the Society records. It is of course possible that he contributed a larger number of experiments than were recorded.

Most significantly, Brown overestimates the *importance* of Charleton's experimental activities. I argue that in order to understand Charleton we should look not only at what medical investigations were performed, but also at how significantly these activities figured in the physician's self–presentation. My disagreement with Brown is not over whether or not the physician did these experiments, or how many he performed, but with Brown's assumptions about what they tell us about *who* Charleton was. I believe that Brown overstates the link between activities and the public presentations (published texts).

Frank argues for the continuity between the Harveian agenda of 1650s Oxford and the Royal Society's experiments. He places Charleton among the 'major scientists' of the era, and indicates his involvement in the Harveian research guided by Oxford physiologists.[30] However the assumption that Charleton's research was directed by principles common to an identifiable community of physiologists is difficult to substantiate. It seems simplistic to argue that a single aim motivated the diverse experimental activities of the time. We must be careful not to distort the range and scope of Charleton's interests for the sake of demonstrating alleged links between the interests of investigative communities.

Charleton did however undertake activities which fitted into the general experimental interests of the Society fellows. His first recorded proposal for anatomical experimentation at the Society was in July 1663. Splenectomy underwent a vogue in the late 1650s, becoming a commonplace among experimenters.[31] The experiment, which revealed that it was possible for animals to live normally without a spleen, suggested the need for reform of Galen's claims for the organ's necessity in sanguification and the attraction of black bile from the liver.[32] Charleton performed several splenectomies in 1663 and 1664.[33] However

[29] Brown, *Animal Oeconomy*, p. 73.

[30] Frank, *Oxford Physiologists*, p. 44.

[31] C. Webster, 'The Helmontian George Thomson & William Harvey: The revival & application of splenectomy to physiological research', *Medical History*, vol. 15, 1971, p. 166.

[32] The first published reference to splenectomies was in Wharton's *Adenographia*, in which he referred to the experiments of Scarburgh, Agricola and others. Webster, 'George Thomson', pp. 159-161. Webster argues that these revisions of the spleen's role opened a crack in the Galenic system, subsequently exploited by opponents of humoral pathology. The Oxford anatomists were also involved in splenectomy trials. Frank, *Oxford Physiologists*, pp. 141-2, 183.

[33] Charleton performed two splenectomies in 1663, (22 July, 5 August), and one similar operation on the kidneys (3 August 1663). See Birch, *History*, vol. 1, pp. 286-92. Records of Charleton's projected and performed splenectomies continue through 1664, including a collaboration with Goddard in February, successful operation on 9 March (Birch, *History*, vol. 1, p. 392), and further experiments with Clarke planned in May. His experiments continued in early 1665, though the plague may have put an end to Charleton's splenectomy trials, as it did to so much.

he published no accounts of the spleen, and thus does not appear to have used this as a means of attacking Galenic philosophy.

Another of Charleton's concerns seems to have been the potential regrowth of organs.[34] This may have been central to Charleton's interest in splenectomy, as it is congruent with several trials in which he was engaged at around this time. Birch's record gives no clue as to the aim of these experiments. Croune and Charleton planned a collaboration with Robert Hooke on skin transplant trials on dogs. Patches of skin were removed from a dog, and then regrafted onto the animal, in order to assess whether the reattached skin would grow again.[35] Another experiment concerned with regeneration of parts by the body was performed by Charleton with Croune and Hooke in 1663, in which the three men attempted to regraft a cock's spur on 28 October.[36] The regrowth of body parts held a peculiar fascination for Charleton, and he developed several related experiments in late 1663. As with the splenectomies, this research was not incorporated in Charleton's published writings. This seems consistent with the nature of Royal Society investigations, which were not focused toward the demonstration of specific doctrines, but were rather framed as non-theorised.

Taxonomy

Charleton worked with Oxford contemporary Christopher Merrett (1615-1695) on several taxonomic projects.[37] Well known for his literary attacks on unlicensed practitioners in London, Merrett made prolific contributions to taxonomy.[38]

[34] Frank, *Oxford Physiologists*, p. 183.

[35] Although Charleton claimed 'that he had tried this experiment formerly', it was resolved by Hooke, who found that the skin shrank too much. Charleton repeated this experiment with Hooke, twice in 1663 (28 October and 4 November), and on his own in 1664 (18 May and 25 May). Charleton's experiments suffered several practical setbacks. On 25 May 1664 the trial was halted because the dog 'a piece of whose skin had been cut off and sewn on again, had got it off'. On 29 June the experiment was ordered to be repeated by Wilkins and Charleton, the 'dog...being run away'. Soon afterwards (6 July) this retrial was projected to be performed by Charleton at Wilkins' house. After this date the skin grafting experiments received no further mention in Birch's record, and attention shifted to other matters.

[36] Charleton repeated this experiment on 4 November. A few weeks earlier, on 14 October 1663, he had contributed his opinions in a discussion about the growth of teeth implanted into the mouth.

[37] Merrett and Charleton obtained their Oxford DM's only a month apart, in early 1643. Wood, *Athenæ Oxonienses*, vol 4, p. 48. Merrett had moved to London around 1645, and joined the Wallis group. He was involved with some respiration trials in January 1663 at the Royal Society. Merrett later worked with a group at Gresham on Boyle's transfusion trials on dogs. Frank, *Oxford Physiologists*, p. 178.

[38] Merrett was called upon on several occasions for his opinions, which were recorded on 17 December 1662. With Charleton, 19 November 1662. See Birch, *History*, vol. 1, p. 166. In December he compiled these findings with those of Merrett and Wilde. 26 November 1662, Birch, *History*, vol. 1, p. 132.

Charleton's observations on wines were published in several editions.[39] He and Merrett also combined their energies in a classification of fishes, and he completed an index of birds on request.[40] Animal taxonomy was a perpetual interest throughout Charleton's career. In 1668 he published an anatomical compendium, which listed the names of all known animals in several languages: *Onomasticon Zoicon, Plerorumque Animalium Differentias et Nomina Propria Pluribus Linguis Exponens*.[41] This volume also contained an anatomical appendix, and some observations on the varieties of fossils.[42] The keen interest in categorisation illustrated by the *Onomasticon* is evident in various of Charleton's projects throughout his years of Royal Society attendance.[43] In this work, Charleton did record some of his own observations. His manuscript notes on his copy of the *Enquiries* illuminate the theological significance of animal taxonomy. The naming and understanding of God's creatures would have been 'the only Religion', according to Charleton, had man not fallen from grace:

> This was the first service that *Adam* perform'd to his Creator, when he obey'd him in mustering, & naming, & looking into the nature of all the Creatures: & this had been the only *Religion*, if men had continued innocent in Paradise, & had not wanted a *Redemption*.[44]

The taxonomic activities at the Society reflect the contemporary interest in discovering 'true' names in the natural world. These can be compared with the Royal Society's sponsored taxonomic publications of Ray, Willoughby (fish, plants, animals), which aimed at knowledge of 'God's second book'. Sometimes

[39] Charleton, *Two Discourses*, London, 1669. See Appendix for further details. Although these were released after Charleton's active involvement in the Society, he probably received encouragement from the other Fellows, after his presentations on the subject.

[40] 22 Oct 1662. Merrett's activities in this area were compiled, along with zoological, botanical and mineralogical classifications, in his *Pinax Rerum Naturalium Britannicarum, continens Vegetabilia, Animalia et Fossilia in hac insula reperta*, London, 1665/6.

[41] London, 1668; reprinted 1671.

[42] See also W. H. Mullens, 'Walter Charleton and his *Onomasticon Zoicon*', *British Birds*, vol. 5, 1911, pp. pp. 64-71.

[43] The preservation and classification of flora were also of interest to Charleton. In March 1666 he was elected to a committee with Wilkins, Cornwall, Hill, Evelyn, Goddard, Cock, Hooke, Harrington and Graunt, to assess the preservation and taxonomy of the Society's inventory of plant specimens. Birch, *History*, vol. 2, p. 73.

[44] Manuscript notes in the dedication to Charleton's *Enquiries*. Charleton's marginalia incorporated a declaration of Wilkins', later echoed by Sprat, that 'praises, when they are offered up to Heaven from the mouth of one, who has well studied what he commends, are doubtless more sutable to the Divine Nature than the blind applauses of the ignorant.' Sprat's *History* reiterated that the study of nature strengthened the principle of natural religion. The wonders of creation revealed God's power and wisdom and encouraged worship: 'to admire the wonderful contrivance of the Creation, and so to apply and direct his praises aright, which, no doubt, when they are offered up to heaven from the mouth of one who has well studied what he commends, will be more suitable to the Divine Nature than the blind applauses of the ignorant.' Sprat, *History*, p. 349. An almost identical phrase was used by Charleton, *Lectures*, pp. 37-8.

mistaken for non-theorised experimentation, taxonomies were often designed to uncover the divine pattern of all things.

The interpretation of structures and differences in the natural world was one of the Society's research agendas, in which Charleton participated enthusiastically. This interest was, in several cases, represented in his publications. These experimental activities on which he published were non-medical, but typical of Society activities. In practice then, he was an experimental natural philosopher. Like many physicians of the Society he was engaged in research unrelated to his profession. He did not participate in a specifically Harveian agenda of medical experimentation. That his publications did not reflect the scope of his activities at the Society is perhaps not surprising. Charleton's performances in the experimental domain did not mean that he identified himself with them textually in medical publications. In many cases experimental practices were not identified in the latter. Reasons for the lack of references, in his publications, to these investigations have already been suggested.

Human anatomies

Charleton's first recorded human anatomy at the Royal Society was the dissection of a boy killed by lightning in July 1662, his observations on which were read on 13 August 1662.[45] The instance later constituted the first of his *Inquisitiones duæ anatomico-physicæ*, published in 1664. The first dissertation discussed Charleton's findings from his autopsy of the boy, probably observed in Cheshire in 1662, as reported to the Society.[46] The text dealt with the 'nature and effects of thunder and lightning', and attacked the prevailing belief that deaths were caused by bolts of thunder. The second dissertation surveyed new discoveries on the properties of the human brain, and included some discussion of Willis' research, founded on the dissection before the Society in response to Willis' *Cerebri anatome*.

However, the lightning autopsy was a response to circumstance, rather than the result of a program of anatomical research, and his main anatomical investigations did not commence until 1664. In January of that year Charleton purchased anatomical instruments, and immediately put them to use in the autopsy of an executed criminal, whose muscles he examined 'after a new method.'[47]

[45] The dissection had involved the boy's head being 'opened . . . by sawing ye skull asunder round about, a little above ye ears', and his brain examined for signs of 'combustion'. The general interest in this question was demonstrated in May 1666, when two Wadham students killed by lightning were dissected by Willis, Lower, Millington and Wallis. Frank, *Oxford physiologists*, p. 183.

[46] Read on 13 August 1662. A copy is preserved in the Royal Society microfilm records, Reel 15, vol. 4 (3).

[47] 17 February 1664, Birch, *History*, vol. 1, p. 385. Sir Peter Wyche later proposed that the lectures given by Charleton on the muscles should be printed, and they were given to the council for further consideration. However the matter seems to have been dropped (as they were never printed). I was unable to find them in the Royal Society archives.

This seems to have been Charleton's first active anatomical investigation on behalf of the Society. Interestingly, his 1680 text on muscular motion referred neither to his own observations, nor to the performance of others' experiments at the Society. Instead he drew his discussion from Jacob Müller.[48] All of his medical writings, as we shall see, were characterised by this absence of experimental evidence.

The following week a committee of the Society's physicians was formed 'to order and manage dissections for the society upon every execution-day'.[49] Jonathan Goddard was given charge. Clarke, Goddard and Charleton continued with their dissections. On 20 April Charleton was officially appointed 'to have the care of dissecting bodies for one year.'[50] He immediately 'proposed an anatomical administration to examine Dr Willis' observations of the brain', as the latter's *Cerebri Anatome* had just been printed.[51] By 4 May Charleton had completed the anatomical examination, and 'found Dr Willis's cuts in the *Cerebri Anatome*, as far as he had gone, true'.[52] However, he also affirmed, that he had discovered many differences between the brain of a man and that of a 'brute.' These findings formed the basis of a larger exercise in comparative anatomy, and Charleton was advised to perform further anatomies on dogs, sheep, and cows.[53] Comparative anatomy was one of his continuing interests.

He gave a lecture on this research, entitled 'Certain differences observable betwixt the brains of a man and the brains of all other animals', on 8 June at the Royal Society.[54] In subject matter it bridged the moral philosophy of his non-medical texts and the use of comparative anatomy to analyse the physical basis of the laws of the passions and the operation of the soul. Charleton's findings were published as *Inquisitiones duæ anatomico-physicæ, altera anatome pueri coelo tacti, altera de proprietatibus cerebri humani* (London, 1665).[55] His conclusions constituted a substantial agreement with the anatomical observations of

[48] Charleton, *Enquiries*, Praelection VI.

[49] 27 January 1664, Birch, *History*, vol. 1, p. 376.

[50] Birch, *History*, vol. 1, p. 415.

[51] This was the first of several instances in which Charleton responded to the writings of Thomas Willis.

[52] 4 May 1664, Birch, *History*, vol. 1, p. 421. *Cerebri Anatome* received imprimatur on 20 January 1663 [i.e. 1664]. It went through nine editions between 1664 and 1683 (the first two in London, the remainder in Amsterdam). The book offered a detailed and systematic account of the anatomy of the brain and nerves, and was founded on a dissections and observations, directed by Willis and aided by Millington and Wren and Lower. Thomas Willis, *The Anatomy of the Brain and Nerves*, ed. W. Feindel, Montreal, McGill University Press, 1965, vol. 1, p. 68.

[53] Birch, *History*, vol. 1, p. 422. Only bovine anatomy, however, was mentioned in the lecture.

[54] The lecture itself survives in the Society's Register book. A copy is held in the British Library, Sloane MS 698, ff. 74r-87r.

[55] Translated, *Two Anatomico-Physical Dissertations; the first concerning a boy killed by lightning; the second, of the properties of the human brain*. The publication claimed to be the transcript of the lecture given, it was printed in Latin (although the original presentation was recorded in the vernacular).

Willis, but a number of philosophical disagreements.[56] The comparison between the human and animal brains, it was hoped, would reveal the origin in the human body of the locus of the rational soul which distinguished man from brute. Charleton aimed to discover the physical location and immediate instruments by which 'the Humane Soul performs even those Actions, which though Organical, are yet more perfect and Excellent, than any actions of the sensitive souls of Brutes'.[57]

The broader program was research on the 'admirable Laws' of the human soul, how it communicated its commands, what parts of the brain it used. He hoped to illustrate the extent to which the soul was bound to the laws of matter. Much debate surrounded the presence or absence of the *rete mirabile*, which Galenic theory identified as the site in which animal spirits were generated from vital spirits.[58] Willis' argument about the lack of *rete mirabile* in the brains of men, according to Charleton, contradicted the 'Wisdom of the Divine Architect'. The evidence lay, he believed, in the 'Maxime of universal truth', that every instrument was 'accommodated both to the nature of the agent that is to use it, and to the effects to be produced by the user of it.'[59].

This application of 'maxims' of classical medicine reflects an important aspect of his epistemology in this publication. Charleton's lecture, although delivered at the peak of his involvement in empirical practices, did not invoke them as demonstrations of veracity. His references to 'universal maxims' indicates that empiricist illustration was not uppermost in his repertoire of persuasion.[60] Although he was performing an anatomical demonstration, for which he was obviously respected at the College, the dissected brain itself was neither the real focus of the lecture, nor the generator of conclusive knowledge. The broader purpose of the lecture was the philosophical question of the relationship between mind and matter. The authority of experiment evidently did not replace that of textual evidence, as Charleton's writings did not refer to his numerous experimental forays. He incorporated several observations on the importance of sound to the human brain,[61] possibly drawn from his own find-

[56] It was decided that communication between Willis and Charleton 'shall not be made public without their consent.' Birch, *History*, vol. 1, p. 436. Harley sees the dispute between the two physicians as the result of a conflict over patronage.

[57] Charleton, *Inquisitiones Duæ*. See also 'Differences', f. 87r.

[58] Porter, *Greatest Benefit*, p. 75. The existence of the *rete mirabile* had been controversial. Berengario da Carpi and Vesalius in the previous century had both questioned its existence. Vesalius claimed that it existed only in oxen. A. Vesalius, *De corporis humani fabrica*, III 14, p. 352. Harvey's *Lectures on the whole of anatomy* conceded its existence, (partly because Bauhin, upon whom Harvey relied, supported its presence in human brains).

[59] Charleton, *Inquisitiones Duæ*. See also 'Differences', 1664, f. 74r. This preformationist undercurrent recurred throughout Charleton's writings and in discussions around his findings. Charleton had engaged in his conversations with Pepys and Brouncker on the topic of preformationism; whether the parts determined the nature, or vice versa. The importance of these debates to Charleton is discussed by Osler, 'Descartes and Charleton'. See also Dew, 'Politics of the Body', pp. 4-5.

[60] He also invoked 'sensory evidence' which was quite external to anatomical exploration.

[61] Charleton, *Inquisitiones Duæ*. See also 'Differences', f. 79.

ings on the subject with Hooke at the Society. However, no mention was made of experiment.

Charleton claimed that Willis' denial of the retardation of the blood, contradicted 'the evidence of Sense'. But this 'sensory' evidence was not physiological demonstration, but a reference to the assurance of *common* sense, 'that Rivers divided, and variously crooked in their Channells, flow more slowly and gently.'[62] 'Evident to sense', did not signify that the subject could be observed in the context of the anatomy theatre, but rather through analogy. His demonstration relied more on analogic understanding than on physical observation in the experimental instance. In fact, Charleton's case for the motion of fluid in the body did not require the presence of a body in the anatomical theatre at all. His argument relied far more upon philosophical logic and the application of reason. The body in the theatre was the focus of attention, but not the source of authoritative demonstration.

Indeed this discussion of the brains of men and brutes illustrated Charleton's healthy eclecticism: 'If you like the hypothesis of the Cartesians better' he noted in his lecture, it would 'not be hard to imagine' the conformity of that explanation to the same empirical findings.[63] He reconciled competing hypotheses on the issue of the *rete mirabile*, in the hope that individual failings might be repaired through judicious combination: 'Each of these Opinions thus failing in some respect to each other; and no one of them proving satisfactory a part, taken single; let us see how neer to probability they may be brought, when their most reasonable parts are conjoyned.'[64]

Inquisitiones Duæ presents Charleton as unwilling to make a final determination on the matter at hand. He concluded with a declaration of impartiality: 'I remain in the same State of Uncertainty, concerning the Principal Seat of the Rational Soul in the Brain, and concerning the Oeconomy, of which I complayned in the beginning . . .' He noted that

> If I have adventured to add my Conjectures concerning the Respective Reasons of those Differences; it was not, I assure you, out of a vain opinion of my being more likely than others to find out the Truth: but out of designe to leave some impressions in you of my being as willing and sedulous as any, in searching after it.[65]

This reflected in part the genre of the lecture, in which the gathered audience acted as the judges of what knowledge ought to be treated as authoritative. The

[62] Charleton, *Inquisitiones Duæ*. See also 'Differences', f. 87.

[63] Charleton, *Inquisitiones Duæ*. See also 'Differences', f. 77.

[64] Charleton, *Inquisitiones Duæ*. See also 'Differences', f. 81r.

[65] Charleton, *Inquisitiones Duæ*. See also 'Differences', f. 87. This same refusal to produce a conclusion is echoed in Nicolaus Steno's lecture on the brain in Paris five years later, which commenced with the declaration that the search for a final answer was pure vanity. 'Instead of promising to satisfy your curiosity concerning the anatomy of the brain, I confess sincerely and publicly here that I know nothing about it.' Nicolaus Steno, *Lecture on the Anatomy of the Brain*, ed. Gustav Scherz, Copenhagen, Arnold Busck, 1965, p. 121.

lecturer was required to avoid appearing to dictate the answer. His task was to illustrate and deepen the question, and to highlight the nature of the debate.[66] The genre of lectures is discussed more fully in Chapter VI.

After the neural dissection Charleton's anatomical exercises before the Society seem to have paused, while he turned his attention to the draining of fens, recommended a manner of preventing rust and experimented with velocity of bullets. Charleton's next contribution on dissection arose from his examination of a buzzard (5 October) in which he had found three testes.[67] When the committee of Society physicians for the development of anatomical researches eventually met, on 5 October 1664 at George Ent's residence, they concluded on the following bases for anatomical enquiries at Gresham: the presence of visible passage of the air into the heart, of a vessel in the brain between the *arteriae carotides* and the *brachia fornicis*, and whether liquor might be expressed out of the stomach immediately into the pancreas.[68] The first of these questions, regarding the visible passage of air into the heart, emerged from Harveian researches on the nature of respiration, which necessitated an enquiry into the role of the air in the lungs and heart.[69]

Charleton and Ent performed a dissection to this purpose at Gresham College on 22 October 1664. Their explorations yielded extraordinary results, as Oldenburg reported to Boyle a few days later: 'Dr Charleton affirmed to me of the last Anatomy, yt ye veines on ye right and left side of ye heart were transposed, so yt ye vena arteriosa was where the arteria venosa useth to be, and vicissim.'[70] Charleton was sufficiently confident to mention his findings widely, citing Ent as his witness. Oldenburg heard the tale 'from [Charleton's] own mouth in ye presence of others, who heard him as well as I'.[71] Scarburgh, when he heard of it, observed that he had also encountered anatomical abnormalities in the pectoral muscles. He urged that the body which was already buried, 'might be unburied again', for further investigation. On 3 November Oldenburg informed Boyle that Charleton had been obliged 'to yield, yt he was mistaken' in his claims, 'both Dr Ent . . . and Dr Scarburgh, having upon ye unburying of

66 This eclecticism is also evident in Francis Glisson's anatomical lectures at the College of Physicians, in the mid-1650s. His lecture on the brain claimed that although he was not able to give 'a clear account' of its operation, he offered instead 'some propositions about it'. Glisson's lectures are in the British Library, Sloane MS, 3306. On Glisson's lectures, see Henry, 'Medicine and Pneumatology', p. 18.

67 This experiment gained the notice of Oldenburg, who concurred with the finding in a letter to Boyle, but declared himself 'afraid of HUDIBRAS' (the 'wit' Samuel Butler, who had mocked Boyle), and thus wary of expressing his confirmation. Oldenburg to Boyle 6 October 1664. Oldenburg, *Correspondence*, vol. 2, p. 247. Boyle, *Works*, vol. 5, p. 310-11. Oldenburg noted on the matter that 'There is certainly something as well in the conformation of the parts, as in the temper of animals, that necessitateth them (if I may say so) to such and such operations.'

68 Birch, *History*, vol. 1, p. 475. It seems they had at last taken up Boyle's suggestion (of February), that the desiderata of anatomy be identified.

69 Frank, *Oxford Physiologists*, p. 158.

70 Oldenburg to Boyle, 3 November 1664. Oldenburg, *Correspondence*, p. 273.

71 Oldenburg to Boyle, 3 November 1664; Oldenburg, *Correspondence*, pp. 281-2.

ye heart, and a strict search and examination, found no such transposition'.[72] Discovery that the claim for anatomical abnormality (which Charleton himself had broadcast) was erroneous, 'addeth but very little to his credit.'[73] Mortified, Charleton disappeared from the Society record between this event and 21 December 1664. The passage of air to the heart seems to have been temporarily forgotten.[74]

On his reappearance at the Thursday meetings, Charleton was granted a licence to print his *Inquisitiones Duæ Anatomico-Physicæ; Altera Anatome Pueri De Coelo Tacti, Altera De Proprietatibus Cerebri Humani* (1664). These were both subjects on which Charleton had contributed at Society meetings. His request for a licence to print them at this time may have been motivated by a desire to reinforce his status, after the incident 'concerning ye Transposition'.[75]

In early 1665 Charleton's anatomical trials continued, as did his comparative anatomical researches.[76] However, shortly afterwards on 15 March 1665 he was disgraced again. The Royal Society had received from Captain Graunt some rare poisonous powder, from Makassar.[77] But an impatient Charleton removed this rare powder to his own residence, where he experimented by poisoning dogs. This behaviour was 'contrary to the order of the Society', for which transgression Charleton was summoned to account for himself at the following meeting.[78] He seems to have escaped punishment. These experiments were satirised by Samuel Butler in 1665, in his 'Occasional Reflection on Dr Charlton's feeling a Dog's Pulse at Gresham College. By R. B. Esq.'[79] Butler's verse illustrated that Charleton's theft, and his subsequent reprimand, were known outside the

[72] Oldenburg to Boyle, 3 November 1664; Oldenburg, *Correspondence*, pp. 281-2.

[73] Scarburgh's observation, however, 'concerning ye defect of ye Pectorall muscle' was confirmed, 'Dr Ent himselfe, upon my particular inquiry asserting it wth all confidence.' Oldenburg to Boyle, 3 November 1664. Oldenburg, *Correspondence*, pp. 281-2.

[74] See Frank, *Oxford Physiologists*, p. 158.

[75] That the incident was to have been widely known, is suggested by reference to a 'fam'd Anatomist' who 'may sometymes slipp into light unaccuracyes', in a letter from Fairfax to Oldenburg in 1669, 30 April 1669, Oldenburg, *Correspondence*, p. 504.

[76] On 8 February 1665 he dissected a diving bird to reveal its respiratory system. Birch, *History*, vol. 2 p. 13. See also 15 February 1665, Birch, *History*, vol. 2, p. 16.

[77] This was to be used in various injection trials on dogs and cats, some of which Pepys witnessed, and recorded in his diary. On 15 March 1665 Pepys saw the poison injected into the veins of a dog. A further trial was made on 19 April, at which both a dog and cat were poisoned, but again no dramatic results ensued. 16 May 1664, 15 March and 19 April 1665. See M. H. Nicolson, *Pepys' Diary and the New Science*, Charlottesville, University Press of Virginia, 1965, pp. 67-8.

[78] That his illicit trials were a failure was duly noted in Birch's *History*. At his hearing Charleton claimed that 'he had opened and tried it' before another witness. 29 March 1665, Birch, *History*, vol. 2, p. 28. The Makassar poison was designated for injection trials, and was to contribute to a series of experiments on 'the infusion of opiates and poisons into the veins of animals'. Charleton must have been up to date with new procedures in transfusion and injection. He exchanged information with Boyle at the meeting on 18 April 1666. Birch, *History*, vol. 2, p. 83. Experiments on transfusion ceased until 20 June 166, and then again until 26 Sept of that year. Birch, *History*, vol. 2, pp. 98, 115.

[79] Butler, *Genuine Remains*, p. 404.

Society. Nicolson suggests that Charleton's misdemeanour was probably an object of humour in the coffee houses of London. The instance indicates his active interest in experiment, and again highlights the absence of such activities (and of conclusions drawn from them) in his texts. He seemed to be writing for an audience for which textual and reasoned knowledge dominated public presentation.

Between 28 June 1665 and 21 Feb 1665/6, Royal Society meetings suffered a long interruption as a result of the plague. When they reconvened Charleton sought reports of the epidemic from those who had remained in London during its ravages.[80] He had himself left London during the plague, spending this period in Cheshire.[81] Indeed in the mid-1660s he spent less and less time in the capital, and was absent for protracted periods from around 1666. One reason for his removal from London might have been the controversy caused by the publication of *Chorea*. Nevertheless, he continued to contribute objects and curiosities to the Society, through intermediaries, after his final recorded appearance at meetings in April 1668.[82]

Charleton maintained his friendship with various notables, and continued to correspond with Aubrey, who informed the isolated physician of recent discoveries and discussions.[83] At some stage during this period Charleton fell out with his long-time friend and patron Lord Brouncker.[84] The physician's absence from London coincides with the later part of Brouncker's fourteen-year presidency of the Royal Society. At this stage, Charleton would have been hoping to accumulate patrons and influence to support his entry to the College.[85]

During his years as an active participant in the experimental life of the Royal Society, Charleton composed several works. These included those works already discussed: *Inquisitiones duæ* (1665); *Onomasticon Zoicon* (1668, 1671) and *Two*

[80] Charleton once again demonstrated his familiarity with recent theoretical literature, introducing the theory of 'vermination of the air', which was 'started in England' by Ent, and subsequently developed in Italy by Kircher. Charleton was to take down the report of 'the masters of the Pest-house', who had 'promised him their observations of the plague'. Birch, *History*, vol. 2, p. 69. However there is no evidence that he tested these theories further.

[81] When the physicians of the Society were ordered to collect observations of the plague, Glisson and Wharton were named as physicians who had remained in London during the epidemic, but Charleton was not. 28 March 1666, Birch, *History*, vol. 2, p. 76.

[82] Austen presented a bird for Charleton in 1666, letter from Hyde. On 29 June 1687 Hooke presented on Charleton's behalf 'some salt shot upon a stick like sugar-candy', the origin of which was said, in the accompanying letter, to be Siberia. Birch, *History*, vol. 4, p. 544.

[83] In February 1671 Charleton was grateful to Aubrey for his bulletin on the contents of the recent *Philosophical Transactions*, 'whereof my present retirement & solitude had made me ignorant.' Charleton to Aubrey, 4 February 1671, British Library, Add. MS Eg. 2231, f. 166.

[84] Pepys recorded dining with both men on 28 July 1666, but their closeness had ended by February 1671, when Charleton sent an angry letter to Aubrey, in which he declaimed Lord Brouncker as 'a rock, on wch I have bin shipwrackd'. Charleton to Aubrey, 4 February 1671, British Library, Add. MS Eg. 2231, f. 166.

[85] His practice in London would have been affected by the plague and the fire. Perhaps this was the reason he moved from Russell to York St.

Philosophical Discourses Concerning the Wits of Men (1669, 1675, 1692). In addition to these pathological, anatomical and taxonomic treatises, Charleton produced *An Imperfect Portraicture of His Sacred Majesty Charles II* (London, 1661)—a propaganda piece for the newly restored monarch, lauding his great piety. Charleton also translated into Latin Margaret Cavendish's biography of her husband, the Duke of Newcastle, published in 1668 as *Guilielmi Ducis Novocastrensis Vita*. Of these only a couple were connected with his Society researches: *Onomasticon Zoicon*, which reflected the taxonomic aspirations of the Royal Society, and *Inquisitiones Duæ*, which drew upon his experimental findings. Both of these works were published in Latin, presumably for a limited readership comprised of members of the 'republic of letters'.

Charleton's actions can be used to situate him as a 'classic' Royal Society investigator. He published some works which reflected his experimental activities, and these might lead us to identify him with the experimental natural philosophy so widely associated with the scientific revolution period. However Charleton's experimental activities were not reflected in his self-construction. Indeed the discrepancy between action and representation in his later medical works suggest a more complex picture of the tensions between the authority of physicians and their activities at the Royal Society. Charleton's medical writings show no evidence of the empiricist epistemology supposedly required for the legitimation of authority in the Royal Society context. I now aim to illustrate the relevance of this tension for Charleton, and the manifestation of this tension across his membership of the two institutions: the Royal Society and the College of Physicians. The Royal Society experiments were certainly prolific, but perhaps, to Charleton, not prominent. While it is relevant to discuss what he did, my primary emphasis is upon what he *said* about what he did—how his practices were reflected publicly in his self-construction.

PUBLICATIONS AFTER THE ROYAL SOCIETY

It is impossible to determine the reasons for Charleton's withdrawal from the Royal Society. His activities there had ceased before he was incorporated into the College of Physicians. Between the height of his Society contributions and the acceptance of his fellowship at the College, Charleton was very busy indeed. He practised among the Cheshire gentry, earning their 'favour and esteem', and returned occasionally to London to deal with personal and financial affairs.[86] He resided with the Crewes of Crewe-Hall. From the house of his new patron

[86] On 4 February 1671 Charleton wrote to Aubrey that he would be returning to London before mid 1671, 'my domestic affairs and cares requiring me to return to my poor home' in Covent Garden, occupied by his eldest daughter Mal. Charleton to a friend of Wood, 20 January 1671; Bodley MS Wood 40. Likewise his reference in 1672 to 'ten weeks' solitude' implies that his visits were interspersed with periods in London. Charleton, *Passions*, preface.

Charleton wrote to many of his acquaintances and attempted to strengthen his patronage network.[87]

Charleton published several works, between his participation at the Society and 1676, possibly to improve his profile at the College. All were heavily derivative of contemporary writings. They were founded on textual rather than experimental knowledge, contrary to the espoused Royal Society commitment to empiricism. In 1672, he published a treatise on scurvy, bound with an attack on quack medicine: *De Scorbuto*.[88] The inclusion of the tirade against unlicensed medical practitioners was probably connected with Charleton's attempts to gain entry into the College, which was concerned with the regulation of medical practice. The volume was dedicated to the 'very Eminent, and very skilled man, Dr George Ent', then President of College of Physicians.[89] Ent was already well known to Charleton, and as College President represented an important contact.

De Scorbuto evidently generated considerable interest, both in England and on the Continent.[90] As I mentioned in Chapter II, the text discussed the nature, origin and cause of scurvy. Charleton believed that scurvy originated in chemical components, and traced its origin in rancid blood, its birth from fixed salt, and its acid genealogy. He examined the diverse names of scurvy, and explained their etymology, the remote and external causes, the question of contagion. In addition his volume explored the issues of particular interest to physicians: the 'Diagnostic indicators', and 'Prognostic signs' as well as therapeutic recommendations for the range of different kinds of the disease. He concluded with a discussion of secondary symptoms, and a diatribe against quacks, whom he saw as incapable of identifying and treating the different forms of scurvy. The work was perhaps composed in response to Willis' 1667 publication on scurvy in *Pathologia Cerebri*.[91] *De Scorbuto* included no experimentally-based material,

[87] Letters from this period exist in the Bodleian Library, MS Smith 13.

[88] Charleton's notes on this subject exist in the British Library, Sloane MS 1532, ff. 41-61.

[89] Charleton referred to Ent as an 'Outstanding Ornament of both his country and his era . . . in which there is no one more precise and more learned yet not more frank and kind in his friendship. And concerning himself and his own efforts deserving the greatest praise, whatsoever of a little book is this the though meagre yet loyal testimony of its esteem and respect and reverence, according to the ancient formula of invocation. WITH GOOD WILL. IN PRAISE [DESERVEDLY] HE HAS GIVEN. HE HAS DEDICATED. Walter Charleton.'

[90] Charleton's *De Scorbuto* was mentioned in correspondence between Henry Oldenburg and René François de Sluse. Oldenburg to Sluse, 21 November 1671, Oldenburg, *Correspondence*, vol. 8 (1671-2), p. 371. Sluse was Canon of Liège. Sluse responded that he would seek the book in Holland, as he was extremely interested: 'For the symptoms of this disease . . . are varied and strange, and I shall willingly learn the reasons accounting for them from this learned author who writes in succession to the distinguished Willis.' Sluse to Oldenburg, 17 December 1671. In Oldenburg, *Correspondence*, vol. 8, p. 411. It was also discussed between Esaie le Bourgeois and Oldenburg. Le Bourgeois to Oldenburg, 31 October 1671, Oldenburg, *Correspondence*, vol. 8, p. 328.

[91] Thomas Willis, *Pathologia Cerebri & Nervosi Generis Specimina, in quo agitur de morbis convulsivis et de Scorbuto*, 1667.

but discussed the chemical bases of scurvy, showing a heavy debt to Willis.[92] Charleton debated the relative significance of different chemical components, rather than drawing from patient reports or experimental procedures. As with his refutation of Willis' cerebral anatomies, Charleton did not assert any competing experimental findings, but offered *disputations* of the philosophical principles.[93] This followed the scholastic method of disputation, rather than a rebuttal through empirical demonstration.[94]

Once again Charleton declared the provisional nature of his own conclusions, and his preface explored the discrepancy between truth and the appearance of truth, in which balance he held that pathology possessed a high degree of certitude. Again he drew extensively from Greek and Latin texts, and quoted from Seneca, whose methods he claimed to imitate.[95] In this medical volume, Charleton's first publication after his removal from Royal Society activities, he invoked not the knowledge generated in the community of natural philosophers, but instead the identity of a solitary scholar. He constructed an identity quite at odds with the experimental philosophy. Charleton indicated his adherence to textual learning, and made evident the origin of his material in the writings of others. The contrast between Charleton and Willis again suggests that there was more that one possible identity for physicians. I do not hope to speak for them as a professional category, but merely to illustrate some characteristics of Charleton, and to suggest that they aided his professional identity in ways not often recognised.

In 1674 Charleton's anonymous *Natural History of the Passions* was published. In the forced seclusion in which he composed the work, he once again engaged with the arguments of Thomas Willis. The latter's *De Anima Brutorum* (1670) was one of the few texts that Charleton had with him in the country.[96] The texts he had with him appear to have included the works of 'those three excellent men, Gassendus, Des Cartes, and our Mr Hobbes', as well as Kenelm Digby.[97] In contrast with the innovative activities of the Royal Society, this book reaffirmed Charleton's commitment to the traditional practice of physic. Professional physicians, according to ancient tradition, emphasised the

[92] Willis' text was more strongly clinical (like the majority of his work) than Charleton's. Willis included detailed analyses of the various available cures, and offered the reader some clinical case studies.

[93] Charleton also used this method in his lecture 'Certain differences'. For example, he invoked the ancient principle that nature acted according to greatest efficiency, to dispute Willis' interpretations.

[94] On scholastic disputations see French, *Harvey's Natural Philosophy*.

[95] 'I have not slavishly delivered to myself anything praiseworthy from those who previously have written concerning the same debate, I have attributed much to the discernment of many men; and I lay claim to something from my own.' Charleton, *De Scorbuto*, p. A6v.

[96] He also had some notes he had previously made 'out of the best Authors'.

[97] In particular, Digby's *Two Treatises*. Thorpe sees the inspiration for Charleton's volume on the passions arising from Hobbes' work on the same subject, by whom the text is strongly influenced. Large sections of Charleton's work were drawn almost directly from Hobbes' *Elements of Law*, and the impact of the *Leviathan* is also evident. The text also demonstrates Charleton's familiarity with Aristotle, the Scholastics and the Cambridge Platonists. Thorpe, 'Two disciples of Hobbes', p. 181.

relationship between good health and virtue, and aimed to treat not only the body but to regulate the regimen of their patients.[98] The *Natural History of the Passions* exemplifies this concern of a practising physician, of providing learned counsel. The principle of self-knowledge as a key to health, prominent in Charleton's *Passions*, referred to an essentially classical definition of self-discipline toward achievement of well-being and the prolongation of life. Charleton concluded the volume with a recommendation for 'the *Ethicks* of *Epicurus*' as second only to the Holy writ as a 'Dispensatory . . . of Natural Medicines for all distempers, incident to the mind of Man.'[99] The orientation of the *History of the Passions* seems to owe more to established ideas of physic than to the new philosophy's epistemology of experiment.[100]

From his publications of this period we can see that Charleton did not attempt to place himself within an experimental context, nor did he associate himself actively with a Royal Society agenda. Rather he continued to publish works founded conspicuously upon textual learning, and which in some cases might be seen as commentaries on recent literature.

CHARLETON IN THE COLLEGE OF PHYSICIANS

Charleton returned to London at some stage during the mid-1670s. His prospects there may have become more promising after the death of his competitor Thomas Willis in 1676, and Brouncker's ejection from Presidency of the Royal Society in 1677. However, his presence at the Society meetings was not noted after his return to London. Instead he seems to have focused upon the College of Physicians. His eventual admission to full status at the College took place on 23 January 1676, when he was 'by a unanimous vote admitted into the order of Fellows'.[101] The College was particularly weak in the 1670s, as a consequence of the Plague of 1665, the Great Fire in the following year, atop challenges from competing organisations.[102] It is difficult to know whether Charleton's admission was affected by the institution's weakness at this time.

Once incorporated he rose rapidly to prominence.[103] He quickly acquired the prestigious position of Censor, and then Elect.[104] As his status in the College hierarchy improved, he became increasingly involved in the regulation of med-

[98] Cook, 'Good advice', p. 14.

[99] The second edition of *Epicurus' Morals* had been printed in 1670, and Hunter & Cuttler suggest that this was almost certainly an added promotion for it, 'Mistaken Identity', p. 91.

[100] While the *Passions* made reference to the physiological and material aspects of the passions, innovations deriving from the rise of the corpuscular philosophy, its emphasis was on the importance and nature of the passions in accordance with traditional physic.

[101] *Annals*, vol. 4, f. 121b.

[102] Cook, 'Institutional Structures', pp. 104-6.

[103] By October of the following year he had achieved the status of second Censor, and in August 1679 he became temporary Registrar. Brown, *Animal Oeconomy*, p. 173. *Annals*, vol. IV, pp. 184-98.

[104] The 'Elects' were an 'inner circle of eight college elders'. Cook, *Trials*, p. 11

ical practice.[105] He eventually took the position of College President, from 1689-1691, a period in which the College 'set about reordering its affairs and trying to regain some unanimity, although matters only quieted for a while'.[106] His profile was high not only in England, but also in the overseas centres of medical learning. Charleton's medical writings in general were evidently known on the Continent. He was cited by Thomas Bartholin, Henry Power, William Croune, Olaus Worm,[107] Antony Deusing[108] and Marcello Malpighi. The fact that he was offered the chair of anatomy in Padua, in 1678, illustrates his high position as a writer and teacher at this time.[109]

During his years at the College of Physicians Charleton composed his main physiological and anatomical works, which have scarcely been analysed or discussed by historians. I will devote the next two chapters to a detailed discussion of two of these medical works: his *Enquiries into Human Nature* (given as lectures in 1678 and printed in 1680) and the *Three Anatomic Lectures* (delivered and printed in 1683). My interest is not in Charleton's lectures as indicative of the philosophical orientation of the College, but in the relationship between his self-construction and epistemology in these works, and the constraints and conditions exerted by performance of anatomical lectures within the College context. Chapter VI examines the intersection between Charleton's personal experience and the lectures as a public expression of his identity as a physician.

Unfortunately we have no reliable indications of Charleton's activities at the College. The College's *Annals*, unlike the Royal Society's records, do not describe research projects, but only detail the College's regulatory activities. Thus we are unable to compare his activities and textual presentation during this vital period in Charleton's later career. However, the lack of information has not deterred scholars from extrapolating. Although there are no eyewitness accounts of the performance of anatomical lectures at the College, there is considerable literature on public dissection, some of which suggests new perspectives on Charleton's practices.

Anatomical lectures and dissections during the seventeenth century were circumscribed within the traditions and strictures of the College of Physicians.[110] Medical lectures were occasionally given at Gresham College,[111] attended by an

[105] In March 1688 Charleton himself was called before the censors for 'disparaging' the practice of a Dr Blackmore. *Annals*, 22 March 1688.

[106] Cook, *Decline*, p. 220.

[107] See Charleton's papers in the Bodleian Library, MS Smith 13.

[108] Although Deusing and Charleton disagreed on some points, Deusing referred to the 'excellent' authority of Charleton, in the epistle dedicatory to *Exercitat Physico-anatomicar. de ultimo Anuimalium nutrimento*. See Charleton's papers in the Bodleian Library, MS Smith 13.

[109] Charleton's letters to Padua are contained in his papers in the Bodleian Library, MS Smith 13.

[110] College statutes permitted no fellow to dissect without its the permission. See Clark, *College of Physicians*. For this reason private anatomies were not generally performed.

[111] Thomas Winston, *Anatomy Lectures at Gresham College*, was published in 1659. Specific topics were left to the discretion of the lecturer, but the presentations were to follow Fernel's tripartite division into physiology, pathology and therapeutics.

audience of London citizens and foreigners.[112] The audience of lectures at the College was often largely comprised of physicians and surgeons. However the performance of public anatomies illustrates the interest of the broader community also.[113] Charleton's lectures, which revolved around the more philosophical dimensions of medical anatomy, would have attracted an audience who were unlikely to apply any of the information he conveyed. In this sense they are a fascinating indicator of how a physician might choose to present his profession before an audience composed of both lay and trained listeners.

While surgical lectures often had a training function, Charleton's 'prælectiones' were clearly designed for theoretical, rather than practical, edification. The publication of these philosophical lectures, which consisted largely of an overview of contemporary medical philosophy, was very unusual. There seem to be no other publications comparable with Charleton's published lectures.

His lectures make a stark contrast with the published surgical lectures of the time, which focused explicitly on the body's physical components. By comparison, the 1659 lectures of surgeon Alexander Read, delivered to barber-surgeons,[114] show an emphasis completely different from Charleton's. Read's text is full of practical directives, detailed specifics and diagnostics. The contrast highlights the philosophical and abstract nature of Charleton's anatomical lectures, and illustrates the great difference between the areas of knowledge of physicians and barber-surgeons.

Broad interest seems to have surrounded medical philosophy in relation to the anatomy of the body in the late seventeenth century. Anatomy bridged the disciplines of medicine, philosophy and theology,[115] and a physician's anatomical lectures thus situated his subject within broader philosophical and moral issues.[116] The wide audience indicated by the list of subscribers to John Browne's 1681 anatomy of the muscles suggests widespread interest in anatomy among groups who had little to do with the practice of physick. Evidently anatomical

[112] Winston complained to the College that his lectures were not particularly popular or well attended. Clark, *College of Physicians*, vol. 1, p. 257. Although there was no grant for the provision of cadavers, evidence suggests that they were used in demonstration. A. Cunningham, 'The kinds of anatomy', *Medical History*, vol. 19, 1975, p. 13.

[113] Cunningham, 'Kinds of anatomy', p. 11.

[114] A. Read, *The Workes of that Famous Physitian Dr Alexander Read, Containing Chirurgical Lectures of Tumors and Ulcers . . .*, London, 1659.

[115] Cunningham, 'Kinds of anatomy'. This point is echoed by C. Lawrence in 'Alexander Monro Primus, and the Edinburgh manner of anatomy', *Bulletin of the History of Medicine*, vol. 62, 1988, p. 213.

[116] As Randall Albury notes, anatomical knowledge, 'addressed issues of theological doctrine and state power as well as questions of bodily form and structure'. W.R. Albury, 'Confessional anatomy in the Counter-Reformation period: "Self-dissecting" anatomical illustrations in the 17th century', in 'Individuals and Institutions in the History of Medicine, Proceedings from the 6th Biennial Conference of the Australian Society of the History of Medicine', 1999. These broader implications are recognised in recent scholarship on the political resonances of Charleton's medical writings, see for example Dew, *Politics of the Body*, and Akihito Suzuki, ' "A Duumvirate of Rulers Within Us": Politics and medical pneumatology in Restoration England', in Marshall (ed.), *The Restoration Mind*.

knowledge was not primarily circulated for the purpose of informing practice.[117] Therefore the audience for these lectures of Charleton's would have been diverse, and not necessarily interested in the practical implications of his arguments. The lectures themselves evince a distinctive eclecticism. The lectures were not oriented toward therapeutic application, but covered contemporary developments in medical philosophy for an interested audience. Charleton does not appear to have attempted to secure consensus for any *particular* theory (although he showed an obvious sympathy for Glisson's hypotheses). The lectures rarely referred to experimental evidence, and far more frequently invoked analogic demonstrations or rational proofs. Again his self-presentation is that of a classicist and scholar, rather than an active empiricist.

In August 1680 Charleton presented the Harveian Oration, in commemoration of the benefits received from William Harvey and other supporters of the College.[118] He was to perform the role of orator again in 1705. Charleton delivered a further series of anatomical lectures on 19, 20 and 21 March 1682/3.[119] These were published as *Three Anatomic Lectures* (London, 1683). They dealt with 'The Motion of the Blood through the Veins and Arteries', 'The Organic Structure of the Heart', and 'The Efficient Cause of the Heart's Pulsation'. These lectures are discussed in more detail in chapter VII, where I consider their epistemological strategies, as well as Charleton's self-construction.

Charleton's only non-medical composition during his fellowship at the College (apart from a translation from Greek in 1684 of *The Life of Marcellus*) was *The Harmony of Natural and Positive Divine Laws* (London, 1680). This was published in the same year as the *Enquiries. Harmony* was published anonymously, and contained no preface from the author. The publisher's letter to the reader presented the work in generic fashion as that consequence of the author's desire to 'confirm his Faith by inquiring into the Reasonableness and Purity of it, and to augment his Piety toward God.' The composition he believed 'worthy a Philosopher and a Christian'. *Harmony* argued that man, as a rational agent within the divine design, was destined to rediscover the principles by which his creation had been effected. The precepts of living which could be discovered through the exercise of reason were 'the very same that are promulgated by the Divine Majesty for the laws of the Kingdom of Heaven'.[120] The *Anatomic Lectures* and *Enquiries* revealed and explained these laws as they affected physiological and anatomical organisation.

[117] The list of subscribers indicates that even strictly descriptive anatomical texts attracted interest from theologians, gentlemen, prominent College physicians, Professors of Physick, Astrology, History, Anatomy, Geometry, Theology. Several doctors of theology purchased copies of the text for themselves and their college. See John Browne, *A Compleat Treatise of the Muscles*, London, 1681.

[118] For further detail on the Harveian Oration, see Munk, *Roll*, vol. 3, p. 360.

[119] Brown, *Animal Oeconomy*, p. 179.

[120] Reason was 'the very law of nature', given by God to every man 'for the rule of his actions'. Charleton, *Harmony of Natural and Positive Divine Laws*, 1682, pp. 8-9.

Charleton's final medical publication examined the causes of 'feminine disorders' and rheumatism of the womb, entitled *Inquisitiones Medico-Physicæ, de Causis Catameniorum, et Uteri Rheumatismo*. This treatise on the physiology and pathology of menstrual flux was published in London with the College's Imprimatur in 1685.[121] The volume was printed and distributed in Leiden, a year later.[122] Its nine chapters recorded the opinions of both ancient and modern authorities. Its structure was typical in that it commenced with a record of the names given to the uterus by Greek and Latin scholars and their associated meanings. He reviewed the nature of the womb (which, he decided, could accurately be described as a 'workshop'), the ultimate cause of catamenia the opinion of the ancients and more recent accounts of causation. He concluded with a genealogy of uterine rheumatism. Theodore Brown argues that the critique of Willis in *Inquisitiones Medico-Physicæ* illustrated the emergence of the debates *within*, rather than *about* iatromechanism, indicating prevailing acceptance of the validity of mechanistic medical explanation.[123] However, it seems viable also to read the significance of the minor refutations in a different way.

Charleton, in line with Royal Society natural philosophers, espoused an aversion to dogma. However, I believe this was for different reasons. Charleton, and perhaps the College, wanted to illustrate his knowledge of contemporary and classical medical theories, and also wanted to promote the continued relevance of traditional knowledge. Eclecticism allowed him to assert the accordance between them, through a number of devices, as we saw in *Natural History*, and as we see again in *Enquiries* and, to a lesser extent, *Three Anatomic Lectures*. He rejected the 'Opinion of the Ancients about the cause of the circuit of Catamenia'. However this did not indicate a rejection of their ancient status, as he promptly dismissed Willis' new theory of uterine ferment. A lengthy 'Digression' on fermentation rejected the other author's assumptions about the centrality of that process.[124] Charleton did not simply protest against the content of the theory, with which he seemed to agree in spirit. Instead he attacked Willis' methods, and the way in which the notion of fermentation had been 'set up as a principle'. He complained that virtually every function or action integral to an animal had been attributed to fermentation.[125] A similar argument

[121] The College edition was formally endorsed by the Censors: Sir Thomas Witherley, Knight, and President; Dr Samuel Collins, Register; Dr Thomas Burwell, Dr Peter Barwick, Sir Thomas Millington and Dr Humphry Brooke.

[122] Charleton, *Inquisitiones Medico-Physicæ, de Causis Catameniorum sive Fluxus Mensui; nec non Uteri Rheumatismo, sive Fluore Albo*, Lugduni Batavorum, 1686.

[123] Brown argues that the volume 'informed the public that the College of Physicians was at the advancing forefront of science.' Brown, *Animal Oeconomy*, p. 180. I disagree with this analysis.

[124] The twenty-six page digression argued against the possibility of fermentation in the blood of living animals. Charleton, *Inquisitiones*, pp. 54-80.

[125] Charleton, *Inquisitiones*, pp. 54-5. He engaged in the typical method of attack, which involved a refutation of the correctness of Willis' terminology, and a claim that his peer's theory had been too dogmatically asserted. For a discussion of traditional methods of disputation, see French, *Harvey's Natural Philosophy*.

appeared in various other texts of Charleton's. He attacked conclusions that he argued were drawn from unquestioning adherence to particular doctrines. In this sense he invoked the identity of an eclectic to attack the apparent dogmatism of an opponent. Thus in this medical text he invoked the avoidance of dogma which characterised an aspect of the Royal Society's self-description.

Inquisitiones advanced Charleton's own theory of causation, a hypothesis based on Ent's theory of nutritive nervous juice.[126] Charleton claimed that uterine disorders were direct mechanical consequences of depraved 'alible juice'. This contradicted Willis' argument for 'fermentative imbalances in superabundant blood'.[127] Charleton's argument invoked Boyle's investigations on the chemical composition of blood.[128]

Charleton seems to have been simultaneously a supporter of authority and an eclectic prepared to seek truth wherever it lay, thus putting him within the Royal Society's methodological ambit.

Charleton's final medical publication was written thirty-five years after the publication of his first. Although prominent in College affairs for a further twenty-one years, until his death in 1706, he composed no more medical texts after 1685. The main purpose of his publications was perhaps to gain him access to the key professional institution of his time, and to attract patrons. In the latter gambit, he seems to have been less than successful, but by gaining the support of the College he seems to have been supported to some extent in his publications.

Because Charleton was involved extensively with experiment, many historians assume that he was a self-styled 'experimental natural philosopher'. However I have argued that to define him according to his experimental activities would be to place his practices above his self-description, as constitutive of his professional identity. His writings did not emphasise the details and performance of experiment, alleged by some historians to define natural philosophical discourse. The meaning he attached to experimentalism reveals that the physician's identity was determined by more complex factors. Medical texts in Charleton's era generated a public profile of the physician as a learned and grave professional. The physician's medical writings were thus designed toward very different ends from that to which his experimental life was oriented. His lectures, even more than the texts that were never publicly delivered, depicted him as eclectic and scholarly. The texts I have mentioned in the present chapter suggest that, during and immediately after his Royal Society involvement, his texts did not cite experimental evidence. Instead he maintained the methodological eclecticism that characterised *Natural History*.

126 Ent's theory was as put forward in *Apologia pro Circuitione Sanguinis*, London, 1641. It was further elaborated and defended in his *Animadversiones in Malachiæ Thrustoni*, London, 1672.

127 For a discussion of their differences see Brown, *Animal Oeconomy*, p. 180

128 He referred to Robert Boyle's publication of the previous year *Natural History of Human Blood*, London, 1684. Charleton, *Inquisitiones Medico-Physicæ*, pp. 72-3.

Classical models of the relationship between practitioners and their patients bound physicians to tradition in a way that did not apply to natural philosophers. Adherence to this classical ideal, rooted in the Hippocratic tradition, was central to a physician's identity. It could not be, nor would it be, discarded, partly because it was implicated in a system of not only philosophical but also practical meaning. These dimensions are sometimes neglected by historians. Physicians were concerned to present themselves not as generators of 'matters of fact', but as responsible, educated providers of medical counsel. Self-depiction as an experimental natural philosopher would not have aided the physician's professional profile. Experimentalism, in contrast to the tried and tested methods of Galenic medicine, must have seemed newfangled and unappealing. Little could be gained by promoting oneself as a proficient experimenter, when one's clientele were unlikely to view such a qualification with favour. Although many physicians were involved in scientific activities, it was becoming evident that association with the new science might do as much harm as good to the learned physician.[129]

The identity of a physician-philosopher was also moulded by the need to gain admission to London's medical regulatory and licensing authority, the College of Physicians. The College leaders were trained with scholastic, rather than practical, emphasis, and the institution valued traditional learning. Ancient authorities also appealed to the aristocratic patrons upon whom physicians were dependent, and because they formed the foundation upon which professional regulation was built. These reasons help us to understand why Charleton's writings emphasised scholastic authority, rather than experimental activities in the Royal Society.

The next two chapters explore in detail Charleton's lectures, delivered at the College of Physicians between 1678 and 1683, and subsequently published. I will illustrate that after the extensive anatomical practices, his sources of authority were consistently scholarly. This highlights many of the issues raised in the present chapter, regarding the influences upon Charleton's identity as a professional physician. The following chapters also explore the distinctive nature of Charleton's anatomical lectures, and their probable audience.

[129] Cook, *Decline*, p. 181.

CHAPTER VI

ENQUIRIES INTO HUMAN NATURE (1680): CHARLETON'S ANATOMY AND PHYSIOLOGY AFTER THE ROYAL SOCIETY

The lectures which constituted the basis of *Enquiries into Human Nature* (hereafter *Enquiries*) were delivered in March 1679. They examined the traditional vital functions in anatomical terms (nutrition, life and voluntary motion), with additional lectures on fevers and anatomy of the stomach. Though the material overlapped with the 1659 *Natural History*, these 1680 lectures were distinguished by their examination of anatomical knowledge of each subject, and their structural, rather than physiological, emphasis. Like many anatomical texts of its period, *Enquiries* was structured around six sections (the prælections).[1] The first three focused on nutrition: 'On nutrition', 'Historia ventriculi' and 'The actions and uses of the ventricle'. The fourth lecture was unusual in its subject matter—'Of life' engaged (inevitably) with theological questions, and also incorporated circulation theory into a corpuscular theory about the nature of life. The fifth lecture, 'Of fevers', was the only one to cover an explicitly clinical and therapeutic question. The final lecture was 'Of motion voluntary', a discussion which echoed *Natural History*'s concluding chapter.[2]

Historians' interpretations of *Enquiries* have been scarce, but with few exceptions they converge on the coherence of Charleton's agenda in terms of experimental practice. Given what we know about the physician's Royal Society activities (as detailed in the previous chapter) we might expect these lectures to endorse the experimental life in which he had been participant. The other main claim made by historians about these lectures is that they illustrate Charleton's turn toward mechanistic philosophy and its application to medicine. The Royal Society's famed emphasis upon philosophical eclecticism might have reduced Charleton's dogmatism, or led him to explicitly reject classical knowledge. I consider whether he consciously identified himself as a virtuoso in the *Enquiries*.

1 Other texts in which the body was divided, for discussion, into six sections, include: Alexander Read, *The Manuall of the Anatomy*, London, 1642; Jean Riolan, *A Sure Guide to the Best and Nearest Way to Physick*, London, 1671; John Browne, *Myographia Nova*, London, 1698.

2 The first, second and third lectures were given on a Friday. All discussed the anatomy and actions of the stomach, and were thus distinct from the later three lectures. The audience had an intermission for a meal between the dissection of the stomach and gullet and the explanation of the actions and uses of the stomach! The fourth was presented the next day, the fifth and sixth the following week. Charleton, *Enquiries*, pp. 384 and 429.

This chapter considers how Charleton balanced the authority of medical knowledge with the importance of new experimental research.

My aim here is to investigate how the author demonstrated his claims in these lectures—his sources of authority, methods of proof, and the differences in both method and content between this work and the earlier *Natural History*. I consider whether the act of experimentation was Charleton's final goal, and whether his emphasis was, as some have argued, consistently mechanistic.

HISTORIOGRAPHICAL PRESENTATIONS OF CHARLETON AND HIS ANATOMY

The most extensive discussions of *Enquiries* can be found in the works of Brown and Nayler.[3] Others mention the medical lectures in passing.[4] Brown sees these lectures as integral to the revolutionary change in the philosophy of the College of Physicians during the 1660s and 1670s. Charleton's 1679 lectures, according to Brown, conveyed the College leaders' explicit support for Willis and for mechanism. He claims that the adoption of mechanical philosophy was founded on the recognition that Collegiate medicine needed to be modernized in order for the institution to withstand the challenge represented by other institutional and professional groups (the Royal Society and apothecaries respectively).[5] He believes that the College adopted mechanical philosophy in order to 'forestall further charges of ignorance, intellectual backwardness, or conceptual confusion.'[6]

Brown argues that the College projected a unified intellectual approach in the 1670s,[7] reaching a crucial point around 1680, when Charleton, amongst others, began to encourage the institution to move in a new direction, aiming to transform it into 'a potentially powerful offensive agency.'[8] Charleton was thus the mouthpiece for the College's declaration of its commitment to 'full public support for the iatromechanical program'.[9] His mechanist argument demonstrated the authority claims of the College as an innovative and competitive institution. Such a view depicts the physician once again as the symbol of transformation, an individual who pushed the College of Physicians to a new frontier of knowledge, changing that intransigent

[3] See Brown, *Animal oeconomy*; and Nayler, 'Insoluble Problem.'

[4] The lectures are mentioned in Jonathan Sawday, *The Body Emblazoned: Dissection and the Human Body in Renaissance Culture*, London, Routledge, 1996. Earlier readings, such as that of Rolleston, describe *Enquiries* as an 'extensively rewritten' version of the *Natural History*. Rolleston, 'Walter Charleton', p. 413. This description is taken up by P. Krivatsy, *Seventeenth Century Printed Books*, Maryland, National Library of Medicine, 1989, p. 236.

[5] Brown, *Animal Oeconomy*, p. 175.

[6] T. Brown, 'The College of Physicians and the acceptance of iatromechanism in England, 1665-1695', *Bulletin of the History of Medicine*, vol. 44, 1970, p. 29.

[7] Brown, *Animal Oeconomy*, pp. 171-8. See also Cook, *Decline*, p. 194.

[8] Brown, *Animal Oeconomy*, p. 175.

[9] Brown, *Animal Oeconomy*, p. 171. This was in the aftermath of the publication of Thomas Willis's *Pharmaceutice Rationalis*.

institution into a vital and authoritative new voice. However, the idea of the Royal College's united presentation has been dismantled by historians such as Cook, who highlights the absence of any 'unified intellectual strategy' of the kind that Brown identifies. The College was far from monolithic, and the writings of prominent physicians cannot simply be understood in relation to the emergence of the mechanist hypothesis. As Cook recognises, many College fellows retained older medical theories, and plenty were uninterested in philosophical debate.[10]

Brown treats one of Charleton's mechanical metaphors (that the body of man is a 'system of innumerable smaller Machines or Engines') as representative of his entire thesis. For him, *Enquiries* constitutes an approach to general physiological problems from this point of view.[11] Thus he claims Charleton was at the forefront of mechanical philosophy. Jonathan Sawday, likewise, sees Charleton in these lectures as representing a profoundly dualistic view of the body, which was 'no more, after all, than a "wonderful engine", designed and constructed according to mechanical principles'.[12] He too quotes Charleton's comment that the body was 'a System of innumerable smaller Machines or Engines, by infinite Wisdom fram'd and compacted into one most beautiful, greater Automaton'.[13] Although Charleton invoked mechanistic explanation in several instances throughout *Enquiries*, I argue that these lectures demonstrate no unified 'agenda' in the promotion of mechanist medicine. Metaphor, in this case the metaphor of a machine, was one of the means by which Charleton was able to negotiate the whole range of available approaches and methods. We cannot afford to treat any single metaphor as a signifier of an overall system.

Brown correctly identifies the stresses under which the College operated in the later seventeenth century, but I suggest that he may have mistaken the institution's response, in depicting Charleton's lectures as mechanistic in intent, and monolithic in emphasis. I concur with Brown that the lectures constituted in part a declaration of the College's public position and how it saw itself. But I argue that the lectures did not adhere to any mechanist portrayal of medicine. Rather, I suggest, these lectures illustrated the consonance between traditional, classical authority (upon which the College's status was founded) while encompassing a broad knowledge of recent anatomical developments, both Continental and local. Charleton's presentation allowed the College to illustrate the consonance between new theories and the bases of authority upon which it had always relied. The lack of monolithic explanation by Charleton echoed the lack of unity within the College itself. *Enquiries* was not a promotion of a single uniform theory on the College's behalf, but rather an illustration that it was not intransigent and ignorant, as some had suggested. *Enquiries* demonstrated the legitimacy of traditional approaches by illustrating their fundamental accordance with the most recent developments. It was perhaps for this reason that

[10] Cook, *Decline*, p. 195.
[11] Brown, *Animal Oeconomy*, p. 173.
[12] Sawday, *Body Emblazoned*, p. 181.
[13] Charleton, 'Preface', *Enquiries*, p. i. See Sawday, *Body Emblazoned*, p. 181.

these lectures were published. The publication of anatomical lectures of this kind was extremely rare. I have found no comparable examples.[14]

Nayler refutes the idea that Charleton's thought was consistent, and her analysis of *Enquiries* is consequently more penetrating than that of Brown. Her interest in the physician is defined by her focus upon muscle physiology, and for this reason she examines in detail only Charleton's lecture on voluntary motion (which I discuss later in this chapter). Her discussion illustrates Charleton's profound debt to the writings of his peers, especially Jacob Müller—from whom, she observes, he plagiarised relentlessly. Nayler sees Charleton's adherence to Müller as a consequence of the fact that the German promoted 'one of the themes of his physiology, namely that "the Omnipotent Creator hath made all things, as in the Greater World, so also in the Lesser Man, in Number, Weight and Measure".'[15] She recognises that Charleton did not aim to promote a systematic view of the animal oeconomy, but simply summarized and reported the opinions of both ancients and peers.[16] By highlighting his close reliance upon the writings of contemporaries, Nayler elucidates Charleton's apparently contradictory statements about his own aims. However, rather than seeing Charleton's eclecticism as a strategy by which he negotiated the contemporary intellectual milieu, she perceives his presentation of both the spirit influx and Glissonian motion as a sign of his 'uncertainty' regarding the cause of muscle contraction.'[17] She believes *Enquiries* illustrates that Charleton did not completely understand Müller's demonstrations and arguments.[18] I believe that Charleton's approach was not founded upon 'uncertainty', but rather upon a commitment to eclectic method.

Dew argues that the *Enquiries*' account of the blood's movement is different from that of *Natural History* in important ways: the removal of the 'pulsifick' faculty, and the greater emphasis on the hypothesis of irritable fibres as causes of motion (thus making the heart entirely without agency).[19] This interpretation is more strongly substantiated by Charleton's writings than is Brown's. The lack of consistent mechanist agenda in *Enquiries* will be demonstrated throughout this present chapter.

Frank places Charleton among the 'Oxford natural philosophers' who resolved various 'physiological problems' through the innovative redefinition of physiological explanation.[20] He describes Charleton as one of the major scientists of the era, along with Robert Boyle, John Wallis, Christopher Wren and

[14] While records of anatomical lectures presented at around this time are in existence, they show a marked difference from Charleton's lectures in the extent of their philosophical emphasis. See for example Thomas Willis, *Oxford Lectures*, ed. and introduced K. Dewhurst, Sandford, Oxford, 1980 and Thomas Winston, *Anatomy Lectures at Gresham College*, London, 1659.

[15] Charleton, *Natural History*, p. 210. Nayler, 'Insoluble Problem', p. 125.

[16] Nayler, 'Insoluble Problem', p. 90.

[17] Nayler, 'Insoluble Problem', p. 134.

[18] Nayler, 'Insoluble Problem', p. 117.

[19] Dew, 'Politics of the Body', p. 31. See Charleton, *Enquiries*, pp. 405-19.

[20] Frank, *Oxford Physiologists*, p. 43.

Thomas Willis. These men are distinguished from 'minor scientists' and 'virtuosi' (the latter 'did little original scientific work themselves, but were eager participants in, and followers of, the activities of their more gifted peers.') Frank depicts Charleton as both innovator and experimental natural philosopher.[21]

Despite his experimental activities Charleton's *Enquiries* drew much of its authority from the texts of contemporaries. These lectures showed a heavy reliance upon the writings of Francis Glisson, as well as Francis Bacon, William Harvey and Jacob Müller. This reliance is largely ignored by scholars who claim innovation for Charleton, especially those who see his definitive activities as the revelation of new knowledge through experiment. This chapter suggests that Brown's portrayal does not depict Charleton accurately. The physician showed few signs of a consistent epistemology, but adopted a *range* of epistemologies in authorising his knowledge of the body. Likewise it suggests that while Frank is correct in identifying Charleton's *behaviour* as empirical, we must look at how he presented himself *in relation to* that empiricist behaviour in order to understand his epistemology.

The teaching role of Charleton's lectures has raised no discussion. Nor has the manner in which they were circumscribed by the traditions associated with learned lectures. The genre within which the lectures were given suggests that we need to be careful about assuming that their subject matter reflected Charleton's particular beliefs. Some of the characteristics shared by the *Natural History* and the *Enquiries* suggest that changes between the two texts were as much the result of a shift in genre as the consequence of his own active participation in new experimentation.

A COMPARISON OF *ENQUIRIES* WITH *NATURAL HISTORY*

A range of reasons can be suggested for the differences between Charleton's *Natural History* of 1659 and his *Enquiries* of 1680. Between the publication of the two works the author underwent a tremendous shift in personal fortunes. In 1653 Charleton had a medical degree and some experience in practice as an assistant to Mayerne. He progressed from private practice and service to the king, to popularity and activity in the key medical and scientific organisations of his time. By the time of the 1680 lectures he held a secure position at the head of the College of Physicians. He had made detailed examinations of various issues, and had been active in an empirical research group. His own experimental experience might be expected to have induced changes in the source of authority of his arguments, and his lectures to contain examples of empirical discoveries, made or witnessed by himself.

We might expect his lectures to the College to reflect the demands of that context. An obvious issue is that of the genre and traditions associated with

[21] Frank, *Oxford Physiologists*, p. 44.

lectures. The 1680 lectures did not attempt a systematic account of the body, as did contemporary anatomical texts,[22] but rather focused on selected insights. Charleton's views in *Enquiries* were tentatively expressed, always open to the judgement of his audience. This difference between his textbook style of 1659, and his lecture style, of 1680, may simply be the result of the genres within which the two texts were written, rather than an intellectual transition. The lectures perhaps required an overview of current developments, rather than the portrayal of a systematic explanation.

Charleton's epistemology might also have been altered by the practices and methodology of the Society. The empirical philosophy by which the latter defined its public image may have diminished his deference to textual authority. Although he was no longer active in the Royal Society by 1679, the extent to which it influenced his self-presentation should not be ignored. In 1676 he was incorporated fully within the College of Physicians, and his desire to accommodate their interests is relevant to our interpretation of the 1680 lectures.

It seems likely there was a common teaching function for both *Natural History*, a textbook, and *Enquiries*, a series of lectures for 'younger anatomists'. Perhaps we should pay more attention to the instructive aims of these two works. His profound eclecticism of approach may reflect the fact that he was educating others in new research and opinions. The institutional context also affected the probable audience for Charleton's anatomical lectures. In 1680 he might expect his works to be read by an extensive range of members of both the Royal Society and College of Physicians. He was known in Padua by this time, for his anatomical expertise. However the text was not designed for an international audience, as it was published in the vernacular. The publication of the *Enquiries* in English supports the contention that it was written as a synthesis of knowledge, for the local audience.[23] The Gulstonian lectures were delivered in English, and various kinds of learned anatomy lecture were either delivered partly in English, or repeated in the vernacular after a Latin presentation.[24]

The changes between Charleton's 1659 and 1680 texts provide only very ambiguous evidence of shifts in his intellectual approach. The 1680 lectures can be used to examine the extent of inclusion of anatomical experiments performed at the Royal Society. However we might assume that contextual factors would have affected the kind of information and manner of verification that he presented to the College audience. The two works provide different perspectives on Charleton's intellectual approach, and any comparisons are complicated by the range of factors identified above. The subsequent chapter will examine the *Three Anatomic Lectures* of 1683, in relation to Charleton's changing views on

[22] See the treatment of anatomy in: Isbrand Diemerbroek, *The Anatomy of Human Bodies*, London, 1689, Johann Vesling, *Syntagma anatomicorum*; Riolan, *Sure Guide*; and English proponents such as John Keill, *The Anatomy of the Humane Body*, London, 1698 and Browne, *Myographia Nova*.

[23] Charleton's comment that some among the audience were 'unaccustomed to any but their Mother tongue', suggests a strong vernacular context. Charleton, *Enquiries*, p. xxx.

[24] Cunningham, 'Kinds of anatomy', p. 12.

the anatomy of the heart. The more eclectic approach that Charleton adopted in the *Enquiries* (compared with the *Natural History*) could be seen either to indicate the influence of the philosophical approach of the Royal Society, or as his response to the demands of a lecture designed to review current medical understandings.

THE GENRE AND AUDIENCE OF ANATOMICAL LECTURES

> Let it . . . at present suffice, if to gratifie the Curiosity of the Yonger [sic] Students of Anatomie, I set before their eyes, not an accurate Map, but a rude Landskip of the Galaxy or Milky way, in which the greater part of the Chyle glides along through the purple Island of the body, to replenish the ocean of blood.[25]

Enquiries did not draw reference to Charleton's own clinical practices. Rather he seems to have drawn upon his knowledge of, and ability to reconcile, both classical texts and new anatomical developments. The lectures emphasised his classical knowledge and scholarly erudition. Dewhurst comments that a physician's literary elegance and ability to demonstrate scholarly prowess through classical allusions was as important as the efficacy of his treatment.[26] In this instance Charleton's classical emphasis reiterated to the gathered students the importance of this aspect of a physician's identity. The College's emphasis upon classical knowledge would thus have been ratified by Charleton's presentation. He exemplified the College's classical expectations, while showing an impressive command of new theories and anatomical discoveries.

Lectures were designed to present an overview of the current thinking on certain topics, for the benefit of the gathered audience. In the case of these lectures, that audience was 'Younger Students in Medicine', 'for whose instruction chiefly it was, that the wise and prudent Authors of the Statutes of this our so worthily renowned College, first instituted and ordained Anatomic Lectures to be therein read by the learned Fellows thereof'.[27] They were not likely to have been designed as an expression of a mechanist agenda but rather, as usual, would be intended to reflect current thinking along a number of different hypotheses.

Students would most likely have been students of surgery, training at the Royal College of Surgeons.[28] Trainee surgeons may have attended these lectures at the College of Physicians. However, if surgeons were present, surely the philosophical emphasis of these lectures would have meant that these men gained little from Charleton's discourse, which was far more philosophical than their training required them to be. Even if the lectures were designed for this audience, they would also have been attended by a basic core of physician Fellows.

[25] Charleton, *Enquiries*, p. 142.
[26] K. Dewhurst, 'Introduction', *Thomas Willis's Oxford Lectures*, Sandford, Oxford, 1980, p. vii.
[27] Charleton, *Enquiries*, pp. 472-3.
[28] It is unlikely that they were students from Oxford or Cambridge.

As attendance at annual lectures had been poor throughout much of the earlier century, the College statutes required senior fellows to attend lectures. This might explain why Charleton addressed an audience of both senior fellows and students.

Charleton's audience consisted also of some junior physicians. His *Three Anatomic Lectures* included visual demonstration in the form of figures, which would 'help both the understanding and memory of younger Students, for whose sake chiefly Lectures were at first instituted in this College.'[29] Charleton declared that he would sooner 'lead my Auditors of the Younger sort by [Borelli's] brighter Torch, than by the Glow-worm light of my own understanding'.[30] He presented his own role as that of an interpreter of recent theoretical developments, offering a synthesis or summary of contemporary thought and writings.

The presence of College luminaries at the lectures would have influenced his presentation. The comparison between the presentation of *Enquiries* and Charleton's later *Lectures* helps us to delineate some features which might have been generic to the lectures format. While the lack of comparable publications makes broader comparisons difficult, we can identify some continuities and differences between *Enquiries* and *Lectures*.

In addition to printing the lectures (two years after they were delivered), the institution probably invited prominent public figures to attend the lectures at the new anatomical theatre.[31] Although the explicitly declared role of the lectures was the education of 'younger students', Charleton was highly conscious of the presence of such luminaries as Hooke, Evelyn, Scarburgh[32] and other notables of the College and the Royal Society. The presence of so many prominent figures indicates that Charleton was addressing a mixture of relatively unlearned and highly learned individuals. The latter would be the ones whom he expected to be able to 'judge' the hypotheses he presented. The former would be encouraged to make judgements upon appropriate principles of reason and logic, in which they had been trained at university.

The lectures commenced with a justification of the utility of anatomy in relation to ancient authors, and provided a careful definition of the scope and foci of anatomical knowledge. This would have reaffirmed to the gathered audience that new anatomical knowledge fitted within the realm of traditional physic. In this sense he claimed anatomy for ancient physic, reiterating the links between the present discoveries and classical learning. Here we can see the influence of the College context upon Charleton's presentation. The lack of practical

[29] Charleton, *Lectures*, p. 42.

[30] Charleton, *Lectures*, p. 49.

[31] Brown, *Animal Oeconomy*, p. 175.

[32] Charleton referred to Scarburgh's presence ('that most excellent Man here present, to whose incomparable Pen Dr Harvey ows half his Glory'). Charleton, *Enquiries*, p. 515. Hooke's *Diary* attests to his presence. Robert Hooke, *The Diary of Robert Hooke, 1672-1680*, ed. H. W. Robinson and W. Adams, London, Wykenham Publications, 1968.

anatomy within these lectures would also have indicated that the physician's part of anatomy was the speculative and philosophical aspect (not the mechanical elements, which were associated with surgeons).[33]

CHARLETON'S METHOD IN *ENQUIRIES*

Enquiries shows the influence of Francis Glisson, sometimes so extensively that Charleton's work almost seems to be a tribute to his understanding of the animal oeconomy. Charleton adopted many of Glisson's ideas, and echoed the elder physician's stylistic approach. The latter was a great reconciler of others' theories. Glisson's inclusion of multiple viewpoints was the consequence of his view that: *Veritas in multis temporis filia est*.[34]

The apparent lack of consistency in method or explanation among the six lectures is in part the consequence of Charleton's heavy reliance on the writings of others. He accepted explanations of particular phenomena, without illustrating any desire to generalise them to account for other functions. Glisson's work on the stomach and intestines was incorporated by Charleton in his own account of the stomach, yet he contradicted other conclusions of Glisson's in a later lecture.[35]

In addition, *Enquiries* illustrated the profound influence of Bacon and Boyle (on fevers, prælectio V); Harvey and Ent (on spirits, prælectio VI) and of course Jacob Müller (prælectio VI). Some of these authorities were acknowledged in the text, but many were not mentioned at all. Though Charleton was interested in the writings of these men, he did not care to demonstrate preference for one over the others, but generally left the decision to his audience, claiming that the issues were too substantial for his own comprehension.

While Charleton made some reductive claims, his emphasis in *Enquiries* remained consonant with the goals of philosophical anatomy. The work was not mechanical in intent, focus or implication. Charleton was likely to accept and reject a principle of explanation (such as the existence of faculties, inherent properties, etc) in immediately adjacent discussions. He juxtaposed conflicting viewpoints, without endeavouring to explain his vacillation. He did not generalise

33 This is immediately evident from a comparison between Charleton's lectures and surgical lectures published at this time. See Alexander Read, *Workes of that Famous Physitian*, and John Browne, *Myographia Nova*.

34 *'Ecquis Philosophorum e tripode dictat; aut spiritu infallibilitatis in rebus naturalibus afflatur? Satis scio de meipso, me facillime errare posse. In arduis naturae, cogitationes primae fere secundis, facta pleniore inductione, cedere coguntur. Cauto est opus: ne praeporpere definias, aut de aliis judices.'* Glisson, *De ventriculo*, p. 272 Giglioni, 'Anatomist Atheist?', p. 134 fn. Glisson attempted a reconciliation of Hippocratic and Galenic medical systems with those of the 'Neoterics.' See Glisson, *De Ventriculo*, p. 389. Giglioni, 'Anatomist Atheist?', p. 127.

35 Likewise the younger physician was ambivalent about the nature of matter throughout his lectures. Glisson's contribution on this issue had been tremendously influential, but Charleton remained skeptical about its application in certain instances, while appearing to embrace it in others.

conclusions from one lecture to another, but took an eclectic approach, avoiding system-building or explanatory doctrines. He was unconstrained by the generation of an exclusively mechanistic, hermetic, chemical or indeed any model of the body. Each instance and topic of discussion was considered on its own merits.[36] Attempts (like Brown's) to discover a systematic theme and explanatory agenda in the lectures of 1680 may be misguided. The range of epistemologies invoked by Charleton suggests problems with interpretations of the *Enquiries* as a declaration of a single preferred method.

Enquiries demonstrated Charleton's fluid movement between different approaches and methods of investigation. One striking aspect of Charleton's lectures is the range of epistemologies that he invoked: experimental, metaphysical (Platonic) and textual. Anatomical knowledge was not equally relevant to all six lectures. The first three lectures, on nutrition, depicted physiological knowledge as possible only through accurate anatomical understanding. However his discussion on the stomach was based not on his own anatomical research, but on the work of Ent and Glisson. The lecture on life combined a Neoplatonic argument about the significance of names, with corpuscular propositions (derived from Bacon) on the nature of heat. These propositions, supposed to build up an irrefutable theoretical basis, were demonstrated analogically, and their speculatively-derived consequences then demonstrated anatomically (invoking experiments taken from Harvey and Lower). The fifth prælection invoked the theory of life expounded in the previous lecture, casting fevers as the perversion of that process. Charleton's lecture on the muscles was informed by textual information drawn from Müller.

CHARLETON'S DEFINITION OF ANATOMY

Understanding the different definitions of anatomy helps us to comprehend the range of epistemologies within Charleton's anatomical argument. The Preface to the *Enquiries* defined anatomy as 'a diligent, accurate and artificial dissection' of the body of animal or man, to acquire knowledge of 'the substance, magnitude, figure, site, structure, connexion, action and use of all and every part thereof.' Two actions were comprehended by the notion of anatomy: a 'work of the hands, Dissections', and 'an exercise of the Mind or Intellect, Speculation.'[37] The dual definition of anatomy had a long and distinguished history,[38] and the

[36] The claim that mathematical propositions would provide the reader with certain truth was cited quite near the argument that there could be no final determination of the truth of a proposition, only its probability. The first of these claims appeared in the middle of the chapter on life, and the other at the end of that chapter.

[37] Charleton, 'Preface', *Enquiries*, pp. xix-xx.

[38] The two varieties of anatomical knowledge had been identified by Matthew Curtius in lectures with Vesalius in Bologna in 1540. See *Andreas Vesalius' first public anatomy at Bologna 1540. An eyewitness report by Baldasar Heseler, together with his notes on Matthaeus Curtius' Lectures on Anatomia Mundini*, ed. and trans. Ruben Eriksson, Uppsala & Stockholm, Almqvist & Wiksells Boktrykkeri, 1959, p. 273.

same division had recently been posited also by Glisson.[39] This combination of observation and meditation was part of the philosophical and practical framework within which anatomy was taught and studied.[40]

Dissections investigated internal motion of body parts, and provided understanding of their structure.[41] This 'Practic' knowledge was 'acquir'd by long use and experience, and natural dexterity', and did not require additional insight. Anatomical 'administrations' provided 'perfect knowledge of the subject, and every the most minute part therof; at least as much Knowledge, as the narrow limits of Human Wit can comprehend.'[42] The manual aspect of anatomy presented complete knowledge as both desirable and achievable, though limited by the extent of human wit. 'Anatomic Administration' required instruments (miscroscopes, syringes), and manual operations, such as close measurements, injection of liquids, 'inflations, extensions, ligations, excarnation' and other excercises.[43] This illustrates the influence of the Royal Society investigations, and of the changing notions of what constituted 'inspection'. Anatomical practice, then, was not merely observation, but active intervention and manipulation.

'Speculation' created 'Theoretic' knowledge, gained 'by reason and sagacity, by hearing the Lectures, reading the Writings of Learned men . . . and by intent meditation.'[44] In this aspect of anatomy individual insight might offer great rewards, and it was here that a physician distinguished himself from a mere surgeon or 'mechanic'. The serious, scholarly and contemplative character of the physician played a role in confirming his ability to discover this kind of knowledge. Charleton's self-construction—as reflective, speculative and bookish—would have enhanced his authority in this aspect of anatomy. The subjects examined under this 'speculative' part of anatomy reflected the aims of traditional scholastic medical knowledge.[45] Aristotelian epistemology placed

[39] Glisson defined anatomy as 'an artificial dissection of [a certayne] object in such maner as may most conduce to the perfect knowledge of the same and all its parts . . . Now this artificial dissection implies not the manuall dissection only but in especiall maner the mentall'. Glisson, original draft of *Anatomia*, London, Du-Gardianis, 1654; British Museum, MS Sloane 3315, ff 165-9.

[40] Andreas Laurentius echoed the division of anatomy into two essential categories: inspection and instruction (or doctrine). A. Laurentius *Historia anatomica*, Frankfurt, 1602, pp. 24-7.

[41] This echoed the Aristotelian definition (invoked by Harvey) that anatomy 'teaches the uses and actions of the parts of the body by ocular inspection and by dissection'. William Harvey, *The Anatomical Lectures of William Harvey*, ed. & trans. G. Whitteridge, London, E. & S. Livingstone, 1964, f. IV, p. 5.

[42] Charleton, 'Preface', *Enquiries*, p. xx.

[43] Charleton, 'Preface', *Enquiries*, p. xx.

[44] Charleton, 'Preface', *Enquiries*, pp. xix-xx.

[45] These included the 'Universal Structure of the whole', together with comparisons within and outside the species; the form of every part, symmetries and sympathies between them; and, finally, their faculties, actions and uses. Charleton, *Enquiries*, p. xx. His 'Historia Gulæ' recounted various names by which it had been known (citing Hippocrates, Aristotle, Cicero, Celsus), progressed through comparative anatomical observations (Willis, Schonfeld's *icthyologia* and Fabricius' *de gulae*), described the similar parts of which the gullet was composed, distinguished the common and proper elements, and outlined contemporary debates (Wharton, Glisson, Hoffman, Steno and Willis). He concluded that its proper action was 'sufficiently manifest' from this structural account. Charleton, *Enquiries*, pp. 25-8.

knowledge of causes above knowledge of structure. The 'final cause or purpose of the parts of the body' could not be known through mere inspection, and thus required the trained insights of a physician.[46]

Philosophical anatomy derived from Aristotelian dictates, and from Galen's panegyric on the final cause in *De usu partium*. The latter described the general importance of knowledge of the uses of the parts, beyond the limited concerns of physic.[47] Despite their differences, Aristotle and Galen agreed that 'true knowledge could only be achieved from knowledge of purpose'.[48] Galen argued that nature did nothing in vain, and all her actions were by definition useful towards her final aims. The study of anatomy was thus the gaining of an understanding of how Nature's 'infinite wisdom' in design operated to fulfil her objectives.

This teleological view, according to some, was one aspect of traditional physic against which iatromechanists rebelled: matter was inert, with no inherent properties. The mechanistic approach to anatomy, according to Cunningham, introduced the notion that the mere construction of the parts made their actions mechanically necessary. The mechanical approach rejected the presence of special properties that determined the actions of body parts. Thus structure was 'the sole criterion of function'.[49] A central aim of mechanist medical thought was therefore removal of purpose from the body: its actions resulted from purely mechanical actions, rather than teleological operations.[50] A 'true' mechanist by this definition would reject the necessity of action-use utility as crucial to anatomical knowledge. *Physiologia*, which presented this view, has been invoked by many as an indication of Charleton's approach to nature. However, as this chapter illustrates, the matter theory outlined in that text had little relationship to his medical ideas—another instance of his theoretical eclecticism.

Natural History often invoked the argument that nature's actions were by definition useful. Nature, Charleton noted, 'always accommodates her instruments to their proper uses and ends, fram'd them of Figures most commodious respectively to their peculiar functions.'[51] He associated anatomical practices with the kind of causal knowledge ratified by centuries of medical philosophy.[52] Not

[46] A. Wear, 'William Harvey and the "way of the anatomists"', *History of Science*, vol. 21 no. 53, 1983, p. 228.

[47] See Galen, *On the Usefulness of the Parts of the Body*, ed. M. T. May, 2 vols, Ithaca, Cornell University Press, 1968, vol. 2, p. 731. Charleton believed himself to have been 'taught, not only by Galen (3. *de usu partium*.) that the discerning ones self, and discovering to others the Perfection of God displaid in His Creatures, is a more acceptable act of Religion, than the burning of Sacrifices or perfumes upon His Altars'. This importance of anatomy was affirmed by Trismegistus: 'the thanks and praises of Men are the most grateful incense that can be offer'd up to God'. Charleton, *Enquiries*, p. xxvii.

[48] R. French, 'The Anatomical Tradition', in W. Bynum & R. Porter (eds), *The Companion Encyclopedia of the History of Medicine*, London, Routledge, 1994, p. 84.

[49] Cunningham, 'Historical context', p. l-li.

[50] French, 'Anatomical Tradition', p. 91.

[51] Charleton, *Enquiries*, p. 133.

[52] Even knowledge of the passions was to be gained through 'frequenting the Scholes of Anatomists'. Charleton, 'Preface', *Enquiries*, p. xi.

only structural knowledge, but understanding and control of the self as a social being belonged to the anatomical realm. He thus broadened the philosophical knowledge of physicians to encompass anatomical expertise. Charleton's anatomy was not restricted to a 'mechanist' perspective (as defined by Cunningham), nor aimed toward a 'mechanist' methodological program (as suggested by Brown). His anatomical works examined purposive issues and structural imperatives which encompassed function, use and purpose. He combined traditional philosophical anatomy seamlessly with recent innovations.

Enquiries stated that there was no function divorced from use. At times the author invoked the centrality of 'mechanic necessity', and at times employed highly reductive arguments. However, his lecture was a search for uses and purposes, according to a long-standing tradition which stated what it meant to give an anatomical lecture. As this chapter demonstrates, he also included essential properties, supposedly rejected by mechanists.

Charleton's definition of anatomy illustrates that his lectures combined a practical guide with subtle philosophical determinations, which had to be balanced according to his status as a university-trained physician. Multiple epistemologies were necessary to comprehend the different aspects of bodily operation. The divergent challenges offered by these two dimensions of the same discipline illustrate how fraught it might be for Charleton to invoke a single method within his lectures. Having examined Charleton's definition of anatomy, let us turn to a consideration of the subject matter of *Enquiries*. As with *Natural History*, space does not permit me to examine every physiological issue in the text. Again I have selected a series of case studies which exemplify Charleton's methods, and which highlight the manner in which he presented himself and his material.

CASE STUDIES IN EPISTEMOLOGY

Materialist physiology—The case of the 'lamp of life'

Charleton's investigation of nutrition in *Enquiries*, in contrast with the *Natural History*, indicates an increasing emphasis on physical structures. Process was less important than structure, in this later explanation of a similar subject. The first three prælections were guided by a focus upon the solid parts of the human anatomy, and incorporated physiological questions only under those divisions. Anatomical understanding was central to this consideration of the digestive process in 1680, in a manner quite different from the *Natural History*.[53] Charleton began with the recognition that putrefactive urgency prioritised the

[53] In 1680 Charleton expressed the notion that physiological explanation required structural knowledge. The same material concerning nutrition was covered here as in 1659, but the categories by which he organised the knowledge had changed.

stomach.[54] However, it is difficult to assess whether this was the result of a change in his own ideas, or simply a reformulation of the material according to the requirements of the anatomical lecture format.

In 1680 the generation, motion, uses and depuration of blood were discussed in relation to the organs responsible, rather than featured as processes in themselves. Where the *Natural History* had focused on the distinctions between the various fluids within the body (blood, milk and chyle), the 1680 discussion emphasised the solid parts, because 'the knowledge of the Fabric of those Organs is necessarily prærequisite to our attaining competent knowledge of their respective Actions and Uses.'[55] The greater materialism of this later work is exemplified by Charleton's use of the 'lamp of life' to explain vital heat. He had stated confidently in 1659 that his beliefs on the decay of the body were shared by 'all Philosophers'. He observed without qualification, in *Natural History*, that: 'Life doth consist in a continuall accension of vital spirits out of the blood, which is the pabulum of the Lamp of life'.[56] However, the parallel discussion in *Enquiries* illustrates that the lamp metaphor was no longer central to Charleton's argument: 'Life consists in and depends upon a continual generation of the Vital Spirits out of the most subtil, active and volatile parts of the bloud'.[57] The lamp of life was replaced with explanation in terms of physical composition.

Charleton's conclusions in *Enquiries* were also far more tentative. The depredator within the body 'seems to be' the Vital Heat of the blood, although understanding of it was 'obscure and inadequate'.[58] *Enquiries* 'supposed' it to be 'an Actual Heat consisting in a certain motion of the various particles of the bloud, and in some degree analogous to fire or flame'.[59] It was however posited by authoritative voices.[60]

Natural History declared that flame, 'is a substance luminous and heating, consisting in a perpetuall *Fieri*', that was, 'accension of the particles of its pabulum, or combustible matter'.[61] In contrast, *Enquiries* questioned the very nature of the heat itself. No longer assumed to have absolute properties, the vital flame nature was refigured through motion:

[54] It implies that *this* lecture was given with a cadaver before him: the stomach and guts 'by reason of impurities contain'd in them, more prone to putrefaction, ought therefore first to be taken out of the cavity of the Abdomen, to prevent noisomness'. Charleton, *Enquiries*, p. 2. Cunningham believes that no cadaver lay before Charleton as he spoke. Cunningham, 'Kinds of anatomy'. The physician made only a few references to a cadaver, and it seems that he was not constrained by the need for physical demonstration. At one point he observed of a certain vessel that 'Tis (ye see) a body round, long, hollow as a pipe.' Charleton, *Enquiries*, p. 24. Evidently not all of the lectures were presented in the presence of a dissected corpse, but only those directly concerned with the stomach.

[55] Charleton, *Enquiries*, p. 20.

[56] Charleton, *Natural History*, pp. 8-9. 'Pabulum' refers to sustenance or food.

[57] Charleton, *Enquiries*, p. 17.

[58] Charleton, *Enquiries*, p. 9.

[59] Charleton, *Enquiries*, p. 9.

[60] He cited Fernel, Heurnius, Descartes, Willis, Aristotle and Bacon. Charleton, *Enquiries*, p. 385.

[61] Charleton, *Natural History*, p. 6.

> Whether we take fire or flame to be a substance luminose and heating; or conceive it to be only a most violent motion of globular parts in its focus: most certain it is, that it consisteth in a . . . continuall agitation or accension of the particles of its . . . fewel.[62]

Although he rejected the explicit comparison, he retained its implications. Thus the lectures illustrate a shift away from the hermetic resonances of the metaphor, toward a more explicitly materialist account.

A question of faculties—The case of the 'ventricle'

The third lecture considered the method, causes and manner of digestion, divided into the operations of the ventricle, 'each of which hath a peculiar Faculty'.[63] Charleton considered the stomach's faculties individually: hunger, thirst, peristalsis, reception, retention, concoction, secretion and excretion. This lecture offers an interesting view of Charleton's approach to the question of the faculties within a medical context.[64] As we saw in relation to *Natural History*, the origin and nature of the faculties dominated many late seventeenth-century medical discussions. This was one of the central issues in reinterpretations of Galenic medicine (in which the faculties were the sentient or at least teleological powers of each organ, by which the proper order of the body was maintained). The following discussion examines two instances in which faculties were explained in *Enquiries*, and considers the approaches that the author adopted regarding their existence and operation.

In 1680 Charleton rejected the idea of attractive faculties as neither necessary (because depression and protrusion could account for the effects), nor probable (as there was no such thing as attraction in nature).[65] He argued that attractive powers of the stomach should be 'wholly expunged' from discussions of the organ, 'not only as fictitious, but absurd also and impossible.'[66] The idea of an Attractive Faculty in the stomach was 'neither necessary, nor explicable either by the hypothesis of Similar attraction, or by mechanic principles, or by the Aristotelean supposition of nature's abhorrence from vacuity'.[67] Its retention as a theory was due to the over-reliance upon authority of those 'servil spirits, who choose rather to err with their Teachers, than to recede from them; as if Philosophical doctrines were, like the principles of Religion, not to be examined, but implicitely believ'd.'[68]

[62] Charleton, *Enquiries*, p. 11.
[63] Charleton, *Enquiries*, p. 65.
[64] In contrast with his explicit rejection of faculties in *Physiologia*.
[65] Charleton, *Enquiries*, p. 99. Galen's claims for attraction organic were also dismissed, as there were 'no instruments fit to effect it'. See p. 100.
[66] Charleton, *Enquiries*, p. 101.
[67] Aristotle's notion of nature's abhorrence of a vacuum was a 'fantastic Sanctuary' that had 'long since been demolish'd'. Charleton, *Enquiries*, pp. 100-101.
[68] Charleton, *Enquiries*, p. 101.

As we have seen, Charleton sustained no opposition to scholastic medical theory himself, but was often reliant upon Aristotelian explanations and Galenic method and terminology.[69] He objected to established authority only where alternative explanations had been provided by a contemporary, and then only sometimes. Thus Charleton's discussion of the stomach's faculties drew upon the work of Glisson and Ent, and echoed their criticisms of the doctrine of attraction. Glisson had recognised the failings of the attractive faculty as an explanation, and had posited instead a 'Receptive faculty', which Charleton described as 'a certain aptitude of the stomach, by which it relaxes its fibres, and inlarges its cavity, to receive meat and drink brought into it'.[70] Although this hypothesis was 'less improbable' than attraction, Charleton expressed doubts about its necessity. A faculty implied an action, whereas the aptitude of the stomach to be filled seemed less a faculty than simply a non-sentient state. A tensile body was not capable of independent extension, because extension was 'repugnant to its nature; and no natural Agent can act towards its own destruction.'[71] In this instance Charleton invoked the Galenic principle of nature's self-preservation, as part of a refutation of Galen's theory of attraction. Clearly his work did not reject Galenic authority *per se*. In contrast with Glisson's view of fibres, Charleton claimed that the 'proper' action of fibres was 'spontaneous Contraction, whereby they restore themselves to their native posture . . . so soon as the stretching they suffer'd ceases.'[72] The stomach changed shape only (and proportionally) as a result of the quantity of food within it. This explanation made the stomach itself entirely non-sentient, allowing only for reaction on a quantitative basis.

However, this statement of mechanical sufficiency in explanation did not mean that Charleton's attitude towards mechanist explanation was consistent. He had doubts about Glisson's hypothesis, which induced him to present arguments for and against the theory, so that 'I may learn from you, which are the more weighty.' He withdrew from any conclusion, declaring that 'Leaving this nice quæstion then to your decision, I proceed . . .'[73] This stance indicates Charleton's own eclecticism, and his sense that his audience was capable of authorising knowledge. He endowed the audience with the status of judges, deferring to them to decide the preferred hypothesis.[74] He constructed himself here as a distinguished conveyor of information. Glisson's theory of natural perception, he lamented, 'seems obscure, intangled with various difficulties, and therefore

[69] Galenic terminology eased the transition to a more physicalistic explanation of the appetitive faculties. The gullet, throat and tongue, for example, were 'cover'd with one and the same membrane, and must therefore sympathize'. Charleton, *Enquiries*, p. 84. He spoke of the membrane's proclivity to dryness, and its 'impatience' of bearing that dryness without 'anxiety', pp. 86-90.

[70] Charleton, *Enquiries*, p. 101.

[71] Charleton, *Enquiries*, p. 102.

[72] Charleton, *Enquiries*, p. 102.

[73] Charleton, *Enquiries*, p. 103.

[74] Perhaps this strategy was peculiar to presentations at authoritative institutions, whose members had power to determine which theories were assimilated into the canon of professional opinion.

doubtful'. Consequently, he would again suspend assent, 'till I have learn'd from you, whether I ought to embrace it, or reject it.'[75] Once again it is clear that the aim of these lectures was not to assert one doctrine (mechanism), but rather to offer an overview of contemporary and ancient theory.

Another example of Charleton's treatment of faculties is the issue of whether the distribution of chyle was the consequence of 'Attraction or Exsuction'. Here he again invoked quantitative and mechanical claims. This time they were taken from Jean Pecquet's *Dissertatio Anatomicæ*.[76] Charleton commenced with a thorough treatment of the question of 'whether there be in Nature any such thing as Attraction, properly so call'd, or not'. He claimed first that it was 'much more probable, that all the motions attributed to Attraction, are really perform'd by Impulsion.' Even if one accepted that some bodies were moved by attraction it would be difficult to explain the motion of the chyle by the same logic.' Charleton cited the failure of the three theories which invoked attraction to account for 'the effect here propos'd'.[77] Instead, he argued, chyle was transmitted into *venae lactae* by way of 'percolation' through the stomach's parenchyma, 'as through a streiner'.[78] This point, borrowed directly from Pecquet, was 'demonstrable even to sense, by various Anatomic experiments made in the bellies of Animals alive or dead'. Manipulations 'by injection of liquors', 'by inflation with air', and 'by the strongest compression' confirmed that nothing could be made to pass from the stomach into the *venae lacteæ*.[79] Interestingly, Charleton did no more than list these failed experimental manipulations, performed by Pecquet. He did not implicate his own involvement, or describe them as if he had been present. It was, he concluded, 'most probable' that the role of the parenchyma was to 'perform the office of streiner to the Chyle'.[80] In this instance a faculty (distribution of the chyle) was explicitly replaced with a corpuscular/mechanical explanation.

These two instances, in which Charleton invoked two different explanations for the attractive faculty, illustrate his eclectic style: he treated the explanations of both Glisson and Pecquet on their merits, and his explanations took into account current research on the topic. The two hypotheses arose from what

75 Charleton, *Enquiries*, pp. 105-6.

76 Jean Pecquet, *Dissertationis Anatomicæ*, 1654.

77 Aristotelian claims for Attraction *ob fugam vacui*, did not explain the distribution of chyle because there was no vacuum, and hence no need of avoidance of it. Attraction Organic was dismissed because the venae lacteæ had 'no hooks, chords, or other instruments, wherewith to take hold of the Chyle, and draw it into their mouths'. Finally, Attraction Similar failed to account for the effect because it supposed Natural Perception, which Charleton saw as 'doubtful', and because there existed no 'affinity or similitude of nature required to be betwixt the Attrahent, and the thing to be attracted'. *Enquiries*, p. 137.

78 Charleton, *Enquiries*, p. 139.

79 Charleton, *Enquiries*, p. 139.

80 As such its purpose was 'receiving and conveying . . . the thinner parts thereof, and excluding the gross and excrementious'. Charleton, *Enquiries*, p. 140. This process was further demonstrated by analogy with 'an Hippocras bag', that 'transmitts the Wine, but retains the Spices infused in it.'

might seem to be irreconcilable views of the nature of matter, yet Charleton presented both, without adhering to either. These were carefully defended as consisting not of any explanatory claims, but simply 'my private Conjectures . . . offer'd rather to your examen, than to your belief'.[81] Thus he took up an impartial stance on the options, like a true eclectic. Clearly his aim was not systematic exclusion of faculties and their replacement with a mechanist account of the body's operations. Rather he echoed contemporary alternative explanations where they existed. Where two theories had been posited, he did not automatically support the more mechanistic option (as Brown's perspective suggests he would). Rather each instance was tackled as if the audience still needed to be persuaded of the need for replacement of faculties.

His redefinition of faculties in purely material terms illustrated a deviation from medical tradition, but simultaneously an adherence to their form, and in many ways to their explanatory power. The faculties were not so much rejected as redefined, and each redefinition carefully explained. Much of the language associated with the traditional doctrine was conveyed to the new definition (for example the inclusion of 'sympathy' between organs, anatomically redefined). This meant that the persuasive impact was retained, and allowed for the argument that traditional authorities were not so much wrong as inadequately informed. Thus new knowledge did not threaten old, but simply allowed his audience to see where ancient notions might be refined. Although Charleton posited redefinitions of the faculties, he did not aim to overthrow the authority attendant upon classical learning. As the previous lecture illustrated, the authority upon which physic was based, as both practice and as philosophy, was that of classical antiquity. He annexed the *authority* of faculties while redefining their operations. I've illustrated two of the ways that Charleton thought about the faculties and their replacement. He ranged over explanations which invoked appetitive, corpuscular and hydraulic processes, without any apparent systematic program with which to replace faculties. While his emphasis on materialist explanations was stronger than in *Natural History*, this work shows no explicit boundaries between the range of available materialist arguments; immaterial agents were often involved (albeit with decreasing importance).

Method and demonstration in Charleton's account of 'life'

The definition of anatomy within which Charleton worked clearly placed the origin of life within the scope of an anatomical treatise. My discussion explores the modes of demonstration he invoked to support his assertions about the origin and nature of life.[82] The starting point in any discussion of life was the asso-

[81] Charleton, *Enquiries*, pp. 141.

[82] His theories on life were related to therapeutic practice only in the 'Epilogue', a discussion that was more a textual analysis than a clinical encounter. Rather than describing his own patients, Charleton used his diagnostic skills on a tale from Philostratus. He linked modern learning with the knowledge of classical antiquity, ratifying newer theories through their ties with tradition.

ciation of life and heat. Contemporary opinion was that heat arose from the motion of particulate matter, and this was how Charleton conceived of life. His lecture represents a range of contemporary perspectives, from Glisson's hylozoistic monism to Boyle's corpuscularism. The immediate prompt for a discussion on this subject might have come from Francis Glisson.[83]

Charleton investigated the effects on the human organs of his proposed theory regarding the vital heat and motion of the blood. He argued that the vitality in the blood was implanted by God. The vitality manifest in the blood's expansive motion was in turn resisted by its grosser elements. The properties of the blood and tendencies toward action were depicted as inherent. The origin of vital motion was the 'vital Spirit regent of the Blood'. This blood was 'naturally agil, active and volatil', and thus 'necessarily contends with the grosse parts that clogg and restrain it.' For this reason it was perpetually 'endevoring to extricate itself'.[84] Clearly the blood's properties were far from mechanistic. In contrast to Aristotelian physiology Charleton attributed no attractive power to the organs, depicting them as purely responsive to the blood. Quantities of blood distended the organs and caused their motion. Despite quantitative elements to his reasoning, and explicitly mechanical logic, he did not propose a materialist explanation. Motion was impossible without 'influent corroboration' from the brain, and from the 'vitality' contained in the blood. Although Charleton at times invoked mechanistic explanations within novel intellectual trends, this lecture illustrates how closely linked they were to classic teleological explanations.

Mechanical analogy in the relationship between soul and matter

Charleton's lecture considered 'original' and 'influent' life in turn. That life was kindled in the blood was affirmed by authorities from Solomon to 'our modern Anatomists'.[85] His next point, that heat and motion were generated by the 'vital spirit' was known by many of the ancients,[86] and had been improved by modern understanding of the circulation.[87] Again, the chief demonstration of this point was the agreement between ancient and contemporary authorities. This is a clear example of his method throughout the work.

The lecture continued with an outline, and then refutation of a range of theories on the nature of life, including those of Cardan, Fernel, Glisson—none of which proved to be 'in all points absolute and Scientific'. However, he retreated from the task of providing 'some new [theory] of my own excogitation, if not

[83] The latter had raised some of these issues in his treatise on the energetic nature of substance.

[84] Charleton, *Enquiries*, p. 394.

[85] Charleton, *Enquiries*, pp. 391-2.

[86] Including Virgil, Suidas, Critias (as reported by Aristotle), Thales, Milesius, Diogenes, Heraclitus and Alcmaeon. Charleton, *Enquiries*, p. 393.

[87] Charleton, *Enquiries*, p. 393. Recent chemical writers however, had 'erred from the Truth' when they argued that life originated from the antagonism of foreign particles. This was perhaps a swipe at Willis, who was influenced by prevailing chemical theories.

more perfect, yet at least less culpable.' For to form 'a true and complete definition' of even the most obvious part in nature, let alone of life itself, he lamented 'would puzzel [sic] a much stronger Brain than mine.' Having justified his refusal to provide another hypothesis for the explanation of life, he left the audience to guess his own beliefs from the ensuing discourse.[88]

Charleton then defined the life of man and the life of brutes. The former consisted in 'the intimate conjunction and union of his Reasonable Soul with his Body' (an assertion with which 'all Divines and natural Philosophers unanimously agree').[89] Animals, by contrast, possessed only the sensitive soul. The substantial difference between these two souls, apart from their function, was that the sensitive soul was comprised of matter.[90] Charleton portrayed the intimate relationship between the (material) souls of brutes, and the organisation of their bodies, which allowed them a power greater than the sum of the material potential.[91] His discussion was supported by textual references, and analogic reasoning.[92]

Mechanical analogies reinforced the corporeal definition of the soul, as they could be used to confirm that bodily actions resulted from the organisation of parts. Charleton asked his reader to contemplate the powers of 'Mechanic Engines'. Although composed from 'gross' and 'ponderose' materials, their 'designe, contrivance and artifice' was such that 'merely from their Figures, positions, and motions of them conjoyn'd into one complex Machine, there do necessarily result certain and constant operations, answerable to the intent and scope of the Artists, and far transcending the forces of their divided ingredients'.[93] In 'vulgar mechanics', the *organisation* of matter was responsible for powers far beyond those of the matter itself. Likewise, in 'a living Automaton or Animal', various powers emerged from the soul, and from 'a conspiracy and cooperation of so many, and so various Organs'.[94] If weak and ignorant man could create compositions that allowed 'weighty, sluggish and unactive' matter

88 Charleton, *Enquiries*, p. 376.

89 Charleton, *Enquiries*, p. 378.

90 This soul was, in turn, dependent upon the plastic virtue, implanted in the seed of animals by the 'Fiat pronounced in the act of Creation'. Charleton, *Enquiries*, pp. 381-2.

91 On the question of how a corporeal soul, composed of matter devoid of sense, might acquire the power of sensation, Charleton bowed out: 'this is indeed the difficulty that remain's here to be solved; but such a difficulty, that I dare not attempt to solve; having more reasons to believe, that it will to the end of the World remain indissoluble.' As for how the 'Faculty of Perceiving or discerning Objects' existed in particles, it 'seems to me far to transcend the capacity of human understanding'. In considering the problem Charleton referred to Gassendi's conclusion, and left the problem, 'as I found it, desperate'. Charleton, *Enquiries*, pp. 390-91. See Gassendi, *Diog. Laertii*, lib. 10.

92 Charleton, *Enquiries*, pp. 386-9. The primary textual authorities were: Lucretius, Thales and Pliny, on the greatness of divine power. Harvey was also cited, on the power of organisation of matter.

93 Charleton, *Enquiries*, p. 387.

94 Charleton, *Enquiries*, p. 388. He seems to have borrowed this comparison from Descartes' *Discourse on Method*: 'many different automata or moving machines the industry of man can devise, using only a very few pieces, by comparison with the great multitude of bones, nerves, arteries, veins and all other parts which are in the body of every animal'. This could only lead to

to produce such effects, 'What ought we to think of the Divine art of the Creator, whose power is infinite, because his Wisdom is so?' The machine metaphor illustrated that the sensitive soul and body were inseparable.

It is often assumed that the division between soul and body was fundamental to the post-Cartesian understanding of life.[95] Charleton's discussion suggests that the division between the rational and sensitive souls was more significant than the division between the sensitive soul and the body. Walker Bynum notes this in her discussion of Christian dualism in relation to the history of the body.[96] This use of mechanical analogies illustrates Charleton's occasional tendency to invoke machines in explanation (a tendency that some historians have latched upon as indicative of his 'mechanist thought'). The analogy was used to make a specific metaphysical point, and was invoked here because of its persuasive explanatory power. Its use did not reflect the presence of a comprehensive mechanist theory underlying his physiology.

Charleton described the blood that circulated as 'impregnate with original Life'. As it moved it endowed the parts with 'the Life Influent'.[97] The influx of blood was the 'general cause' of the 'noble Actions' of the body. However the blood's vital heat was not the sole cause; it could not effect results unless corroborated by the particular temperament, or '*spiritus insitus*', of each member. The temperaments of the organs themselves were crucial in assisting the faculties, and explained the individual operations of the liver, spleen, stomach, kidneys, etc.[98] The general vital heat from the circulating blood was a non-specific benevolent influence, for which Charleton offered a barrage of terminologies. Just as the sun's rays were essential to the growth of seeds in the soil, so this vital heat was to all animals, 'the Sun within them, their Vesta, perpetual Fire, familiar Lar, calidum innatum, Platonic Spark'.[99] The presentation of this range of terminologies asserted the equivalence of a variety of philosophies, in accordance with his eclectic avoidance of adherence to any specific explanation.

Varieties of demonstration in the 'uses and acts of the blood'

Charleton's lecture on life outlined the 'uses and acts of the blood' in its circuit throughout the body. These explanatory discussions offer an extraordinary

---

a view of the body of man as 'a machine, which, having been made by the hands of God, is incomparably better ordered, and has in it more admirable movements than any of those which can be invented by men'. R. Descartes, *Discourse on the Method of Properly Conducting One's Reason and of Seeking the Truth in the Sciences*, Penguin, Middlesex, 1968, part 5.

[95] See Stafford, *Body Criticism*; Drew Leder, 'A tale of two bodies: The Cartesian corpse and the lived body', in *The Body in Medical Thought and Practice*, Netherlands, Kluwer, 1992, pp. 17-35.

[96] C. Walker Bynum, 'Why all the fuss about the body? A medievalist's perspective', *Critical Inquiry*, vol. 22, 1995.

[97] Charleton, *Enquiries*, p. 416.

[98] Charleton, *Enquiries*, p. 417.

[99] Charleton, *Enquiries*, p. 417. This exemplifies the interchangeable nature of terminology within Charleton's vocabulary.

insight into the range of epistemologies invoked by Charleton's anatomical lectures. My discussion treats several of these as case studies, considering the kinds of demonstration invoked by the author.

Reconciling doctrines—The case of the generation of life

Charleton invoked two different theories to explain life. The first of these I have already outlined. Drawn from Descartes and Gassendi, it stated that human life consisted in the union of the rational soul and the body.[100] Bacon and Glisson's theory, in contrast, stated that the life of all animals (man included) consisted in 'the expansive motion of the Spirits in their Blood'. Charleton, in characteristic form, concluded that 'these two positions, though seemingly opposite, are yet really capable of reconciliation each to the other; and by consequence, both may be true'.[101] Jewish commentaries, he declared, indicated that God endowed man with 'a double Life', one immortal, which was 'essential to the rational Soul' the other mortal 'common to Brutes also, and extinguishable by death'.[102] This combination of doctrines had not been, to Charleton's knowledge 'rejected by the Christian Scholes', as heretical, or even unsound, and therefore he claimed that it was 'not unlawful for me to embrace it'.[103] The existence of two souls was, after all, 'very antient, highly consentaneous to reason' and had been defended 'not only by many eminent Philosophers as well antique as modern, but even by some Divines of great learning, Piety and Fame'.[104] In claiming that the two theories could be reconciled, Charleton invoked a range of textual agreements, and the lack of specific refutation.

This attitude to competing theories, about the relationship between material and immaterial entities, parallels his ambivalence regarding the wider implications of the competing theories of Descartes and Glisson on muscular motion. His final lecture on motion did not attempt to combine the alternative theories, but he was not concerned that they asserted a different relationship between material and immaterial aspects of the body.

The role of experiment—The case of 'excitation of the pulse'

The 'excitation of the pulse' was presented by Charleton as the cause of the heart's motion. As blood filled the body parts, its expansive motion distended those parts, causing 'irritation' which incited the retraction of the fibres.[105] This contraction (through natural restitution) squeezed the contained blood out into the ventricles,

[100] Charleton, Enquiries, p. 403.
[101] Charleton, *Enquiries*, p. 403.
[102] The latter only relied upon the expansive motion of the spirits of the blood. Charleton, *Enquiries*, p. 403.
[103] Charleton, *Enquiries*, pp. 404-5. He invoked St Paul's division of the man into body, sensitive or vital soul, and spirit/rational soul. The apostle's testimony was affirmed by reference to 'many Ethnic Philosophers, and some antient Fathers'. Charleton also cited his own *Natural History of the Passions*, as a more thorough discussion of this matter.
[104] These included the Catholic Gassendi and the Anglican Dr Hammond. Charleton, *Enquiries*, p. 404.
[105] Charleton, *Enquiries*, pp. 405-6.

and thence onward as circulation. This systole and diastole was 'the grand cause of the Perpetual Circuition of the Blood'.[106] Thus the diastole of the heart, claimed Charleton, was caused not by the blood's properties, but simply by 'the Quantity of Blood flowing into and distending them'.[107] This quantitative theory countered the role of 'ebullition', as Aristotle argued, or 'rarifaction', as Descartes claimed.[108] The substance of the heart itself was further confirmation—its fibres and membranes would assist its restitutive contraction. Further evidence lay in the analogy between the filling and emptying of the ventricles and the action of the stomach, bladder, womb and other bodily cavities.[109]

'But why do I insist upon these Reasons', Charleton asked, 'when an easie Experiment offers itself to determine the Question?' Here, uncharacteristically, he promoted experimental observation as conclusive. However, there is no suggestion that this was his own experimental research. He detailed Richard Lower's vivisection experiments on the heart of a dog.[110] When ligatures prevented the motion of the ventricles, pulsation ceased. Release of the ligatures immediately restored pulsation. Thus, the link between blood volume and pulsation was demonstrated by experimental manipulation.

Evidence of the distention of the ventricle by a volume of blood was provided as if from Charleton's personal experience. Distention was evident 'even to Sense', by touching the turgid diastolic ventricle. Vivisection of an eel or viper would also verify the loss of blood from the heart during systole.[111] Proof of the ventricle's distention adequately demonstrated, for Charleton, that the same was the cause of circulation.

Charleton copied much of his demonstration of the heart's motion from Harvey's *De Motu Cordis*.[112] This source accounts for the greater prominence of experimental evidence in this lecture. His audience would have been familiar with Harvey's text, and would probably have recognised these references. It seems likely Charleton gained authority from the textual reference, from knowledge of such an important text, and from imitation of its style on this specific topic. He recounted experiments in the future subjunctive: 'if the two Vessels . . . be girt with ligatures . . . their orderly pulsation will cease.'[113] These experimental descriptions were presented almost as a guide for readers to use in their own dissections ('If you open an Eel or Viper alive, you may observe the Heart to become white in the Systole'.[114] This style can be found in *De Motu Cordis*, and specific demonstrations described in *Enquiries* are easily traceable.

[106] Charleton, *Enquiries*, pp. 406-7.
[107] Charleton, *Enquiries*, p. 411.
[108] Charleton, *Enquiries*, pp. 407-8.
[109] Charleton, *Enquiries*, p. 411.
[110] Charleton, *Enquiries*, p. 412.
[111] Charleton, *Enquiries*, p. 410.
[112] See esp. Harvey, *De Motu Cordis*, Chapters 4 and 10.
[113] Charleton, *Enquiries*, p. 412.
[114] Charleton, *Enquiries*, p. 410. See corresponding instance in W. Harvey, *De Motu Cordis*, ed. and trans. G. Whitteridge, Oxford, Blackwell Scientific, 1976, Chapter 10, p. 289.

Charleton's personal endorsement of the experimental method here was a close echo of Harvey. After making these quantitative and empirical declarations, he moved on to his next source, Glisson. He presented the latter's explanation, and adopted its associated method as his own. Charleton did describe the experiments in such a way as to give the reader an indication of how to perform them. However, the authority of prominent experimenters such as Lower and Harvey was not invoked through reference to any experimental context. He did not seem to feel any compulsion to locate the experimental instance in a specific time and place, or to construct a 'virtual witnessing' of the kind Dear and Shapin portray. Experiment was clearly an important mode of demonstration. However, Charleton did not cite experiments as a matter of course, and did not link himself to them, even when he had been present at, or involved in, relevant examples. Quite the contrary, a great number of the experimental instances he cited were in fact drawn directly from textual authorities. Where experiment was a useful means of demonstration, its specificity was not (as some suggest), central to its persuasive power, at least in Charleton's mind.

Charleton invoked a range of methods of demonstration. The divine origin of the vitality of the blood was proven through reference to Scriptural authority and classical doctrine. The nature of that vitality was figured through the corpuscular hypothesis, and the actions of corpuscles demonstrated by analogic instances. These analogically demonstrated actions were fitted into a quantitative framework, which explained the causes of observable motions. The force of observable motions was demonstrated by reference to observed actions of the body, and invoked as proof of the quantitative causal hypothesis.

His epistemological manner varied from text to text, according to the source from which he drew. The epistemology of Lower, the demonstrations of Harvey, and the method of Glisson were all brought into play. There was no single, uniform mode of identity within which all physicians were bound, but rather a range, among which Charleton seems to have roamed extensively. Clearly many epistemological methods were interchangeable for Charleton. He ranged between qualitative and quantitative explanations, invoking empirical demonstration and speculative analogy. He both rejected and adhered to ancient pathological principles, without apparent sense that the ancients were to be symbolically rejected or uniformly embraced. Discrete epistemological categories, created by historians to make our comprehension of the past easier, seem instead to make Charleton more impenetrable.

The continuity of Galenic therapeutics—The clinical episode

Despite his claims, the main body of Charleton's lecture made no reference to clinical experiences. The Epilogue engaged with therapeutics in the form of comments upon Philostratus' tale concerning the apparent death and miraculous recovery to life of a maiden, under the physic of Apollonius. The story itself was omitted (perhaps assumed to be familiar to the audience) but Charleton's comments covered several pages. Charleton used this epilogue to illustrate his own classical knowledge, making a pronouncement upon the translation from the Greek, and emphasising his own interpretation of the original text.[115] Philostratus' expression for original life, he noted, echoed exactly that definition that he himself had shown was 'consentaneous to Truth.'[116] Thus Charleton's own definition was ratified.[117] His conclusions on this episode reflect the continuity of Galenic therapeutics.[118] The epilogue illustrates that, despite his considerable authority as a physician, Charleton did not refer to his own clinical experience, or promote himself as a practitioner before this audience. He constructed his authority before this audience as a man of scholarly training, able to interpret classical texts and with broad knowledge of contemporary publications.

Reconciling ancient knowledge—The case of fevers

Charleton's discussion of a clinical issue, such as fevers, reflects an intersection between his fundamentally traditional pathological ideas and the new ideas and discoveries that he had set himself to outline for his audience. The ultimate aim of his discourse, he claimed, was to increase knowledge in order to facilitate cure. To find out 'the most probable Causes, and reason of curing great Diseases, is the principal scope and end of all our Enquiries, as well Physiological, as Anatomical.'[119] This reaffirmed the therapeutic status of the physician's identity. It was important in this discussion to reconcile traditional therapy with innovations. Treatment of fevers was one of the more lucrative aspects of clinical practice for physicians.[120] Here was an important juncture at which to reassert the legitimacy of the traditional treatments that he himself offered.[121]

Charleton's epistemological approach was stated in his opening paragraph: it paralleled that by which mathematicians inferred 'Theorems or consentaneous Speculations' upon a series of propositions. The verity of such theories, he argued, was equal to that of the propositions upon which they were based. His own method in the discussion of fevers was an imitation of that method. He had

[115] Again Charleton was concerned with the importance of naming and definition: Philostratus' definition of life was identical to that of Salomon. He also cited Ecclesiastes, and claimed that the coincidence between their definitions 'is alone sufficient, as to give credit to the Expression itself'. Charleton, *Enquiries*, p. 424.

[116] Charleton, *Enquiries*, p. 424.

[117] The maiden, he claimed, had suffered from a temporary 'eclipse' of the vital spark. Charleton, *Enquiries*, p. 425.

[118] He diagnosed the maiden as suffering from a 'fit of the mother' (caused by a wandering uterus) and as possessing the 'frigid and delicate Constitution' common to women. Charleton, *Enquiries*, p. 424.

[119] Charleton, *Enquiries*, p. 430.

[120] Epidemic fevers were the most lucrative area of seventeenth-century treatment. Dewhurst, *Willis's Oxford Lectures*, p. 12.

[121] Charleton's recipe books, indicating the traditional nature of his curative therapies, are held at the British Library.

already laid down 'the Primordia, perpetual Source, and circular race of Life', from which he wished to make some pathological deductions which might 'afford some glimses of Light toward the discovery, and nature and causes of a certain malady'.[122] The lecture on fevers was therefore a logical progression from the lecture on life. Blood was the origin of life, and fever was an interruption or perversion of the spirits ruling the blood.

Brown sees Charleton's lecture as a 'public endorsement of the corpuscular approach to medicine', which provided an 'unquestioning' or uncritical outline of the basis of corpuscular iatromechanism. Charleton's lecture on fevers was an 'impressive testimony' to the extent of Willis' influence.[123] He claims that Charleton spoke highly of Willis in the *Enquiries*.[124] However, I argue that Brown overestimates the extent and importance of Charleton's adherence to Willis. He cites Charleton's terminology as evidence of his consistently mechanistic approach,[125] ignoring the multiple terminologies invoked by the physician. Brown argues further that Charleton's disquisition on fevers was a 'formal announcement' of the College's 'total support for the iatro-mechanical approach to medicine.'[126]

Brown believes that the lecture on fevers constituted 'a public advertisement' for Willis' *Diatribæ Duæ*.[127] He makes no mention of the important figure by whom Charleton was undoubtedly also influenced, Francis Glisson. Charleton put forward 'Dr. Glisson's new Doctrine, concerning the most general and obvious differences of Crudities apt to produce Fevers', and depicted himself as attempting to 'assert and explain' what he termed the 'Glissonian Hypothesis'.[128] Brown is keen to associate *Enquiries* with the College's mechanist agenda, for which he sees Willis as the primary vehicle. However Charleton's debt to Glisson might equally be read as an expression of support for a hylozoistic monism, in contrast with Willis. Evidence for Charleton's strict adherence to Willis is slender, and we must be circumspect about any such adherence, given Charleton's consistent tendency toward eclecticism. In this lecture Charleton aimed to illustrate convergence between traditional and innovative theories and treatments.

Charleton's discussion of the fevers was also influenced largely by Galenic and chemical theories, and Hippocratic notions of cause and effect. This issue, close to his own therapeutic and clinical practices, doubtless would have inclined the author to invoke traditional explanations. New treatments had not emerged from new theories of causation, and therapeutics tended to remain consistent with traditional clinical practices. Although there was much innovation in his discussion of particulate processes within the body, the effective causes and recom-

[122] Charleton, *Enquiries*, p. 429.
[123] Brown, *Animal Oeconomy*, pp. 173-4.
[124] Brown cites *Enquiries*, pp. 26-7. Brown, *Animal Oeconomy*, p. 173.
[125] Brown, *Animal Oeconomy*, p. 173, cites Charleton's language on pp. 451-2.
[126] Brown, *Animal Oeconomy*, p. 175.
[127] Brown, *Animal Oeconomy*, p. 175.
[128] Charleton, *Enquiries*, pp. 467-72.

mendations for treatment were consistent with traditional therapeutics. Febrile corruptions were generated by the incorrect use of the six non-naturals.[129]

The great variety of the crudities responsible for fevers could not possibly be known by any physician 'however curiose', and Charleton therefore reduced them 'to the Analogous Humors of the Antients'.[130] The commotions caused by crude humours could be termed a 'Ferment, according to the Name given to it by all Modern Physicians.'[131] Charleton went to great lengths to illustrate that the fermentation he described was 'the same thing that the Antient Physicians meant by the Putrefaction of the Blood in Fevers'.[132] This exemplifies Charleton's continual attempts to integrate ancient principles with modern concepts. His main emphasis here was upon the correspondence between modern notions about fevers and those ancient principles upon which therapeutics were based.

The 'disorder of the State of the Body' was caused partly by nature, which was responsible for the motion of the spirits. But insofar as that 'mictation' was violent and hostile, it must result from corrupt or alien matter.[133] Some matter contained spirits that were 'hostile' and 'pernicious to the Vital Spirits, and incapable of being tamed' (*Semina heterogenea*).[134] This illustrates a combination between this fundamentally new theory of a destructive agent and the 'Doctrine of the Ancients', which attributed tertian fevers to 'Choler, Quotidian to Phlegm, and Quartan to Melancholy'.[135] Charleton answered this by resort to the theories of Glisson, 'who in all arguments endevor'd, as far as his devotion to Truth would permit him, to sustain the authority of the Antients'. Glisson had reconciled ancient and modern theories through the claim that the humors described by the ancients 'were, or at least might be taken, either for the reliques of the stale and vapid Blood, or for Humors analogous to them.'[136]

[129] 'intemperance, too full Diet, Surfeits, Compotations, and other Debauches and Disorders; or for want of exercise to correct and dissipate the crudities they have congested'. Charleton, *Enquiries*, p. 462.

[130] Charleton, *Enquiries*, p. 466. He detailed these similarities again, on pp. 466-7.

[131] Fermentation was 'not indeed such as the Leven of Bread, or as the Yest [sic] of Ale and Beer; but such, that being in our Bodies mixt with the Blood which perpetually conceives new vital Heat in itself, produceth the like commotions therein, that those domestic Ferments do in their respective Subjects'. Charleton simultaneously rejected and ratified the analogy of yeast fermentation. Charleton, *Enquiries*, p. 432.

[132] The ancients' term 'Putredo, is the very motion of the matter of a Fever tending to purulency: and this Motion is the very same, that most of the Neoterics Name Fermentation'. Charleton, *Enquiries*, pp. 433-4.

[133] Charleton, *Enquiries*, p. 433.

[134] While Nature had 'taken care to preserve the Blood pure and undefiled' through the use of secretory organs, she had no mechanism to deal with crude humours still imbued with spiritous vitality. Charleton, *Enquiries*, pp. 437-8. This was first observed by Cardan.

[135] Charleton, *Enquiries*, p. 439.

[136] Charleton, *Enquiries*, p. 439. The only humor Charleton did not accept was melancholy, and here he diverged from the doctrines of the ancients. Notwithstanding 'the respects and veneration due to those Fathers of our Art', he observed, 'the interest of Truth, which is still more sacred and venerable, obliges us to affirm, that they erred most egregiously' regarding the production of melancholy in the spleen. Charleton, *Enquiries*, p. 440. Charleton claimed that 'in the whole Body we find no peculiar organ provided by Nature for the secretion, reception, and exclusion of any such Humor', p. 440.

Analogy had a significant role in verification. Charleton took some license, in calling melancholy analogous, because 'nothing that holds any the least resemblance or analogie, can be any where in the whole Body found'. Nevertheless it was 'lawful' to claim

> that the Analogie that some parts of the Blood seem to have to that fictitious Humor, which the Antients imagined to be separated and received by the Spleen; may serve to excuse us, if, out of compliance with custom and the vulgar Doctrine of the Schools, we retain the denomination, while we rectify the Notion of Melancholy.[137]

Glisson's 'reliques of the blood' were therefore reconciled with classical 'humors', and encompassed by the same nomenclature.[138]

Charleton here claimed that a common term could be used while its meaning was revised. This illustrates the importance of retention of classical knowledge to clinical/therapeutic questions. He further implied that an imagined analogy was adequate resemblance to justify the use of the same name: 'For, though the Analogie be wanting; yet if in reality a thing respondent thereto, hath existence in Nature; the supposed Analogy is enough to justify the appellation.'[139] He concluded that 'there seems to be no difficulty in reconciling the Doctrine of the Ancients concerning the Humors contained in the Blood, with the constituent parts of it now observed by us'.[140] Charleton reconciled modern knowledge with established doctrines, claiming to have illustrated the material cause of putrid fevers 'without demolishing what the Antients have delivered of the same'.[141]

Yet at other times Charleton attacked perceived adherence to authority. The question over the existence of a 'Quotidian' fever, had generated 'a kind of Civil War among the Sons of Aesculapius'.[142] Quotidian fevers had been incorrectly rejected by some authors. These men, he argued, had been 'carried away by prejudice, and the authority of their Predecessors, rather than conducted by the light either of certain Experience, or of right Reason'.[143] Such opposition to ancient learning was not grounded upon profound epistemological resistance, but on disagreement with particular doctrines. The charge of over-adherence to classical knowledge appeared sporadically. Yet such criticisms were outnumbered by proofs or demonstrations verified by ancient doctrines. Thus the physician's criticisms of adherence to doctrines did not constitute a rejection of classical knowledge *per se*.

Charleton's conclusions did not attempt to promote any theory as the final answer to the question. Instead he claimed that his audience 'may easily judge

[137] Charleton, *Enquiries*, p. 440.

[138] 'The Humors analogous to the newly enumerated Reliques of the Blood, are signified by the same names', Charleton noted: 'in particular, the viscid, insipid, and white part of the Blood, is called Pituita, or Phlegm; the hot, drie, acrimonious, and pungent or corroding, Bilis or Choler; the cold, drie, blackish, and adust, Melancholy (if at least any such Humor may be admitted to lye concealed in the Mass of Blood.)' Charleton, *Enquiries*, p. 440.

[139] Charleton, *Enquiries*, p. 440.

[140] Charleton, *Enquiries*, pp. 441-2.

[141] Charleton, *Enquiries*, p. 442.

[142] Charleton, *Enquiries*, p. 470.

[143] Charleton, *Enquiries*, pp. 470-71.

of the reasonableness of the precedent Hypothesis', and left it to them to decide 'how far the same may deserve your approbation, or dislike.'[144] This depicted the audience as men of reason, rather than men of experiment. The strength of the rational proofs were of course the most important persuasive elements. Thus his purpose seems to have been largely the reconciliation of ancient with contemporary theories.

The continuity of inherent properties in explanation

Charleton's lectures appeared to refute claims for (scholastic) inherent properties in matter. Yet he argued that the poisonous nature of the *fermentum febrile* 'doth consist, not in any manifest quality, but in some Seminal Nature',[145] which was 'secretly repugnant and pernicious to human Nature'.[146] This resort to essential qualities in explanation was one of many.[147] He several times invoked the idea that differences between objects were implanted in an essential form. For example, he argued that all food possessed 'certain indelible characters, or insuperable qualities', which could be 'tamed or kept under, but can never be totally destroy'd'.[148] In the process of digestion the body needed a powerful capacity to tame 'the reliques of the former seminal impressions of the aliments: in all which there certainly remain some vestiges or prints as it were of their pristine form, which we may properly call their seminal impressions'.[149] These impressions had to be 'subjugated' to admit new impressions.[150] This was an animated struggle, in which active principles were subjugated, so that a dominating force could impose a new regimen. There is no evidence of a corpuscular or mechanist framework in this explanation.

The 'Therapeutic Corollary'

The 'Therapeutic Corollary' was designed to be 'useful to Younger Students in Medicine', for whose instruction the anatomic lectures were designed.[151] Charleton considered the best remedies, citing Hippocrates, Seneca, Livy and Galen. The most appropriate advice was always preventive: correct regulation of diet, and lifestyle. Therapy focused on expulsion of harmful matter from the body (cupping, bloodletting, diaphoretics). Having outlined the proper remedies,

[144] Charleton, *Enquiries*, p. 451.
[145] Charleton, *Enquiries*, p. 442.
[146] It was noxious, regardless of dose, because of its 'seminal Malignity'. Charleton, *Enquiries*, p. 443. The resonances are clear here with the 'real qualities' of scholastic doctrine.
[147] Charleton identified cause through effect when he described the origin of various fevers: the cause of malignant fevers was more 'malicious and greivously hostile' than that of putrid fevers, because the severity of the fever itself was worse. Charleton, *Enquiries*, pp. 442-6.
[148] Charleton, *Enquiries*, p. 117.
[149] Charleton, *Enquiries*, p. 117.
[150] Charleton, *Enquiries*, p. 117.
[151] Charleton, *Enquiries*, p. 472.

Charleton cited 'a certain Rule of Hippocrates',[152] concerning the over-use of purgative medicines, which seemed, he acknowledged, 'plainly repugnant' to his former claims for their importance. Charleton set out the situations in which it was not only lawful, but expedient, for a learned physician 'to institute convenient Evacuations'. This contemporary debate, over the extent to which physicians should prescribe purgatives, was concluded by Charleton with the claim that he could 'reconcile these two equally true aphorisms'.[153] This discussion of fevers owes more, in its eclectic method, to Glisson than to Willis.

Brown argues that Charleton simply justified traditional therapy through reference to 'modern, mechanical theory'.[154] However, it is arguable whether Charleton followed a mechanistic theory, and far from evident that he saw its development as defining his own role in the presentation of the fevers lecture. It seems rather that he did not promote any hypothesis as a definitive solution, but instead wished to illustrate the reconcilability of ancient doctrines and modern ideas, and considered physical hypotheses without concluding which was the most satisfactory.

In this sense I understand the consequences of Charleton's lecture on fevers to be broader than Brown concludes. The lecture allowed the reconciliation of ancient with modern knowledges. Rather than presenting a conversion to a fully mechanistic approach to medicine, this lecture exemplifies Charleton's eclecticism, which allowed him to bring together traditional and innovative ideas. In so doing, he reaffirmed the status of the approach by which the College had traditionally defined itself. This was also the basis upon which his practice as a physician was founded. He would hardly have been acting in his own interests in disparaging it. However, while supporting the legitimacy of that knowledge, the content of the lecture confirmed that he as a College fellow was fully conversant with the new theories of the day. This, in all promoted his identity as well-read, conscientious and in tune with new developments. From this position of authority he was able to justify the traditional physic that was the cornerstone of his professional status.

Charleton's textual reliance—The case of 'voluntary motion'

Charleton's final lecture raises questions about the sources of his arguments, and the influence of his experimental activities within the Royal Society. Charles Webster situates Charleton as an active participant in experimental procedures upon muscular motion, in collaboration with his country's experts, Charles Scarburgh and Christopher Wren.[155] This text on voluntary motion has been the

[152] Charleton, *Enquiries*, pp. 482-3.
[153] Charleton, *Enquiries*, p. 487.
[154] Brown, *Animal Oeconomy*, p. 174.
[155] Webster, *Great Instauration*, p. 318. On Wren's activities, see W. C. Gibson, 'The biomedical pursuits of Christopher Wren', *Medical History*, vol. 14, 1970, pp. 331-41. Webster cites Charleton as the 'first to give a mechanical analysis of the action of the voluntary muscles'. Webster, 'Solomon's House', p. 400, fn. 20. Goodall argues for Scarburgh's contribution as the 'first who introduced Geometrical and Mechanical Speculations into Anatomy'. Goodall, *College of Physicians*, p. xi.

most frequently noted aspect of his *Enquiries*. The most insightful discussion of Charleton's work on muscle physiology is that of Margaret Nayler. She suggests that Charleton's presentation of the ideas of Jacob Müller was a result of his wish to apply geometrical and mechanical principles to muscular motion.[156]

Nayler illustrates Charleton's profound debt to contemporary textual sources on muscular motion, both English and Continental. She maintains that Charleton was not directly involved with those contemporaries whose contributions to muscle physiology have become famous.[157] Charleton was not at the forefront of experiment, but drew his authority from the College's adoption of the geometrical hypothesis proposed by Jacob Müller. *Enquiries* reiterated the fundamental details of Müller's explanation as presented in *Natural History*. Charleton had performed a muscle dissection himself, at the Royal Society in January 1664,[158] and had probably also witnessed Croune's experiments at the Society on 6 November 1661.[159] Clearly he could have presented his own experience in muscular dissection, annexing the authority of the experimental philosopher. Although since 1659 he had witnessed numerous muscle dissections, he used the same textual source, and the same essential doctrine, in 1680 as in 1659. Despite the experimental demonstrations he had performed or witnessed, he chose instead to present a textual extract, describing and contrasting the account of Müller with the theory of Glisson. This suggests that experimental authority did not constitute the most powerful or desirable epistemological device for a physician in Charleton's position in a public lecture. His Royal Society experimental activities illuminate only part of his self-definition.

Alternative metaphors and competing explanations in muscular physiology

Two theories existed to explain 'animal' or voluntary motion. The first, that of Galen, saw the nerves as conduits through which animal spirits passed, on their

[156] Nayler, 'Insoluble Problem', p. 143.

[157] The physician's attacks on Willis, for example, do not indicate familiarity with the details of Willis' theory. Nayler, 'Insoluble Problem', p. 143.

[158] At the meeting on 20 January 1664, Charleton offered to 'open the muscles after a new method.' He performed the dissection following the next execution, and it was later suggested that his lectures on the muscles might be printed [17 Feb 1664]. See Birch, *History*, vol. 1, pp. 373-85. Apparently this came to nothing, as no such lectures exist within the Royal Society archives.

[159] Birch notes the demonstration of the lifting of a weight through water pressure in a bladder: *History*, vol. 1, p. 53. Croune's work influenced Steno, whose *Elementorum myologiæ specimen* was published in Florence in 1667. Croune's theory of chemical explosion, and Steno's suggestion of multiple fluids within the muscles, influenced Thomas Willis and John Mayow. Willis believed the reaction within the muscles was akin to the explosion of gunpowder. T. Willis, *The Remaining Medical Works of . . . Thomas Willis*, trans. S. Pordage, London, 1681, pp. 40-42. Mayow replaced spirits with his nitro-aerial particles. J. Mayow, *Medico-Physical Works*, London, 1674, pp. 244-64. Although Borelli's work was not published by the time Charleton gave his lectures, he later added a note on Borelli, in the margins of his own copy of the *Enquiries* (now in the British Library), p. 542.

way from the brain (where they had been prepared and purified out of blood) to the extremities. In this model, the animal spirits received sense impressions from the external world, on behalf of the brain, and then acted as the brain's agents for the control of the muscles.[160] The second hypothesis stated that the nerves were under perpetual tension, and that communication from the brain consisted of 'vibrations', which were transmitted along the nerve like vibrations along 'the tightened string of a musical instrument.'[161]

Glisson's hypothesis replaced spirits in conduits with the contracting motion of fibres. The soul's communication to the muscle was performed 'by a mere contraction of such Fibres of the Brain as are continued to that Nerve.'[162] The nerves originated in the brain, and were attached at the other end to muscles, so that a single uninterrupted cord linked brain and muscle. Thus a motion commenced in the fibres of the brain, must be simultaneously conveyed to the nerve and thence the muscle.[163] While this theory avoided the necessity of spirits as agents of muscular motion, it offered no real suggestion on the problematic relationship between incorporeal soul and corporeal matter. Its focus was the nature of the communication to the muscles, through what Glisson termed 'natural perception'.

In 1659 Charleton had enthusiastically supported the ancient doctrine that the soul acted on the muscles through the transmission of animal spirits.[164] The muscles were moved when they were distended by a 'gale of spirits', issued out of the brain.[165] The following discussion examines Charleton's 1680 exploration of the properties and processes of each theory of muscular motion. Despite the similarities in source and subject matter between *Natural History* and Charleton's 1680 lecture on voluntary motion, the later text discussed the issue quite differently. Whereas the 1659 text accepted traditional Galenic explanation through spirits, in 1680 Charleton declined to show his support for either hypothesis. Instead he claimed that he would 'put into the Scales the principal Arguments alleaged on both Sides, together with their respective difficulties not yet sufficiently solved'. The audience could then judge 'which of the two Opinions is the more probable'.[166] He presented cases for and against the exis-

160 Frank, *Oxford Physiologists*, p. 4.

161 Wilson, 'Croone's theory of muscular contraction', p. 161. This view, held by Scarburgh, later influenced Glisson, who incorporated his own distinctive matter theory. Croune described a reaction between a materialised spirit within the nerves and blood passing through them, causing a 'commotion and agitation of the particles'. W. Croune, *De ratione motus musculorum*, London, 1664, p. 21, sect 27. His significant addition, however, was that muscular motion resulted from vibration in the nerves, rather than merely movement of fluid. Wilson, 'Croone's theory of muscular motion', p. 164.

162 Charleton, *Enquiries*, p. 495.

163 Charleton, *Enquiries*, p. 496.

164 Charleton, *Enquiries*, p. 495 [mis-numbered, duplicated no's on pp. 494-5].

165 Charleton, *Natural History*, p. 188. This process could be demonstrated anatomically by the observation that a severed nerve effectively prevented motion: 'because the intercourse betwixt the brain and that particular Muscle is wholly destroyed'. Charleton, *Natural History*, p. 199.

166 Charleton, *Enquiries*, p. 496 [proper].

tence and necessity of animal spirits, claiming that he himself had no determined preference. His own position was that of a neutral presenter of debate.[167]

The contested question was whether the soul required intermediary instruments. The theory that communication between soul and limbs was performed by animal spirits had been 'without any dispute embraced and asserted through a long train of Ages'.[168] Yet, as he observed in 1680, that 'antique Hypothesis' no longer possessed the same authority. Recent theories had suggested that the spirits were 'both improbable and unnecessary'. Charleton conceded that this left 'great uncertainty', not only of the origin, nature and qualities of animal spirits; but 'of their very existence'. He found himself 'unable to determine what I ought to conclude, of the Antient, and at this day vulgar opinion, of their being absolutely requisite, both to sensation, and to all voluntary motion'.[169] Consequently, he called upon the 'more discerning judgement' of his audience, for a decision of this so difficult controversie'. Again he acknowledged his ignorance, and declared that his subject was 'wrapt up in clouds of impenetrable darkness'.[170]

Charleton's negotiation of these alternative hypotheses in his 1680 publication illustrates a different self-construction from that of the *Natural History*. His eclecticism was perhaps increased by Royal Society emphasis upon this mode of presentation. However, the difference in genre between the 1659 and 1680 texts is also important. A learned lecture required a different presentation from that of a textbook. Thus it is quite possible that his avoidance of preference for one hypothesis over the other was founded on the requirement that he present both, with their difficulties, for the 'Younger students' to determine their veracity.

Arguments for *the animal spirits*

Beginning, 'out of respect to the Antients', with the conduit theory, Charleton commenced his examination of the question of voluntary motion. His arguments *for* the spirit-conduit theory invoked: logical deduction based on analogy and theological principles; traditional anatomical theories (structure indicates purpose; comparative anatomy—the idea that nature doesn't vary methods between species; experimental observation recorded by others (Lower, Ent, etc). The following discussion considers some of these demonstrations, which offer an insight into how Charleton reached his conclusions.

One argument for the animal spirits theory was that, as motive power was not inherent in the muscles or nerves, it must therefore be 'immitted into them from the Brain'.[171] Glisson's natural perception hypothesis was derided as incongruous

[167] Although in 1659 he had frequently presented hypotheses as conditional, his stance in 1680 made that avoidance of doctrines more explicit.

[168] Charleton, *Enquiries*, p. 495 [mis-numbered, duplicated no's on pp. 494-5]. He had embraced it himself in 1659.

[169] Charleton, *Enquiries*, p. 518.

[170] Charleton, *Enquiries*, p. 518. Yet he claimed that 'this my ignorance must not deter me from proceeding in the administration of my province.' Charleton, *Enquiries*, p. 519.

[171] Charleton, *Enquiries*, p. 498.

with the 'logic' of anatomy. The nerves Charleton claimed, were not strings by which the parts could be moved, but 'conduit pipes' through which racing spirits communicated. For if the nerves were intended to act as strings, why were they composed 'of many slender Filaments or Threads?'[172] The anatomy of the nerves confirmed that their role was in conveying spirits. This could be further verified by the observation that prevention of spiritous access through the nerves caused a palsy. However, the anatomical observations on which this argument was based were themselves founded on the assumed necessity of spirits. If the animal spirits were agents of rapid communication, the structure of the body must contain pathways for their travel. Examination of the nerves therefore revealed precisely these findings, and Charleton found himself 'almost obliged to acknowledge, that the Fibres were so disposed in Parallels, with small Canales running along betwixt them', so as to allow 'quick passage' to the spirits transmitted from the brain to the muscles.[173] If anatomy did not verify the presence of such pathways, this was a failing of anatomy.[174] Here we can see clearly that for Charleton argument held greater significance and import as a form of demonstration than did empirical evidence.

Further proof of the conduit-nature of the nerves was found in the fact that Galen himself, 'compares the Nerves, not to Strings or Cords, but to rivuletts, or Conduit-Pipes'.[175] Charleton relied not upon an account of Galenic anatomy, but on the ratification of Galenic *analogy*. This indicates his respect for the results of physiological and anatomical insights gained through compelling analogies. Another argument in favour of the theory derived from comparative anatomy. The voluntary motion of snails and earthworms was aided by a 'certain aerial or spirituose Substance, issuing from their Heads'.[176] Charleton made explicit the analogical power of comparative anatomy: just as an 'Aerial Substance' aided the motion of other animals, likewise 'our muscular Motion is made by the help of a spirituose Influx from the Brain into the Nerves and Musc[le]s.'[177] Though Nature varied the instruments of voluntary motion in animals according to their respective forms and functions, nevertheless 'she seldom

[172] 'And why were not these small Threads closely twisted together into a strong Cord, but extended, as parallels, side by side, and intercepted with many narrow Canales or Passages betwixt them, and all inclosed within a membraneous Coat extremely thin and tender?' Charleton, *Enquiries*, p. 499.

[173] Charleton, *Enquiries*, p. 499.

[174] This point is noted by Clarke in relation to Riolan. Clarke, 'Neural circulation', p. 302. As Riolan claimed: 'A Nerve is a Channel made to carry animal Spirit: and because the spirit is most subtil, therefore the Cavity is so smal, that it is not discernable'. Riolan, *A Sure Guide*, p. 27.

[175] Charleton, *Enquiries*, p. 499.

[176] Charleton, *Enquiries*, p. 500. This analogy was borrowed from Regius' observations on slugs: *Fundamenta physices*, Amsterdam, L. Elzevir, 1646, see pp. 225-6. However, Charleton did not cite Regius, but rather a M. Bochart, who had seen a similar phenomenon in the motion of a 'Horse-leech'. In the Latin version of *Natural History*, Charleton had cited Regius in relation to this demonstration. Charleton, *Oeconomia Animalis*, p. 21. However, *Natural History* contained no mention of Regius as the source of this observation.

[177] Charleton, *Enquiries*, pp. 500-501.

varies her more general ways of formation, but upon necessity, and when she doth, her defections from her common Method are never so wide, but still they carry some Analogie to them'.[178] This principle ratified further analogically-derived insights.

Experimental proof was not the pinnacle of knowledge in anatomy. A compelling analogy could constitute a stronger proof than could an anatomical demonstration. Analogy held a powerful explanatory role, and observations made in one instance often sufficed as the proof in other (analogous) dimensions of physiology.

Arguments against *animal spirits*

Having presented arguments for the animal spirits, Charleton proceeded to defend the alternative position. Here he attacked the 'fanciful' notions of Willis. Although heavily reliant upon Müller, Charleton did not acknowledge this source.[179] Many of Charleton's refutations of the spirit hypothesis were drawn directly from Harvey and Ent. He criticised supporters of spirit theory for their failure to generate an appropriate analogy by which the action of the spirits could be understood. He argued that their proposed explanation did not stand up to logical analysis, and claimed that it was inconsistent with the circulation model (which was assumed to apply to all fluid motion). Perhaps Charleton's most telling criticism of the spirit-theorists was that their theory failed because it invoked spirits as a *factotum*.

Physicians, complained Charleton, invoked animal spirits as 'the plenipotent and immediate instrument of the Soul in all her operations upon the Body'.[180] Spirits were thus endowed with the status of a total explanation of physiological process. The virtuoso emphasis upon non-dogmatic research dictated against over-fondness for particular explanations. Accusations of over-use constituted an attack on the failings of authors to maintain the properly skeptical stance. Charleton censured the 'omnipotence' of spirits in medical explanation. He ridiculed Willis, for having 'enrich'd the Commonwealth of Philosophy with a whole Legend of the empire of Animal Spirits'.[181]

Charleton also attacked the method by which experiments had been generated to support the hypothesis. Thomas Bartholin, having been 'desirous to prop up the antique opinion of Animal spirits' was 'first so ingenuous to suspect, and later so lucky as to find certain holes and open passages' for spirits in the petrified brain of an ox.[182] Visual verification of Bartholin's theory was not

[178] Charleton, *Enquiries*, p. 500.

[179] Nayler, 'Insoluble Problem', pp. 133-5 and 401. *Enquiries*, unlike *Natural History*, omitted the geometrical demonstrations. *Enquiries* also contained illustrations drawn from Fabricius, *De musculi utilitatibus*, pp. 135, 150, 151.

[180] Charleton, *Enquiries*, p. 501.

[181] Charleton, *Enquiries*, p. 502. That the author under attack was Thomas Willis is confirmed by Charleton's notes in the margin of his personal copy of the *Enquiries*, in the British Library (on pp. 56-7).

[182] Charleton, *Enquiries*, p. 517.

satisfactory proof, as his method was fundamentally flawed because based on prior expectations.

To question the truth of this empirical finding 'would be Incivility; to believe all of it, shameful credulity; and to conclude from thence that there are Animal spirits, down right folly.'[183] Charleton rejected the experimental account as highly improbable, in contrast with his deference to the analogic demonstration.[184] Empirical knowledge did not automatically supplant rational technique or analogy.

In a culture which valued skepticism concerning dogmatic claims, Charleton's criticism of others' reliance on theories as universal explanations constituted a deliberate and damaging attack. He was less concerned with the spirits themselves than with the manner in which they were invoked. Such a refutation did not prevent his own use of spirit terminology. As the centre of his presentation were the redefinition of spiritualist terminology, and the definition of himself as an 'equal arbiter'.[185] Thus I conclude that experimental demonstration was not essential to illustration of argument. All the theories considered here were textually-based (and strongly derivative of Müller's work). This case study required the provisions of proofs/dis-proofs, but experiment did not in any of these instances prove decisive, nor did mechanist explanation automatically triumph.

Testing the analogy—Glisson's alternative explanation

These points concluded Charleton's attack on the traditional theory in which the animal spirits were required for voluntary motion. The question remained as to how the 'Ruling Faculty in the Brain' communicated with the muscular fibres.[186] Having demonstrated the flaws of the conduit hypothesis, Charleton presented the alternative, based on the theory of his late friend and colleague Francis Glisson. The latter, discerning the problems with the spirit hypothesis, had 'not only wholly rejected it, but excogitated another plainly contrary thereto.'[187] This theory was that in all voluntary motion the fibres of the muscles contracted themselves 'by their own proper vital Motion', therefore 'needing no copiose and suddain afflux of Spirits'.[188] The hypothesis depicted the

[183] Charleton, *Enquiries*, p. 518.

[184] Through his analogue between spirit and wine, he demonstrated that the spirit could not be rectified in the climate of the brain. 'What Chymist hath at any time attempted to rectifie spirit of wine in a vessel whose head was stuffed with damp raw silk, or other like villose matter, more apt to repercuss and condense, than to refine it?' Charleton, *Enquiries*, p. 516.

[185] He deliberately contrasted this eclectic self-presentation with the writings of Thomas Willis. A similar attack upon Willis can be found in Charleton's *Inquisitiones Medico-Physicae*, in which he attacked Willis for over-reliance upon the notion of fermentation.

[186] Charleton, *Enquiries*, p. 508.

[187] Charleton, *Enquiries*, p. 507.

[188] Glisson's theory did not remove the spirits, but relegated them to a minor role, in which they were not essential to the immediate processes of motion. He argued that the spirits moved extremely slowly, in a manner analogous to the way in which 'the nourishing Juice of Plants is believed to creep up from the Roots along their Fibres'. Charleton, *Enquiries*, p. 508.

communication as an 'invigoration, running along the Threads of the Brain, toward the beginning of the Nerve to be excited, and requiring a Motion of the Nerve conform thereunto'.[189] Glisson posited a system of strings, which drew the parts into motion, and vibrated with resonances from the brain.

Glisson had delivered a series of anatomical lectures on the brain in the 1650s, at the College of Physicians.[190] It seems likely that Charleton attended these, or discussed the subject in depth with the older physician.[191] Like Charleton, Glisson questioned the ability of an immaterial soul to act upon the corporeal body.[192] Both were concerned with the interaction between soul and body, and the laws by which the actions of the soul could be characterised. Glisson did not claim to provide 'a clear account', but merely offered 'some propositions about it'.[193] He proffered an array of ideas illustrating the eclecticism and modesty which seem to have characterised learned lectures. Neither attempted to achieve certain knowledge, but discussed the arguments openly, in recognition of the provisional nature of knowledge.

Charleton examined Glisson's hypothesis in detail, as he had that of spirit-conduits. He seemed to hold a fondness for his elder colleague's explanation, referring to it as 'perhaps coming neerer to Truth, than any other hitherto excogitated.'[194] However, out of his own 'honest desire of learning' Charleton felt obliged to raise 'one or two Questions'.[195] He accepted that the brain might be the source of natural perception. But it could not explain voluntary motion, which depended on a principle 'incomparably more sublime and energetic; namely the Soul'.[196] Only the soul could respond to the appearance of good or evil. Otherwise, Glisson's hypothesis removed the need for a soul at all, because the brain endowed with natural perception 'would alone serve to do all the Offices of a sensitive Soul.'[197] Though the late physician's hypothesis did not

[189] Charleton, *Enquiries*, p. 509.

[190] Henry, 'Medicine and Pneumatology', p. 18. See Glisson's text on the brain in the British Library, Sloane MS 3306. I am indebted to John Henry's 'Medicine and Pneumatology', for the excerpts from Glisson's lecture. Henry cites the lectures as taking place shortly after the publication of *Anatomia hepatis*.

[191] Charleton's entire text was highly derivative of Glisson's lecture, and of the elder physician's writings on energetic matter (*Tractatus*, 1672 and *De Ventriculo*, 1677).

[192] The brain, Glisson noted, 'cannot move the nerves, either by its perception or appetite, because those are both immanent actions & cannot reache any thinge without their owne subject.' Henry, 'Medicine and Pneumatology', p. 19. Glisson, Sloane MS 3306, f. 163.

[193] Henry, 'Medicine and Pneumatology', p. 18.

[194] Charleton, *Enquiries*, p. 509. Charleton summarised Glisson's hypothesis as follows: 'The motion of the Brain from within outward, by which it rules the Fibres of the Muscles, is made known to the Fibres to be moved, not by sense (for the Intellect hath no notice or cognizance of it at all) but only through Natural perception: and consequently that the Brain, by mediation of this Perception Natural, doth, at the command of the Phansie, excite the Fibres of the Muscles to Motion, and recompose them, at pleasure.' Charleton, *Enquiries*, p. 508.

[195] Charleton, *Enquiries*, p. 509.

[196] Charleton, *Enquiries*, p. 511.

[197] Charleton, *Enquiries*, p. 511.

promote the nerves above the soul, it attributed motion to two different sources within the body.[198] While Charleton's first argument played on an imbalance in the hierarchy of authority in the body, his next related to a disproportion in physical action.

If communication between brain and limb were dependent upon motion at the source, it would require the motion in the brain to be far smaller than the motion it caused in the part. Yet, he argued, this defied the laws governing fibres, as proven by 'mechanic experiments'.[199] Given this, he concluded, it was quite impossible to conceive how a small contraction in the brain's fibrils could cause a greater contraction in the nerves.[200] Charleton invoked the analogy of a lute-string to explain the action of the nerves in the body.[201] He did not illustrate his point anatomically, but rested upon the substantiating power of the lute-string analogy. Once again, evidence provided by analogy was more fundamental to his argument (and sufficient to his purposes) than was anatomical proof.

Charleton's conclusions on the 'reason and manner' of motion of the muscles, allowed that natural motion operated only in contraction, having no capacity for extension, nor sentience.[202] Charleton's refutation of natural perception retained many of Glisson's ideas, and indeed the explanatory metaphor, but shifted the attribution of causal power, located by the elder physician in the inherent sentience of fibres.

The topics and perspectives considered in Charleton's final lecture, on muscle physiology, seem to have been guided by Müller's writings on the muscles and Glisson's lecture on the brain. Once again, his debt to contemporaries is more considerable than his own original contribution. Charleton's textual reliance is a striking characteristic of his lecture on muscular motion, despite his own activities in

[198] Charleton refuted Glisson's theory using the body politic metaphor. Glisson had 'set up in the Brain a new Government by a kind of Duumvirate, the Regimen being divided betwixt the Will and natural Perception, so that neither of them can, without the assent of the other, excite any the least animal Motion. And how unstable, how obnoxious to Divisions must that State be, that hath two Heads?' Charleton, *Enquiries*, p. 514.

[199] 'all Cords, when pull'd or stretch'd at one end more or less, are equally extended all along quite home to the other; and this by reason of the continuity of their Parts.' Charleton, *Enquiries*, pp. 511-12.

[200] Charleton, *Enquiries*, p. 511. He left out the crucial reason why, according to Glisson, the strings of the body did not act mechanically - their inherent sentience. Therefore in his refutation Charleton [deliberately?] neglected a fundamental aspect of Glisson's hypothesis.

[201] 'that a Lute-String stretcht by a Plummet hung perpendicularly at one end, and held by a Mans finger at the other, is extended in all parts alike, is beyond all controversie. And this is sufficient to my purpose. For from hence it is most evident, that howsoever, and in what part soever a Nerve is contracted, the contraction must be equal in all Parts of it: not only a Hairs bredth at one end, and two or three Inches perhaps at the other.' Charleton, *Enquiries*, p. 512. For a discussion of musical instruments and metaphors of bodily operation, see J. Kassler, *Inner Music: Hobbes, Hooke and North on Internal Character*, London, Athlone, 1995. The analogy of a lute string was used by Borelli in relation to the muscles, in *De motu animalium*.

[202] As he argued, 'the Fibres spontaneously retracting themselves, return to their native Position, after the cessation of that force that extended them; as all other Tensil Bodies do by the motion of Restitution'. Charleton, *Enquiries*, p. 527.

this area. He worked within an intellectual world circumscribed by texts, rather than by experiments. He examined the potential difficulties with these hypotheses through rational and analogic rather than experimental criteria.

This concluding lecture juxtaposed the two metaphors for the body. The first focused on spaces within the nerves, which allowed the transmission of spirits, and on the composition of those spirits. The second invoked a body moved by nerves as vibrating cords. Both hypotheses had their origins in external metaphors or analogies, projected into the body to make sense of unseen physiological processes. Charleton's contrasting of alternative metaphors suggests he used them to explore and evaluate different possible hypothetical explanations. He assessed their plausibility through practical or logical manifestations outside the body. Where experimental demonstrations were used, they were proofs or disproofs of the logic of the analogy, rather than demonstrations of the anatomies of parts in question. The Bartholin example illustrates how an anatomical demonstration could be dismissed merely because it was not methodologically sound.[203]

Charleton avoided decisive approval of either Glisson's theory or the doctrine of the hollow nerve. In closing he observed that there were 'knots . . . in this most intricate Hypothesis, which my weak Reason is not able to untie.' His own researches had merely demonstrated 'how uncertain that knowledge is, which even the greatest Wits have attained' concerning the animal spirits and voluntary motion.[204] As with all the major issues discussed in the *Enquiries*, Charleton confirmed the provisional and incomplete state of knowledge about muscular motion, leaving the conclusion 'to the consideration of wiser Heads than mine'.[205] His presentation, he declared, had been 'designed only as an Essay, rather to excite the Industry, than to prescribe to the curiosity of the more perspicacious and more learned, to whose judgement mine shall be always ready to Conform.'[206]

In conclusion, *Enquiries* does not appear to have been a public declaration of support for the mechanical philosophy, as expressed by Willis or anyone else. Although his text contained some mechanistic metaphors, and explanations, he

[203] Earlier in *Enquiries* Charleton discussed an 'Anatomic Experiment' performed by Richard Lower. Although it was 'true and ingeniose', nevertheless Lower's inference was 'more than can be, according to the Laws of right ratiocination, from thence deduced.' Lower's failure to record certain details of the experimental findings allowed for the possibility of 'other causes', to which the effects 'may be with more verisimilitude ascrib'd.' This illustrates that the status of anatomical demonstration was not sufficient to guarantee the veracity of the inference. Experimental results were subject to speculative inference, and these could most certainly be flawed. Charleton's rebuttal of Lower illustrates the power of reason in response to anatomical experimentation. *Enquiries*, pp. 145-7. The original experiment is recorded in Richard Lower, *Tractatus de corde*, cap. 5.

[204] Charleton, *Enquiries*, p. 514.

[205] 'sitting down content with my Ignorance lest I should farther expose it, I hast to a Conclusion.' Charleton, *Enquiries*, p. 543. The avoidance of final determinations also characterised Glisson's lecture on the brain.

[206] Charleton, *Enquiries*, p. 544.

made no consistent attempt to demonstrate the superiority of a mechanical explanation of the animal oeconomy. Although iatromechanical theories were distinctively different from traditional pathology, medicine's requirements were practical, and in this realm the contributions of iatromechanism were minimal.[207] As far as patients were concerned, there was nothing new or preferable about therapy administered by an iatromechanist. Therefore this new philosophy did not affect patient perceptions of elite physicians and academic physic.[208] In addition, an innovative mechanist philosophy might potentially contradict the traditions on which Charleton's status as a physician, and the College's authority as an institution were founded.

Enquiries does not indicate the kind of consistent intellectual agenda identified by Theodore Brown. Indeed the lectures did not promote any particular explanations, but covered a range of hypotheses, and left the final determination of preference to the audience. The lectures illustrated that traditionally defined processes of the animal oeconomy were explicable within a contemporary anatomical framework.

The first important step toward coming to understand the various epistemological strategies employed by Charleton in this work is to understand that his definition of anatomy encompassed not only empirical observation of the physical structures of the body, but also theoretical understanding of causes and uses. His very definition of anatomy encompassed knowledges that we define as philosophical, rather than anatomical. This broader definition of anatomy meant that Charleton's text included speculative or theoretical methods of demonstration. Analogic and comparative devices were as vital to demonstrations as were experimental manipulations. In many cases analogic demonstrations seemed to possess a higher status in explanation. This understanding of Charleton's definition of anatomy helps to account for one of the most striking characteristics of the *Enquiries*.

While Charleton was clearly present at many experiments, and had an active experimental life himself at the Royal Society, his writings do not refer to these as the chief source of authority. Indeed, the *Enquiries* indicates a deliberate self-portrayal as a classical authority, and an ongoing attempt to reconcile ancient knowledge with recent developments. His self-presentation was affected less by his involvement within the Royal Society, than by his professional status as a physician.

Enquiries balanced the textual, traditional, doctrinal nature of medical practice with some of the empirical, non-dogmatic values associated with the virtuosi. But the physician did not appear to view these epistemologies as incompatible. His own experimental experience did not apparently change the source of authority of his arguments. His lectures did not emphasise the role of empirical discoveries, made or witnessed by himself. Although the text did

207 Henry, 'Doctors and Healers', pp. 210-11.
208 Henry, 'Doctors and Healers', p. 212.

include more accounts of experimental findings than had the *Natural History*, this is partly because the texts from which Charleton drew many of his theories were more substantially founded on empiricism. The *Enquiries* include references to his own experimental observations, but they do not possess primary epistemological status.

The eclecticism of the *Enquiries* might also be partly the consequence of the teaching function of the lectures, which required an overview of contemporary literature on various topics. Differences between his texts of 1659 and 1680 may illustrate a shift of genre, rather than an intellectual transition. This genre in turn affected Charleton's treatment of new anatomical discoveries. He integrated them into his text, but made no attempt to confront the discrepancies between the explanations they were used to promote.

The changes in Charleton's self-presentation, between his 1659 text and that of 1680, seem to be overshadowed by the persistence of classical and textual authority, analogic reasoning, and his devotion to rational, rather than experimental, knowledge. The important continuities in his self-presentation and epistemology through these substantial changes in his personal circumstances and experience suggest that there were important reasons for the retention of an identity which did *not* emphasise experiment.

CHAPTER VII

THREE ANATOMIC LECTURES (1683): WAYS OF KNOWING AND THE ANATOMICAL BODY

Having considered *Enquiries*, as an example of Charleton's writing after his experimental experience at the Royal Society, I now turn to the final of the three medical texts under scrutiny. *Three Anatomic Lectures* (hereafter *Lectures*) was the physician's penultimate publication.[1] *Lectures* responded to the Neapolitan anatomist Giovanni Alfonso Borelli (1608-1679), whose *De motu animalium* had recently been published.[2] As a modified translation of Borelli, *Lectures* allows us a special insight into how the English physician selected and altered material for his audience. By examining where and how he deviated from the Neapolitan's presentation, we gain insight into Charleton's own epistemological emphases in the investigation of the heart. For example, *Lectures* deviated from Borelli's text in the addition of extended analogies, and the omission of numerous physical and mechanical demonstrations. Such alterations can suggest what Charleton saw as persuasive presentation of the material to his English audience, and also his preference for analogic forms of demonstration over others. In addition, of course, Borelli had mathematical training that Charleton lacked.

This chapter discusses further the questions of identity and epistemology developed in the preceding chapters. It explores the differences between *Lectures* and *Enquiries*, as comparable texts with different scope and foci, and considers what *Lectures* can reveal about Charleton's self-construction within the College context. Charleton, we have seen, had performed anatomies of the heart within his Royal Society investigations. When he presented the work of the Italian anatomist he could have done so with reference to his own experimental activities, if empiricism had been central to his public presentation. I consider the bases of authority upon which Charleton relied, and the extent to which he invoked empirical evidence and the experimental context. The present chapter looks at how he weighed the empirical evidence offered in Borelli's text, and whether he afforded epistemic primacy to these demonstrations. If not, what other forms of proof did he invoke?

Enquiries revealed the physician's self-construction as a consummate scholar, able to penetrate difficult questions by meditation and reference to authoritative texts. While the *Enquiries* showed a strong tendency toward reconciliation of ancient and contemporary medical theories, *Lectures* suggests a different

[1] His final publication was the 1685 *Inquisitiones Medico-Physicæ*.

[2] Throughout this chapter I refer to Paul Maquet's English translation: Giovanni Alfonso Borelli, *On the Movement of Animals*, Berlin, Springer-Verlag, 1989.

emphasis. In the lectures of 1678 (*Enquiries*) he depicted his findings as tentative, rather than conclusive, and deferred any final conclusions to his audience.

The *Lectures* were delivered in the anatomical theatre at the College of Physicians, on 19, 20 and 21 of March 1683. They covered, respectively, the circulation of the blood, and the effects of circulation, the heart's pulsation, and lastly, the efficient causes of the heart's pulsation. College President Thomas Coxe ordered the lectures to be printed, and they were published by Walter Kettilby, appearing for purchase in the same year.[3] It was unusual to have lectures published in this period, and the fact that Charleton had not one but two volumes published, with the College's encouragement, is remarkable. As the location, genre and likely audience all share considerable overlap with those of *Enquiries*, this comparison is particularly instructive. Given the similarities in audience and genre for the *Lectures*, it is important to consider what characteristics were shared by the two works. The lack of comparable publications makes it difficult to gauge which characteristics of Charleton's work arose from the subject, and which from the genre. Here a comparison between the two offers important insights. While the *Enquiries* responded to a number of theories on a range of subjects within the animal oeconomy, the *Lectures*' task was more specific: to explain the final and efficient causes of the heart's motion by a hypothesis that did not invoke fermentation. All three lectures were directed toward this end.

Borelli's text had been published posthumously, and only in Latin. Charleton was thus the first to bring Borelli's theories into the English language. The extent of his reliance upon Borelli was never fully acknowledged in the work, and has rarely been recognised by scholars (with the exception of Nayler).[4] The only acknowledgment of Borelli in the text was in the 'Præloquium', in which Charleton stated that he would 'make use of some mechanical demonstrations particularly to my purposes out of Borelli, the most skilful of all mathematicians.'[5] Although he expressed admiration for Borelli's intelligence, learning and perspicacity, Charleton's *Lectures* aimed to revise the former's theory regarding the motion of the heart. Yet the English physician's work appears at times to be little more than a translation of Borelli's text, with occasional differences.

HISTORIOGRAPHY

Scholars have paid little attention to Charleton's *Lectures*. Brown argues that this work 'was trying to work out an attitude to physiology and pathology

[3] Charleton's *Three Anatomic Lectures* was prefaced merely by a Latin 'Præloquium'.

[4] Nayler, 'Insoluble Problem'.

[5] Charleton claimed that he would 'borrow from his writings omitting all the others who have written their opinions after our Harvey, about the circulation of the blood, because I do not want to waste your time or lest I seem to distrust the accuracy of your memories.' Charleton, 'Præloquium', *Lectures*.

that combined criticism of certain of Willis' ideas with a basic loyalty to Willis' general approach.'[6] The lectures were, he claims, Charleton's first public declaration of this view. In my opinion, Brown misinterprets both Charleton's words, and the motivation behind them. He states that Charleton's overt attacks upon Willis signalled the depth of 'his commitment to iatromechanism'. Extensive criticism of detail, claims Brown, indicates acceptance of the basic theory: 'and for Charleton iatromechanism, clearly, was firmly established.' Because of his conviction regarding the basic mechanistic principles, 'he therefore felt free to innovate within the broad scope of iatromechanical possibilities.'[7]

Lectures, according to Brown, replaced Willis with Borelli, as 'the principal spokesman for iatromechanism'.[8] This assumes that Charleton had a mechanist agenda and that between 1679 and 1683 he swapped one mechanist authority for another. Yet a consistently mechanistic approach is not evident in the *Enquiries*, and while *Lectures* promoted a mechanist explanation of the heart, this was not generalised to the whole body. *Enquiries* illustrated Charleton's tendency to apply particular hypotheses to specific instances, without attempting to synthesize a total system of explanation. This eclectic method is equally relevant to understanding *Lectures*, with their specific focus on the heart's motion. Charleton's adherence to Borelli's explanation of the heart did not mean that he explained other motions of the animal oeconomy by the same principles. There is no evidence that he intended to build a system upon this basis.[9] *Lectures* indicate that Charleton's aim was to 'interpret' Borelli's theories for his audience. This indicates a need for caution in considering how it reflected Charleton's personal beliefs. We must examine the complex interactions between his and Borelli's works, rather than taking the Englishman's words at face value. In suggesting these qualifications concerning the extent to which Charleton's presentation of Borelli entailed his own conversion to mechanism, I do not wish to suggest that Borelli's ideas should not be seen as 'mechanist'. I simply question the reductive basis of Brown's assertions about what this meant for Charleton and its ramifications concerning how we should read this work.

Charleton's aim, according to Brown, was to 'keep up with the latest developments in the mechanical philosophy', in order to gain himself 'that precious commodity—a progressive reputation.'[10] The rejection of Willis is alleged to have been part of his strategy to gain such esteem. However, Brown presents no evidence that Charleton desired a progressive reputation, nor that he might have

[6] Brown, *Animal oeconomy*, p. 178.

[7] Brown, *Animal oeconomy*, pp. 180-81.

[8] Brown, *Animal oeconomy*, p. 179.

[9] Nayler acknowledges that while Charleton's conclusion in the *Lectures* ('the heart is an automaton, moved by mechanical necessity'), could be used to portray him as 'a true Cartesian mechanist', the physician in fact viewed cardiac motion as a special case. While he tackled this specific subject in an apparently Cartesian manner, he did not advocate broader understanding of the animal oeconomy within this same framework. Nayler, 'Insoluble Problem', p. 141.

[10] Brown, *Animal oeconomy*, pp. 180-81.

achieved it by rejecting Willis.[11] He does not establish that a 'progressive reputation' was the highest form of prestige attainable. While Brown sees Borellian mechanism as the 'advancing forefront of science', this does not mean that Charleton agreed, or even that if he did agree, he wanted to be seen at the forefront.[12] Brown's presentist assumption is that 'progress' through mechanical philosophy was automatically equivalent to high status. Throughout Charleton's works we have found that his self-construction emphasised attributes of modesty, scholarly education and eclecticism. These were more highly valued than an identity associated with theoretical innovation. Brown's interest in institutional gambits for authority leads him to neglect the complexities of individual identity.

BORELLI'S TEXT

Borelli, a key member of the Accademia del Cimento, had studied muscle physiology, respiration, the nervous system and cardiac motion. *De motu animalium* was published (posthumously) in two parts. The first, published in 1680, concerned the movement of limbs and locomotion of man and animals. The second volume, which was directly influential upon Charleton's *Lectures*, examined the internal motions of animals and their causes. It was printed in 1681. Borelli's claims were explicitly mechanistic. He explained the problems of human and animal locomotion through reference to 'demonstrations based on Mechanics'. He analysed the structure of the muscles to reveal 'which forces and which mechanical organs' were responsible for movement of the limbs, and for motions that were involuntary or unconscious (such as the heart's pulsation). He wished 'to illustrate and enrich the part devoted to Physics by mathematical demonstrations', and to 'enlist Anatomy into Physics and Mathematics not less than Astronomy.'[13] This reflected his personal priorities, and his career as a lecturer in mathematics at Pisa.

Borelli followed the principle that nature acted by the most direct, and most economical methods.[14] Mechanical necessity dictated the differences between entities: 'the operations of Nature are easy, simple, and follow the mechanical laws which are the laws of necessity.'[15] Followers of this approach generally accused their opponents of presenting cases 'contrary to nature'.[16] The English

[11] Indeed Brown fails to perceive Charleton's frequent attacks on Willis in his earlier works.

[12] Brown, *Animal oeconomy*, pp. 180-81.

[13] Borelli, *Movement*, p. 6.

[14] 'Nature regularly takes the simplest ways, and dislikes roundabouts and changes.' Borelli, *Movement*, Bk II, prop. 75, p. 279.

[15] Borelli, *Movement*, p. 240. This standard aspect of Aristotelian teleology was echoed by Charleton.

[16] His understanding of the operation of the valves of the heart was verified on the basis that, by the alternative suggestion, 'Nature would provoke half the effect by a stupid double effort.' Borelli, *Movement*, Bk II, prop. 55, p. 261. This was echoed almost exactly by Charleton, 'Nature would foolishly by a double endeavour attain but half her end'. Charleton, *Lectures*, p. 65.

physician even turned this claim against Borelli, censuring the latter's hypothesis on cardiac motion as 'incongruous to the Wisdom of Nature'.[17]

Charleton, rather comically, compared his motivation to reform the Borellian Hypothesis with that of Tycho Brahe, who set out to 'excogitate' a system 'of more probability and neatness' than the Ptolemaic.[18] As Brahe had attempted to repair some of the 'unhansom' characteristics of the Ptolemaic, likewise he set out to invent a hypothesis free from Borelli's failings. Despite the grandeur of the English physician's claim, his deviation from Borelli was incredibly minor. The rather ludicrous comparison with Brahe may have been invoked by Charleton to present himself as avoiding the traps of dogmatism and blind adherence to doctrines.[19] It was certainly far from the modest presentation that characterised *Enquiries*. However, although he claimed to generate an improved hypothesis, he was worried that 'anyone here should believe, either that I have usurped to my self this whole System of the Motion of the Heart, from that most excellent Mathematician Alphonsus Borellus . . . or that I lay claim to more than a Candid attempt to reform it.'[20] Thus he defended himself against claims of plagiarism and censure for overconfidence. His limited claim for a 'candid attempt' reinforced that his stance should be seen as impartial.

The main dispute between the two men was over the direct cause of the motion of the heart. Borelli argued that it was 'the swelling of the vesicles of its pores resulting from fermentative ebullition of elements of tartar of the blood by some spirituous juice instilled from the orifices of the nerves.'[21] Charleton rejected the presence of any 'fermentive ebullition', and instead explained the heart's motion in purely mechanistic terms. However, this issue was not raised until the final pages of his third lecture. Prior to this, Charleton's reliance upon Borelli's propositions, experimental reports, analogies and illustrations was consistent.

Charleton wished to create an explanation 'agreeable to the Organical Structure of the Heart, to which above all things it was requisite I should endeavour to adjust it.'[22] This highlights the distinctly materialist methodology of *Lectures*. Throughout the text his reference point was the mechanical 'conformation' (structure) of the heart. He argued that the function of the organ derived directly from

[17] Charleton, *Lectures*, p. 105.

[18] Charleton, *Lectures*, p. 104.

[19] The circumstances and genre associated with lecturing may have been the reason for this inflated claim. While the speaker made substantial claims to innovation, to justify his authority as a presenter, he was largely in agreement with the source he claimed to rebut. *Enquiries* illustrated Charleton's fear of having to create new theories to replace rejected hypotheses. Although he pretended in the *Lectures* to create a new hypothesis, he did not deviate substantially, and his epistemology seems identical.

[20] Charleton, *Lectures*, p. 105.

[21] This was similar to that by which all muscles were moved. Borelli, *Movement*, Bk II, Prop. 77, p. 282. Quoted in Borelli's Latin by Charleton, on p. 98 of *Lectures*, as the source of his disagreement.

[22] Charleton, *Lectures*, p. 105.

its structure, itself a manifestation of divine intelligence. *Lectures* is striking for this consistent emphasis upon structure. The surest way to gain knowledge, it claimed, was through the observation of structure and the deduction of cause from effect. However, I argue that while this emphasis was consistent, it did not make Charleton a 'mechanical philosopher' *per se*. He simply explored a mechanistic physiological explanation in this work. While the *Lectures* was strongly mechanistic in emphasis, this work should not be treated as definitive evidence of Charleton's late life conversion to a mechanist understanding of the body.

The physician depicted himself to this audience as 'an Interpreter' of Borelli's theory for the purpose of explication,[23] rather than as an innovator or experimentalist. His claim was for a theoretical reform to be more consistent with physical evidence. Yet he did not locate experimental demonstration, or laboratory authority, as the centrepiece of his claims about physical structures. This resonates with the humanist notion of compilation and interpretation, rather than with some newly developed experimentalist identity.[24] More in keeping with the construction of a modest identity, Charleton emphasised the difficulty of his task, and the uncertainty associated with his subject. His progress through the topic of the heart's movement, he claimed, 'is yet dark, and rocky, with Precipices on both sides: and all the light I can expect, must be from a few Sparks stricken out of my Flinty subject by the force of conjecture.'[25] Borelli had likewise attributed only modest success to his own endeavour.[26]

Charleton's modesty was also displayed in his continual deference to his peers, such as Glisson and Ent.[27] He down-played his own status, perhaps not wishing to be seen to be pushing his own agenda. By so under-emphasising his own role he retained the necessary deference to his peers which, paradoxically, served to assert his own authority.

[23] Charleton, *Lectures*, p. 49.

[24] Charleton, *Lectures*, pp. 75, 76. He provided a 'summary of what Borellus hath from a long chain of most ingeniose propositions and Theorems inferred.'

[25] Charleton, *Lectures*, p. 80. He went on to complain that 'to explore wherein this nice difference may most probably consist, is a work of so great difficulty, that I wish it were possible for me to revoke the temerarious promise I made to attempt it'.

[26] 'If my endeavour does not succeed completely, at least cleverer and more learned men, under my stimulus, will be able to improve and complete this science by firmer reasoning, and better methods.' Borelli, *Movement*, Bk I, p. 6. This permitted great civility between the originator and later revisers of any theory. Borelli was described in the preface to his book, as 'such an example of modesty, sobriety and equanimity that he looked like Socrates, Plato, or some other of the ancient Founders of the schools'. p. 4.

Borelli claimed that 'those who admit their ignorance of natural matters' were 'more commendable than that of those who take the liberty of being daring in Philosophy.' Yet both stances were wrong: 'Neither must we accept hypotheses which do not conform to the behaviour of Nature and to common sense, nor must we be deterred by any objection before having examined its importance carefully.' Borelli, *Movement*, Bk II, prop. 28, p. 237.

[27] The claim that 'the Brain, is the fountain of life' had been shown by 'experiments Anatomical' and by reason, 'by Sir George Ent, Dr. Glisson, and (if it be lawful for me to put my self in the same period with such excellent men) my self.' Charleton, *Lectures*, p. 82.

Charleton hoped that his theories would be 'consentaneous to right reason, agreeable to the Animal Oeconomy, congruous to the Organical structure of the Heart (to all which I have been careful to adjust them) and in fine consistent among themselves'.[28] Borelli's modest aim, as declared in his dedication to Queen Christina, was to illuminate some of the 'geometrical reasonings which are expressed by divine wisdom in the structure and operations of the animals.'[29] He did not, like Charleton, declare the parameters of proper natural knowledge. The Englishman was more explicit about knowledge criteria.

As I argued in the previous chapter, *Enquiries* did not follow Willis, or indeed any authority, consistently. Consequently his adherence in these lectures to Borelli (rather than Willis) is less remarkable than Brown believes. The latter assumes that Charleton's ridiculing of Willis in *Lectures* illustrates his departure from Willis' authority.[30] However, these refutations illustrate a more complex negotiation of knowledge than Brown recognises. Rather than merely playing off Borelli against Willis, his derision of Willisian fermentation attacked its *use* as a factotum. As we have seen this was a trenchant criticism in the seventeenth-century context. His rejection of its use in this manner reinforces my sense that Charleton chose to present himself as eclectic in manner and preference.

He complained that, 'so plausible and favourable hath the Hypothesis of various Ferments . . . seemed to many of the Virtuosi of this our inquisitive age; that they have not doubted to ascribe to them a powerful energy and necessary influence'.[31] Charleton accused the virtuosi of credulity. The theory of fermentation was attacked not merely for its explanatory inadequacy, but because of its adoption as a factotum.[32] The attacks on Willis in the *Enquiries* are not noted by Brown, who also seems unaware of the extent of Charleton's plagiarism of Borelli. Nayler, by contrast, recognises the English physician's reliance upon Borelli, and the limitations of his own mechanism.[33] While Charleton generally ignored the contradictions between theories he presented, he chose occasionally to reflect on his changes of direction. Where he did deviate from Borelli, Charleton cited the reasons which had led him 'to withdraw my judgement from the conduct of Borellus, whom before I had so closely followed'.[34] Perhaps such

[28] If they were assessed as such, then he would 'with assurance conclude, that the Heart is, as all Automata are, moved by Mechanic necessity'. This very principle was 'what I proposed to demonstrate . . . as necessary to be supposed, in order to the Solution of this great Probleme [sic] of the Pulsation of the Heart, and that of the motion of the rest of the Muscles.' Charleton, *Lectures*, p. 97.

[29] Borelli, *Movement*, p. 4.

[30] Brown notes that Charleton specifically ridiculed the theory of fermentative agitations in the blood. Brown, *Animal Oeconomy*, p. 179. Brown fails to note Charleton's frequent attacks on Willis in his earlier works.

[31] Charleton, *Lectures*, p. 1.

[32] Finding the theory of ferments 'lying, like a block, in my way: I thought it concern'd me rather to remove than to leap over it'. Charleton, *Lectures*, p. 4.

[33] See Nayler, 'Insoluble Problem', pp. 138-42.

[34] Charleton, *Lectures*, p. 100.

an explicit statement was intended to illustrate that he was no dogmatic adherent of a single hypothesis. His interest in the discovery of truth, he indicated, was more important than his wish to appear consistent.[35]

After providing three possible explanations why pulsation occurred in the heart but not in other muscles of the body, Charleton suggested that if no single theory was sufficient, 'yet if ye please to conjoyn and twist them all together into a triple chord, ye may then perhaps find them strong enough to pluck up the proposed Difficulty by the roots.'[36] Thus he affirmed his identity as a properly 'disinterested' and eclectic enquirer, with no particular investment in one theory to the exclusion of the others. Already we can see emerging the author's self-presentation in *Lectures* as one pursuing modest aims in relation to the topic at hand. This was the most focused in subject matter of his English medical works, and it exhibited less of the eclectic tendency than the other works I have examined. But we can see the elements of a continuous identity: modest, textually-reliant and eclectic, with a strong emphasis on the reconciliation of philosophies.

THE AUDIENCE FOR *LECTURES*

His modesty and eclectic manner might have been influenced by the genre of public lectures. The audience for these lectures would have been similar to that of his 1678 lectures at the College. As I argued in the previous chapter, the College context, and the nature of the audience, would certainly have influenced the manner of Charleton's presentation. The author's depictions of the audience at the 1683 lectures suggest a mixed crowd. Charleton referred to the presence of young men, students of anatomy, and also to 'this venerable Assembly of most Learned men, where I have as many Judges as Auditors'.[37] The College audience was depicted as capable of authorising and determining the status of his claims. The physician humbly claimed that he was only able to proceed with his philosophising because of 'the well known Candor and benignity of my most Learned Auditors.'[38] Having outlined his own hypothesis, in refutation of Borelli's, he declared that he would not 'commend [it] to others, unless your approbation shall encourage me'.[39] This standard strategy constructed Charleton as disinterested and willing to be corrected.[40] It identified the speaker as rejecting dogmatism, excessive confidence or self-aggrandising rhetoric. If he failed to present a convincing case, 'humanity will oblige you, rather to put forth your hands to support, or guide me, than to deride my blindness.'[41]

[35] The importance of this conceit is noted by Shapin, *Social History of Truth*, p. 222.
[36] Charleton, *Lectures*, p. 91.
[37] Charleton, *Lectures*, p. 3.
[38] Charleton, *Lectures*, p. 80.
[39] Charleton, *Lectures*, p. 100.
[40] For a discussion of this approach, see Shapin, *Social History of Truth*, p. 276.
[41] Charleton, *Lectures*, p. 80.

Throughout *Lectures* Charleton maintained this strategy. He retreated from any apparent confidence in the persuasive power of his own theory: 'But this Chair doth not make me a Judge. To hear and determine, most Excellent President, and my most Learned Collegues [sic], is your right; which I ought not to usurp.'[42] He thus humbly, and without reserve, submitted 'all parts of my Disquisition (for I pretend not to know, but only to inquire after truth) to your examen and judgement.'[43] Through such strategies he reinforced the idea that he did not intend to provide certain and definitive answers to these questions.

Charleton's audience consisted also of some junior physicians. He included visual demonstration in the form of figures, which would 'help both the understanding and memory of younger Students, for whose sake chiefly Lectures were at first instituted in this College.'[44] Charleton declared that he would sooner 'lead my Auditors of the Younger sort by [Borelli's] brighter Torch, than by the Glow-worm light of my own understanding'.[45] He presented his own role as that of an interpreter of recent theoretical developments, offering a synthesis or summary of contemporary thought and writings.

The comparison between the presentation of *Lectures* and *Enquiries* helps us to delineate some features which might have been generic to the lecture format. While the lack of comparable publications makes broader comparisons difficult, we can identify some continuities and differences between *Enquiries* and *Lectures*. Although these lectures were more focused and less philosophical in the range of their discussion, they were still far more meditative and abstract than contemporary surgical lectures. It seems that in this instance Charleton was presenting and commenting on a foreign theory for the benefit of his educated English audience. Within his discourse he added comments on his country's contributions to knowledge on the subject. These lectures did not include significant input from empirical demonstrations, nor did they attempt to suggest that their author was a master of empirical manipulation. Instead Charleton drew extensive reference to textual sources, and it seems that this is where he expected his audience's interest to lie.

CHARLETON'S PRESENTATION OF BORELLI

Although he copied the majority of his claims from Borelli's propositions,[46] Charleton reshuffled them to fit his own thematic discussion. Charleton

[42] Charleton, *Lectures*, p. 96.

[43] Charleton, *Lectures*, p. 96. When he detailed the competing claims to the discovery of blood's circulation, he quoted extensively from Fabricius ab Aquapendente and Father Paul of Venice, and then, characteristically, retreated: having 'faithfully recited the Pleas of these two great men, I leave it to you to decide the controversy'. Charleton, *Lectures*, pp. 18-20.

[44] Charleton, *Lectures*, p. 42.

[45] Charleton, *Lectures*, p. 49.

[46] Charleton lifted all of his stated propositions from *Movement*: The first lecture included propositions: 31, 33, 68, 69, 75, 34, 35, 36 and 37. Charleton's second lecture incorporated Borelli's propositions: 38, 48, 49, 50, 51, 52, 53, 54 and 55. The third and final lecture contained propositions: 66, 77, 78 and 80.

declared that he would 'try, with great effort, to take the kernels separately for the sake of clearer teaching'.[47] Thus the Englishman's rearrangement was partly the consequence of dividing the material up for three lectures of set duration. *De motu animalium* presented Borelli's points as a series of 'propositions'. The Italian's propositions would have made a very stilted verbal presentation, and the need for greater fluidity may have motivated Charleton's changes. Although Charleton's structure was less rigid and formulaic, his presentation of material echoed Borelli's in many ways.[48]

While the Italian presented a chain of reasoning, in which each proposition led to the next, Charleton omitted many demonstrations, selecting his points from Borelli's. The Englishman's emphasis was more philosophical. He abandoned much of the mathematical method which characterised Borelli's text. He wished to illustrate the probability of mechanical operation, but did not incorporate the mechanical and mathematical proofs which followed Borelli's precise style of demonstration.[49] Evidently there was a difference between the expectations of the two men, and also a disparity in their mathematical knowledge—Charleton possibly simply did not have sufficient background in mathematics to understand some of Borelli's proofs.

Authorities cited in the *Lectures* were often merely echoes of those in Borelli's text.[50] However, in some instances, Charleton made additional references to fellows of the College of Physicians (particularly Lower).[51] He also deviated from Borelli in the provision of an extended digression on the original discovery of the valves.[52] This consisted of an energetic defence of Harvey, but also detailed the competing claims of Fabricius ab Aquapendente and Padre Paolo (Father Paul) of Venice.[53]

Whereas *Enquiries* invoked ancient authority, *Lectures* were more concerned with recent investigations. Charleton referred to Lower's work and Ent's *Antidiatribæ*.[54] He traced Ent's influences to Galen, Erasistratus, Hippon and Censorinus.[55] This suggests that the veracity of claims could be increased by

[47] Charleton, *Enquiries*, 'Præloquium'.

[48] Echoes of scholastic presentation in Charleton's text, for example, generally seem to have been taken directly from Borelli.

[49] Charleton did refer the reader to Borelli's text at one point, where he would find the point made 'more mathematico'.

[50] Charleton's references to Bellini, Steno, Lower and Malpighi were all copied from Borelli, *Movement*, p. 249.

[51] Gerardus Blasius of Amsterdam had accused Lower of copying Steno but, according to Charleton, without providing evidence. Charleton, *Lectures*, pp. 40-41. Although he defended College fellows Harvey and Lower against their detractors, the purpose of his lectures was not 'to wipe off the dirt that either envy or ill nature has thrown upon any member of this most learned Society'.

[52] Borelli simply mentioned the partial discovery by Cesalpino, and the conclusive findings by Harvey. Borelli, *Movement*, Bk II, prop. 30, p. 242.

[53] Charleton, *Lectures*, pp. 18-19.

[54] Charleton, *Lectures*, p. 81.

[55] Charleton, *Lectures*, p. 82.

showing their classical origins. Thus as with *Enquiries*, albeit to a far less significant extent, he integrated ancient authority and new insight. Nevertheless Charleton concluded his discussion with a quote from Cicero (absent from Borelli's text), and also referred to Seneca. In association with a mechanical insight, Charleton invoked Seneca's description of passing through the *Crypta Neapolitana* (a reference to the Neapolitan Borelli) and his delight upon returning to the light:

> The same surprising alacrity, methink, I now feel within my self, after my passage through the no less darkness, in which Nature had, through a long Series of ages, involved her great secret of the Motion of the Heart, made more obscure by the dust of mens various opinions[56]

The conclusion to his search was his arrival 'at the light of knowledge', regarding 'by what Mechanic necessity' cardiac action was performed.[57] A revelatory search concluded in a mechanical exposition.

Charleton's presentation in these lectures did not emphasise the reconciliation of ancient knowledge with modern to the same extent as *Enquiries*. This was probably because he presented an interpretation of a single, new text (rather than an overview of medical issues). In this case, he was putting forward an Italian contemporary's theory on single issue, and showing its links with the English context in which his audience operated. In *Enquiries*, reconciliation with ancient authorities might have represented a means of justifying and contextualising his choice of topics. Given this circumscribed subject matter it was not necessary to establish the classical origins of his subject in the same way.

CONTINUITIES AND DEVIATIONS FROM BORELLI'S *DE MOTU ANIMALIUM*

Charleton's omissions

To ascertain exactly how Charleton responded to Borelli's influence it is necessary to look at what he omitted as well as what he included from *De motu animalium*. He neglected many of the 'lemmas', or practical demonstrations provided by Borelli.[58] Like Borelli, Charleton rejected the agency of animal spirits. However, he did not call upon the Italian's demonstrations, which examined in detail the supposed nature of these agents, and revealed the practical impossibility of their supposed actions. Charleton's refutation, by contrast, focused on Willis' and Lower's argument that the heads of animals required additional

[56] Charleton, *Lectures*, p. 62.
[57] Charleton, *Lectures*, p. 62.
[58] Charleton omitted the detailed mechanical demonstrations contained in Borelli's propositions 56-65.

nerves in order to convey animal spirits to their brains.[59] Willis and Lower, according to Charleton, had not 'sufficiently proved, that there are such things as Animal Spirits in *rerum natura*'. Therefore 'till I see their Existence otherwise than precariously asserted, I am justly excusable if I doubt thereof.[60] He did not provide evidence to disqualify their claims, but railed against the inadequacy of their demonstrations to satisfy his criteria.

While the Italian provided experimental disproofs that Charleton could have copied, he chose instead to ridicule the excessive *use* of the spirits in Willis and Lower's explanations: 'In some books indeed whole Commonwealths of them are found, so that ye can hardly pass along without meeting crouds of them.'[61] Again, it seems his attack was founded on their use as factotum, rather than simply their inadequacy as an explanation. At stake then was method, as well as hypothesis. There is a clear continuity with *Enquiries* in terms of Charleton's opposition to Willis, and of his criticism of excessive reliance upon a single doctrine.

Charleton's appropriations from De Motu Animalium

Of the empirical examples in Charleton's work, the vast majority were taken from Borelli. Charleton appropriated the Italian's experiments without acknowledgment. He did not, however, attribute epistemological primacy to experiment. His annexation of Borellian experiment seems to have been equivalent in authority to his use of Borelli's textual references. Charleton did not promote his own empiricism as a means of affirming his opposition to Borelli's theory. *Lectures* did not situate individual sensory experience as the core of authoritative knowledge, did not cite the laboratory context and ultimately, did not promote an empiricist methodology. This does not mean that experiment had no role to play in Charleton's epistemology, but rather that Shapin's formulation does not accurately depict the manner in which this physician, and many of his peers, gained authority within their public presentations.

Charleton was content to refer his audience to authoritative texts. Those whose 'curiosity is not urgent enough' to induce them to unravel the muscles of the heart anatomically were referred to the 'testimonies of those learned Anatomists of this age, who have professedly written singular books of the structure of the Heart': Marcello Malpighi, Richard Lower, Lorenzo Bellini and Nicholaus Steno.[62]

The relatively minor importance of direct experience for his authority in these *Lectures* is demonstrated by the fact that Charleton chose to perform a textual

59 Willis and Lower argued that the horizontal position of the heads of brutes meant a slower supply of animal spirits. See Nayler, 'Insoluble Problem', pp. 137-8. The issue was discussed in Charleton's lecture at the Royal Society, 8 June 1664, recorded as 'Certain differences between the brains of brutes and men'.

60 Charleton, *Lectures*, p. 45.

61 Charleton, *Lectures*, p. 45.

62 Charleton, *Lectures*, p. 40. Interestingly, the investigation of the muscles had not been performed by Borelli, but by Lower and Bellini. *Movement*, prop 37, p. 249.

analysis and purely hypothetical rebuttal of Borelli. He did not pit experiment against theory, but rather accepted the inevitability of theory in making and performing experiments (he was not, as Shapin would have it, dividing experiment from meditation). Charleton ascribed an important role to contemplation and meditation. He attributed Harvey's discovery to 'admirable Sagacity of Spirit, by numerose Experiments and Observations Anatomical', and 'assiduous Meditation', and suggested also that 'the secret Manuduction of Fate . . . had reserved the secret for his knowledge'.[63] This characterisation was not derived from Borelli's text.

In contrast to *Enquiries*, *Lectures* seems not to have referred extensively to traditional medical authorities. There may be a number of reasons. The more specific nature of the subject meant that fewer authorities were relevant. *Lectures* was an interpretation of Borelli, rather than an overview of medical thought on a range of issues, as *Enquiries* was. However there are many continuities with *Enquiries*. These include the construction of a modest and impartial identity, the use of textual references (though here contemporary, rather than classical) and the lack of emphasis upon empirical demonstration. Another continuity is the role that the physician ascribed to meditation and contemplation. Again Charleton's presentation reiterates some of the distinguishing features of a physician's identity. Here again the College context and its identity are relevant to our consideration of Charleton's lectures.

THE EPISTEMOLOGY OF *THREE ANATOMIC LECTURES* AND *DE MOTU ANIMALIUM*

Having examined some overlaps between Borelli's and Charleton's presentations, the following discussion explores the extent to which they shared epistemological principles. One key question for the purposes of this discussion is the priority that Borelli and Charleton ascribed to the discovery of final causes, and the means by which they aimed to discover them. A second crucial issue, both for the comparison between the two men, and for a deeper understanding of Charleton's epistemology, is the degree of certainty expected from natural knowledge.

The use of inductive reasoning

Borelli invoked inductive reasoning in his explanation of method: 'The simplest way of analysing a motive force consists of knowing its most notable effect.'[64] Knowledge of the effect produced by the pulsation of the heart, he noted,

[63] Charleton, *Lectures*, p. 23.
[64] Borelli, *Movement*, p. 271.

allowed one to 'more easily investigate its cause which is the movement of the heart, its motive force and its mechanical action.'[65] Charleton echoed Borelli's method of natural knowledge: 'The natural method of acquiring Science, ye know, is to begin from things more known, and then to advance to things less known; to proceed from effects to their causes.'[66] Thus both works shared an inductive frame.

Charleton's second lecture acknowledged the accusations of atheism levelled at mathematical and mechanistic methods: 'To measure the Divine Wisdom elucent in every Organ of an Animal, by the short line of human Reason, is indeed extreme folly', he recognised, 'yet I doubt not to applaud and follow the counsel of Erasistratus, who (as Galen relates) advised Physicians to solve all the actions naturally done in the body of an Animal, by Mechanic Principles.'[67] Here he linked contemporary mechanist understanding with the classical version of the same venture.

Like Borelli's text, *Lectures* cast mechanical understanding as the highest form of knowledge, in both practical and theological terms. One could not possibly achieve philosophical knowledge of the final and efficient causes of the heart's pulsation and the blood's motion 'without enquiring the Mechanical reason of the continual motion of the Bloud through the Veins.'[68] Understanding of the anatomical structure of the body was thus elevated to the highest kind of natural knowledge.[69] By 'Mechanical' Charleton meant 'necessarily consequent from the structure, conformation, situation, disposition, and motion of the parts, by which they are respectively performed.'[70] In addition to his explicit statements of the importance of 'mechanical knowledge', Charleton invoked mechanical principles (motion, volume, velocity, quantity) as explanatory.[71] Charleton, like Borelli, claimed that natural knowledge could be deduced by mechanical reasoning, drawing upon the divine plan of God. Yet of course man could never comprehend divine intention. To say otherwise would be to stray into the claim that man was capable of usurping the divine role of creation.

Borelli emphasised the piety of a mechanist approach. His preface portrayed the natural world as:

[65] Borelli, *Movement*, Book II, p. 248.

[66] Charleton, *Lectures*, p. 31.

[67] Charleton, *Lectures*, p. 37.

[68] Charleton, *Lectures*, p. 12.

[69] Mechanical understanding was the criterion by which Charleton assessed alternative hypotheses. Fabricius of Aquapendens, like Father Paul, 'understood no more the Mechanic reason of [the valves'] conformation, than if he had never heard of or seen them.' Therefore neither could claim discovery of 'the more noble and more difficult invention of the CIRCULATION of the bloud', because it was 'morally impossible for any man to deduce from their absurd opinions concerning the use of Valves'. Charleton, *Lectures*, pp. 21-3.

[70] Charleton, *Lectures*, p. 69.

[71] For examples, see pp. 27-8. Charleton argued that 'whenever the bloud is quiet or ceases from motion . . . the constitution or contexture of it is dissolved and corrupted'. *Lectures*, p. 31. This argument too was lifted from Borelli, whose proposition 34 was 'Normal composition of the blood is maintained by its circulation.' Borelli, *Movement*, Bk II, p. 246.

> the treasure of His infinite Wisdom which appear in the open book, constituted by the heavens, the earth and all the visible creatures and, in the first place, in the anatomy of animals and man. Through them, as by degrees, the invisible intelligence of God appears.[72]

The English physician cited Plato, 'that God always works by Geometry', and Aristotle, who cast God as 'the Mechanic of the world'.[73] If these authorities were not sufficient, 'we have the greater authority of the Sacred Scripture itself, that God hath framed all things in number, weight, and measure.'[74] The individual who intended to study any part of God's works, 'must therein chiefly consider number, weight, and measure, i.e. the Mechanism of it: otherwise, in the end he will find his mind rather swell'd with opinion, than fill'd with knowledge.'[75]

Charleton's *Lectures*, after Borelli, situated mechanical understanding as the core of possible knowledge, and he was determined to defend its theological validity: 'Why then may not we, who are Christians as well as Natural Philosophers, take those parts of an Animal to be Machines or Engines, which evident reason, and chiefly sense shew to be such?'[76] The defence of those who examined the body's mechanisms as 'Christians as well as natural philosophers', indicates the presence of detractors.

Enquiries was influenced by Glisson's theory of fibrous irritation (consequent upon 'natural perception'). Charleton's discussion of fibrous action in the *Lectures* bypassed the role of natural perception, replacing sentience with structural properties: 'all Fibres even in their natural posture are somewhat upon the stretch'. When cut, they instantly retracted, which would not occur, 'if they had been constituted in a middle state betwixt laxity and extention, as a Bow unbent is quiet, suffering neither contraction nor distraction of its parts'.[77] These analogies, which omitted Glissonian sentience, were copied exactly from Borelli. Charleton's adherence to the theory of natural perception in earlier works was simply replaced by Borelli's argument, which affirmed the non-sentient nature of fibrous composition.[78] Charleton did not necessarily reject Glisson, but

[72] Borelli, *Movement*, Bk I, p. 1.

[73] Charleton, *Lectures*, p. 37. This directly imitated the preface to *Movement*. Borelli declared that 'The language in which the Creator speaks in His works is geometry. This was expressed most plainly by divine Plato. When asked what God did, he answered: "God exerts geometry".' Borelli, *Movement,* Part I, p. 1.

[74] Charleton, *Lectures*, p. 37. This principle, although stated in *Enquiries*, did not carry the epistemological primacy in that work, that it did in the *Lectures*.

[75] Charleton, *Lectures*, p. 37.

[76] Charleton, *Lectures*, p. 37.

[77] Charleton, *Lectures*, p. 30. He stated that the fibres would 'exercise, by natural necessity, that mechanic power they have of contracting themselves'. Charleton, *Lectures*, p. 30. 'Whence it appears . . . that the spontaneous constriction of the distended Arteries is the cause of the expulsion of the bloud out of them into the substance of the parts'. Charleton, *Lectures*, p. 30.

[78] He echoed Borelli's image of the fibres 'as Webs of Linnen exposed to the Sun are kept upon the stretch by many small Cords tied on each side'. Charleton, *Lectures*, p. 67.

copied Borelli, who happened to disagree with Glisson. This deviation from Glisson's view is interesting, as we saw in *Enquiries* how fascinated Charleton was with this theory of his friend and colleague. This ambivalence toward Glisson does not necessarily indicate that Charleton shifted authorities because he disagreed explicitly. After all, his declared aim was to simply interpret Borelli for his audience.

Despite the centrality of 'mechanism' in *Lectures*, Charleton did not at any point suggest that the operations of the entire body were explicable along the same universal principle. He referred only to 'those parts' which operated as machines. He did not depict the whole body mechanistically. This is relevant in considering claims that his understanding was comprehensively mechanistic. His application of mechanist philosophy referred only to the subject in question, the heart. Some comments that appear to illustrate comprehensive mechanism might instead have aimed merely to persuade his audience of the plausibility of the mechanical explanation of the heart's motion. This therefore does not simply support Brown's claim for Charleton's conversion to mechanism, but suggests that he was interested in exploring it (within a broader eclectic framework) as a possible explanation in the question at hand. Charleton's method echoed that of Borelli's in significant ways throughout *Lectures*, and both followed inductive reasoning at times. He seems to have aimed to reaffirm the classical and theological legitimacy of mechanist reasoning, in the face of contemporary criticisms.

The role of empirical knowledge in Charleton's Lectures

This section examines the role of empirically derived knowledge in *Three Anatomic Lectures*, and considers how Charleton invoked Borelli's experimental demonstrations. *Enquiries*, despite its author's Royal Society investigations (which were in some cases directly relevant to his discussion), did not promote the experimental context, nor refer to his own involvement in it. However, Borelli's *De motu animalium* cited a number of empirical demonstrations, which Charleton's presentation in *Lectures* could have incorporated, if empiricism offered a superior form of demonstration. Yet Charleton did not incorporate specific details of experiments, and it seems contextual verisimilitude was not relevant to the authority of the demonstration. When he did report experimental observations he did not attempt to indicate the precise context of their performance, nor did he imply that he himself had performed them (as the Shapin model suggests that ought).

Instead he relied heavily on the textual reports of others' experiments, the authors of which were not always mentioned. This would have been normal practice for many natural philosophers (in contrast with the detailed experimental site descriptions hypothesised by Shapin *et al*). Many of the empirical findings reported in *Lectures* were drawn directly from Borelli's text. However, Charleton tended not to acknowledge his debt to the Italian. For example, with a demonstration of the contractive power of the heart's fibres: by placing a finger into an incision made into the heart of a live animal, he noted, 'in every Systole of the heart, you shall feel your finger pinched all round about, as by a pair of pincers, by the swoln and indurated flesh of the Heart.'[79] Although he may indeed have performed this same demonstration, his account of the procedure was taken directly from Proposition 14 of Book II of *De Motu Animalium*.[80] While he may have been happy to imply that he himself had performed the experiment, he was not excessively concerned that this was so, and gave no circumstantial account to increase its persuasive power.

On the question of whether arteries were always full, or whether they only contained blood during pulsation, Charleton repeated Borelli's observations: if empty of blood during intervals between pulses, then when 'laid naked to the sight' the arteries would appear loose and lank. However 'our eyes assure us, that on the contrary they retain their round and plump figure, and being press'd by the finger resist the pressure'.[81] This observation was drawn verbatim from Borelli's text: 'If they were bloodless when at rest, then, when exposed, they would appear constricted and flat like ropes or ribbons. This is contradicted by autopsy. They retain their wide and round shape and their cavity is full of liquid. The fact that their cavity is full can be palpated by the fingers.'[82] Examples of such mimicry abound.[83] Charleton's presentation of the experiments was influenced significantly by the manner in which they were presented in his source.

[79] Charleton, *Lectures*, p. 48. Borelli's claim was 'If a finger is introduced into an incision in the apex of the heart of a live animal, during each pulsation the finger undergoes a violent constriction and compression, as if it were in a press.' Borelli, *Movement*, Bk II, Prop. 14, pp. 218-9. Again: 'if a finger is introduced into a ventricle of the heart of a live animal . . . constriction of the lateral parts if perceived'. Borelli, *Movement*, Bk II, Prop 50, p. 257.

[80] Borelli, *Movement*, Bk II, Prop. 19, p. 227. This empirical example reappeared in Proposition 19. It was only when Borelli used it a third time, that he noted that the experiment had not been performed by himself, but by Bellini and Lower. Borelli's third mention was in *Movement*, Bk II, Prop. 38, p. 250. He did not refer to a specific text.

[81] Charleton, *Lectures*, p. 25.

[82] Borelli, *Movement*, Bk II, Prop. 68, p. 273.

[83] That the heart was a muscle like others was also demonstrable by boiling its 'Carnose Fibres', upon which 'we may plainly perceive, that it is composed of robust fleshy fibres of the same Prismatical Figure, of the same colour, and consistence and tenacity, as the Fibres of other Muscles have'. Charleton, *Lectures*, p. 39. This was a quote from Borelli, *Movement*, Prop. 37, p. 249. That the heart's pulsation was not controlled by the Will, was showed through vivisection. If the heart of a viper was removed, and placed in warm water it continued to pulse for several hours. By severing the nerves 'all commerce betwixt the Brain and the Heart' was prevented. Therefore 'no sensation, or election can be imagined to ordain and command that motion.' Consequently 'we are obliged to confess, that the first and mediate cause of the hearts Pulsation is in some respect of other, divers from that whereby other Muscles are incited to motion, at the command of the Will.' Charleton, *Lectures*, p. 79. This example was lifted directly from *Movement*, Bk II, Prop. 78, p. 282. Similar demonstrations were drawn from examples of 'Animals whose Arteries are transparent, as in Snakes, Vipers, Eels, Froggs &c.' Charleton, *Lectures*, p. 25. This was also from Borelli, although the latter only mentioned frogs. Borelli, *Movement*, Bk II, Prop. 68, p. 273.

Charleton stated experimental findings as universal consequences of the laws of nature, rather than as specific findings in particular instances. His form was generally: 'whenever x then y', rather than 'Borelli performed x and found y', or even 'I performed x and found y'. Charleton appropriated not only Borelli's theories but his experiments, and did not care to acknowledge his source.

De Motu Animalium demonstrated a similar lack of specificity in relation to demonstrations. Although he presented more specific mechanical proofs than did Charleton, Borelli's anatomical accounts were not particularly detailed, and did not imply his own activity. He declared at one point that 'All these facts were demonstrated by many experiments carried out by Harvey and others'.[84] Although Borelli seemed to claim status as a direct observer when he described himself as 'the first to see' the fibres of the heart dissected 'in Pisa in the presence of the famous Malpighi in 1657', he then remarked that 'Afterwards, more recently, others made the same observation.'[85]

Charleton did not cite Borelli as the source of the experimental report (nor did the latter cite Bellini). It is quite possible that in many of these cases Charleton had performed comparable experiments. However it is significant that his written presentation of them (although not acknowledged) relied on Borelli. Textual accounts of experiments were perhaps a source shared between physicians (as indubitably among natural philosophers). As texts were the traditional source of physicians' professional knowledge, it would perhaps be remarkable for them to be replaced in the manner that some historians describe in their profile of natural philosophers' identity. Textual authority seems, for them, to have been sufficient without the ratification of experimental report.

Experimental observations appear to have been treated as a kind of common reservoir of experiential observation: a collective experimental understanding.[86] These demonstrations lacked precisely that contextual detail that Shapin argues was so central to their authority. Rather than implying specificity to gain authority, Charleton did the opposite, drawing upon experiment as a source of universal observational knowledge. He depicted physical knowledge arising from nature's revelation of her secrets. For example, to demonstrate the brain's influence on the heart he first invoked Borelli, but then declared,

> why have I recourse to the authority of men, when Nature herself, in her constant process of forming the parts of an Embryon, seems to teach us, that some influence (whatever it be) derived from the Brain to the Heart, is absolutely necessary to the incitement and perpetuation of the Motion of the Heart.[87]

[84] Borelli, *Movement*, Bk II, Prop. 54, p. 261. He did not refer to a specific text.

[85] 'Additionally, the famous Lower and Lorenzo Bellini investigated the exact texture of the heart . . . I missed observing that, after boiling, the fibres are easily separated . . .' Borelli, *Movement*, Bk II, Prop. 37, p. 249. Borelli's passive presentation of experimental knowledge was echoed throughout the text (eg. 'This is contradicted by autopsy', Prop. 68, p. 273).

[86] For example, it was 'universally acknowledged by Anatomists, that the Motive force of the heart depends upon some influence from the Brain.' Charleton, *Lectures*, pp. 80-81.

[87] Charleton, *Lectures*, p. 81.

This kind of empirical knowledge belonged to no man, but resulted from universal natural laws. As such, medical insight derived from nature's institutes about her own work of art, the idea of personalising and contextualising experiment was perhaps inappropriate. If the physician gained his knowledge directly from nature, then the notion of individual possession or performance of experimental knowledge was irrelevant.[88] It seems likely that this is true also of natural philosophers, though that question is beyond the scope of this discussion.

Some of Charleton's observations were not drawn from Borelli's text, and perhaps derived from his own experience. Yet he did not provide circumstantial details to establish them irrefutably as his own experience. He treated knowledge of empirical experience much as he treated textual authority. Regarding the continual flow of the blood, he questioned, 'why do I mis-spend time in alledging reasons to prove a truth that is manifest to sense in Phlebotomy?' Physicians in the audience would themselves have observed that as soon as a vein was opened, 'the Bloud flows forth in a swift stream . . . without pauses or interruption, which is a demonstration of the thing proposed'.[89]

That the two sides of the Coronary Artery were linked, Charleton suggested, was evident 'if you take the heart of a Calf or any other very young Animal . . . and with the back of a pen-knife gently impel the bloud from one side of the heart toward the other'. This illustrated the connection between the two sides.[90] This experiment did not appear in Borelli's text. Again this contrasts with the reconstruction of experimental instances as depicted by Shapin. The means by which physicians supported their claims as authoritative were clearly divergent from those depicted by Shapin as characteristic of natural philosophers. The discrepancy between the Shapin model and Charleton's behaviour could suggest larger problems with application of the former's model to natural philosophical activity and authority.

On the question of the immediate cause of heart's motion, Charleton cited a combination of 'common experience', presumably shared by his audience, and evidence given in Richard Lower's *Tractatus de Corde*.[91] This added Lower's authority to that of Borelli. His source of authority was unmistakably textual.

[88] For example, Borelli noted that 'Nature presents examples of such operations'. Borelli, *Movement*, Bk II, Prop. 79, p. 283. Empirical proofs, when given, were frequently in the form of common experience, drawing upon observations from the domestic context. It seems the laboratory was not invoked as a site of authority for these demonstrations.

[89] Charleton, *Lectures*, p. 12. Phlebotomy was a common medical practice: this clinical reference would have been understood by an audience at these lectures. However the blood's continuous circulation was also proved by reference to experimental manipulation 'as is commonly done in Frogs and Vipers'. Charleton, *Lectures*, p. 11.

[90] Charleton, *Lectures*, p. 46.

[91] If the nerves of the eighth conjugation in the neck of any animal were compressed by a ligature, or severed, 'there suddenly will ensue a visible change in the Motion of his Heart; witness the memorable experiment made by Dr. Lower, and recorded in his excellent book *de Corde*'. Charleton, *Lectures*, p. 80. If the nerve was cut, the 'power of motion in that Muscle is presently intercepted, or totally destroy'd: as common experience witnesseth'.

It is as if he was presenting material to an audience who did not have the chance to experiment, and therefore provided a guide to the literature in which key experiments were described.

These characteristics of Charleton's text suggest an attitude towards experimental findings that is not recognised in the secondary literature influenced by Shapin's hypothesis: one in which the performance of the experiment is not important, but instead the findings are treated as general principles. It seems that this variety of presentation would have suited a physician such as Charleton, who did not wish to emphasise his own involvement in manual procedures.

The role of analogy in explanation

Borelli stated that when using inductive reasoning, if the primary cause of an action was 'unknown and imperceptible', the enquirer would be 'led to the action of the heart by analogy with other actions of Nature (as required by correct philosophic method)'.[92] Where it was impossible to 'perceive the cause from the effect', he claimed, 'We are thus forced to guess the cause itself from some other sign from which we find the magnitude of the effect. Such a sign is the similarity and analogy of the heart muscles with other muscles of the same animal.'[93] Thus analogy had a vital role in correct method. This principle was adopted avidly by Charleton, who repeated Borelli's doctrine verbatim.[94]

Analogy was a crucial means by which to demonstrate mechanistic principles. The invisibility of the minute constituents of matter dictated that they themselves could not be observed, but the assumption of structural homogeneity of all physical entities meant they could be understood by analogy. Therefore analogies and models from the macroscopic world could be used to represent microscopic and invisible operations.[95] Analogy was thus the principle underlying many experiments. Empirical demonstration involved manipulation of some apparatus, after which its equivalence to the human body had to be established. As Nadler notes, the best form of persuasion was to postulate the existence and composition of a causal mechanism, demonstrate its plausibility and then illustrate that it was preferable to alternative hypotheses.[96] The assumption of the uniform operation of nature supported the validity of such analogic experiments.

[92] Borelli, *Movement*, Bk II, Prop. 79, p. 283. This principle underlay much of the empirical information presented by both men. He stated this principle twice, see also Book II, Prop. 66, p. 271.

[93] Borelli, *Movement*, Bk II, Prop. 66, p. 271.

[94] The shape of the heart determined that 'we are compelled, from some other Sign, to raise a probable conjecture, whence we may investigate the greatness of the effects.' This sign 'shall be the Similitude and Analogy, which the Muscule [sic] of the heart seems to hold to other Muscles of the same Animal.' Charleton, *Lectures*, p. 74.

[95] Charleton invoked this principle extensively in *Physiologia*.

[96] Nadler, 'Doctrines of Explanation', pp. 520-21.

The most highly developed analogy shared by Borelli and Charleton compared the fibres of the heart with ropes or cables. 'Let us represent to our imagination', declared Charleton, 'a glome or bottom of small twine or thread.' He asked his audience to imagine the reactions of thread to various stimuli, and then to comprehend the similitude between its action and that of the heart. In order to account for the movement of the heart, Charleton, after Borelli, invoked an experiment on what he argued was a comparable apparatus.[97] If 'a crooked gut tied about with a thred' was partially filled with water, and the water impelled toward either end by compression, then 'the gut becomes crooked . . . and the other pendulous extremity will be erected, and strike against your hand held a little over it.' Having established the action from this analogous observation, he linked it to the physiological issue in question—the movement of the heart—which 'plainly follows from the curve figure of the membrane of the gut'.[98]

Some of the more interesting deviations from Borelli are the analogies added by Charleton in *Lectures*. Although incidental analogies from domestic experience were often lifted from the Latin text, Charleton added several extended analogies which were vital in his demonstrations. It seems that although he adopted Borelli's philosophical justification of analogy, he wished to extend its power. Presumably the extended analogies used by Charleton carried some authority within the College context. He may have omitted Borelli's empirical 'lemmas' because he thought they would not carry persuasive power for this audience. Instead he chose to illustrate the probability of Borelli's hypothesis through analogic examples.

For Charleton, analogy enabled the transition from possible to probable statements of knowledge. Visual and structural similarities indicated the law by which more obscure truths could be deduced. These progressed from knowledge of similar causes to deductions about similar effects. Charleton used analogy to make Borelli's hypothesis about the action of the heart persuasive. A persuasive analogy or similitude could elevate the theory to a higher status of knowledge or degree of certainty. The Borellian hypothesis was that drops of liquor instilled into the heart's fibres would, by dilating their pores, cause their abbreviation such that the heart was constricted.[99] Charleton aimed to make the theory 'appear not only possible, but also probable'.[100] While the argument, and the founding metaphor of fluid in fibrous cord, was initiated by Borelli, Charleton went further in his use of analogy to demonstrate its explanatory power within the body:[101]

[97] Charleton, *Lectures*, p. 61.

[98] Charleton, *Lectures*, p. 61. Another example was the demonstration of the composition of the vena cava. Its trunk was 'furnished with fleshy circular fibres, by which it is constringed' was understood to operate 'in the same manner as when a gut or bladder is outwardly constringed by the hand, the liquor therein contained is expressed'. Charleton, *Lectures*, p. 47. Borelli, *Movement*, Prop. 37, p. 249.

[99] Charleton, *Lectures*, p. 92.

[100] Charleton, *Lectures*, p. 92.

[101] Borelli had perhaps been influenced by Fabricius and Harvey, who also drew the comparison between muscle contraction and wet rope or cord. Harvey, *De motu locali animalium*, p. 101. Nayler discusses this link, 'Insoluble Problem', p. 140.

> the Possibility of so great an effect from a cause that seems to be so weak and inconsiderable . . . may be without much difficulty proved from the just Analogy or Similitude of this effect to many other as great, if not greater effects commonly observed to arise from the like Causes.[102]

Analogy held an important role, in the recognition of the possibility of similar instances. Like Boyle, Charleton affirmed the plausibility of his hypothesis by using analogy to provide feasible examples of theorised actions within the body. The extension of the power of the analogy and the certitude of the knowledge, was the Englishman's addition to Borelli's framework.

Charleton wanted to demonstrate Borelli's hypothesis on cardiac motion by comparison with the reaction of string when wetted: its cavities filled by fibrous expansion, while its external bulk remained unchanged.[103] The analogy had been linked with the dilation of cardiac muscles by 'internal humectation'.[104] To establish this point with greater certainty, Charleton invoked a series of 'Mechanic Examples', not used by Borelli. The Englishman perceived that they would illustrate the probability of his theorem. He selected two situations in which wetted ropes shifted enormous weights. The possibility of this was mentioned in Borelli's 29th proposition, which asserted that weak ebullition could cause tremendous effects.[105]

The first 'Analogy or Similitude', was that of 'a new Cable, which upon wetting will very much swell or become thicker, shrink, and shorten itself'. The extent of its retraction was 'beyond belief of any but a Mariner',[106] indicating a source of authority beyond the sphere of natural philosophy. 'This ye will confess to be admirable . . . Yet common experience testifies this to be true'.[107] This variety of experience referred to knowledge gained by common folk (derived perhaps from the Royal Society's study of trades). The specific details were taken from Galileo, who had observed that the abbreviation of ropes was 'of so great efficacy, that the violence of a Tempest, the weight and jerks of a loaden ship of 1000 Tuns burden, and the current of the Sea, cannot by their united forces extend the Cable to its former length'. The second analogy would, Charleton believed, 'raise your admiration to a higher degree', being even more

[102] Charleton, *Lectures*, p. 92.
[103] Charleton, *Lectures*, p. 51.
[104] Charleton, *Lectures*, pp. 49-51.
[105] Borelli, *Movement*, Bk II, p. 241. Ebullition was weak, but the power of the muscles was very great. This had to be explained. Proposition 29 discussed the 'Mechanism by which the weak ebullition which occurs in the muscles is able to exert a huge force'. The 'example of a wet rope shows that such an action is possible, and actually occurs. The rope contracts and raises heavy weights for no other cause than the small weight of the particles of water which, when driven into the fibres, like wedges, provoke the contraction of the rope.' Borelli, *Movement*, Bk II, prop. 29, p. 241.
[106] Charleton, *Lectures*, p. 93.
[107] Charleton, *Lectures*, p. 93.

memorable.[108] This example, drawn from the writings of Monantholius, was that of the transportation of an enormous obelisk.[109] This task baffled the engineer in charge, but a simple carter realised that the obelisk could be successfully raised by shrinking its holding ropes with water.

These analogies were included to convince the audience of the tremendous contractive power of fibres, and of the possibility that a small cause might be responsible for great effects. Charleton had not yet discussed the relationship between the historic examples and the issue under discussion, the heart. However, after his description of the second similitude, Charleton asked the reader to 'reflect upon this example', and upon the fact that the use of water to dilate the pores of the cables had 'swell'd and shortned them with force great enough to overcome the immense gravity both of the Obelisk, and of themselves'. Consequently, he argued, 'I am confident, you will no longer think it impossible for a few little drops of liquor diffused through the Fibres of the Heart, and like wedges dilating their little Meshes or Pores, so to swell and abbreviate them, as to cause a constriction of the Ventricles.'[110]

The persuasive power of the analogy revolved around the plausibility of such powerful motion from this cause. The analogy established the plausibility of the physical principle, and then how it might be applied. The probability of a proposition 'cannot be too obscure to any man of common sense' who considered first 'the near similitude that is between the threds of a chord, and the Fibres of the heart, in Figure, in tenacity and strength, in aptness to surcle, and consequently to shorten themselves upon humectation'. Further persuasion was provided by the similarity between the two fluids under discussion:[111]

> For since the two Agents, viz. water and the Succus Nervosus, are so alike in their efficacy, as to the dilatation of the Pores of tensile bodies; and since the two Patients also, viz. the threds of a chord, and the Fibres of the heart, have so full a resemblance in their nature: it is highly probable, if not necessary, that like effects should be produced by them.[112]

Thus the degree of truth was increased by the analogy, to the point of being almost a 'necessity' (possessing the status of a law). The probability was greater yet, because 'of all the other Efficient Causes hitherto excogitated by Learned men', none could be suggested 'which is either so intelligible, or so congruous to the whole Mechanism of the Heart, as this which in this Lecture I have endeavour'd to assert.'[113] Probability was partly a product of the extent to which

[108] Charleton, *Lectures*, p. 93.

[109] Allegedly 170 feet high and 12 feet wide, weighing 9,586,148 pounds. Charleton, *Lectures*, p. 94.

[110] Charleton, *Lectures*, p. 95. This was taken from Borelli, 'Muscles contract with considerable force because their fibres are swollen by an additional substance as if it were by wedges.' Borelli, *Movement*, Bk II, Prop. 15, p. 219.

[111] That is, 'the little or no difference betwixt water and the Succus Nervosus, as to the power of insinuating into, and dilating the Pores'. Charleton, *Lectures*, p. 98.

[112] Charleton, *Lectures*, p. 98.

[113] Charleton, *Lectures*, p. 96.

explanations were 'intelligible', 'congruous to the mechanism', and 'facile'.[114] Their intelligibility was one of the functions increased by persuasive analogy, which highlighted the congruity of analogy with subject.

These examples presented Charleton as an individual who read and observed widely, rather than one involved in the active manipulation of nature and the *creation* of confirmation for his theories. Thus, his understanding of the natural world was presented to, rather than sought after by, him. He presented himself here as an armchair theorist, considering hypotheses and seeking confirmation or disconfirmation through the consideration of instances in textual sources. This self-description fits the circumstances of his life during much of the 1670s, in which he spent time in scholarly solitude in exile from London (following his active experimental commitments at the Royal Society in the 1660s). His knowledge of textual sources was reiterated to this audience as a vital part of his identity. While the analogy was Borelli's, many of the proofs were Charleton's own. Although he omitted Borelli's mathematical and mechanical lemmas, presumably because he did not see them as essential to proof of the theory's probability, he added these extended analogies. The demonstrative power of analogies derived from wide reading was apparently greater than the repetition of experiments from Borelli. Their inclusion was presumably based on his perception of the most persuasive means to convince a College audience of the theory's veracity.

Charleton's refutation of others' analogies illustrates the central role of analogy in theory confirmation. Attacks upon the founding analogy of a hypothesis constituted profound rebuttal. Charleton devoted considerable energy to refutation of a competing analogy used by Lower in *Tractatus de Corde*.[115] Blood, he argued, was *not* expelled out of heart by 'a Spiral contortion or twisting of the heart, such as that by which water is commonly squeez'd out of a wet napkin, as some late Writers have thought'.[116] This refutation involved a detailed discussion of the implications of the analogy, and a series of reasons why the process described did not accurately represent the heart's action.[117] Likewise in *Enquiries* one of the bases upon which he had rejected the presence of animal spirits was that its supporters could not find an appropriate analogy for the spirits.[118] Thus analogy was the basis of major debates between authorities. The applicability of competing

[114] Charleton, *Lectures*, p. 87.

[115] Borelli also rebutted this model, though he had not named its author. See R. Lower, *Tractatus de Corde*, London, 1669.

[116] Charleton, *Lectures*, p. 59.

[117] Charleton accepted that expulsion was performed by 'constriction of the cavities and Pores, which were filled by the fluid', as it is in 'wringing of water out of a wet cloth', but 'at the same time I deny, that such a constriction is made in the heart, and such expression of the bloud thence, by the same cause, the same Organs, and the same Mechanic action, by which water is squeez'd out of wreath'd Linnen.' Charleton, *Lectures*, p. 59

[118] The doctrine of spirits was unacceptable because it had not yet been 'by certain reasons or Experiments, taught us, to which of all the Fluids that are known to us, that is like'. By implication, the discovery of a compelling analogy could be a crucial form of demonstration. Wine had been proposed as the fluid most analogous. Charleton did not refute the fact that wine was rich in spirits,

analogies was critically assessed by non-experimental theorists. A healthy debate over competing hypotheses occurred quite outside the experimental domain.

A persuasive analogy: The Segovian mint

Charleton's third prælection, 'Of the Efficient Causes of the Pulsation of the Heart', opened with a description of his search for appropriate analogies to understand the heart. He had 'revolved the Books of the most Celebrated Authors, who have professedly written of Architecture, and of Hydraulic Engines; in search of some example of a Machine, that might be, at least in a few respects, compared with it.'[119] The secret motions of the heart were to be understood by profound meditation upon appropriate analogies or similitudes, not by probing the body itself. This declaration situated his search for proof in the scholarly context, not in the laboratory, or in the anatomy theatre. It prioritised analogic method in gaining natural knowledge. Of the many that he found, that 'nearest in similitude to this inimitable Prototype of Nature, was the Hydraulic Mint at Segovia'. The Segovian mint had been a focus of discussion for both Hobbes and Digby previously.[120] Charleton's knowledge of the Segovian mint seems to have been drawn from Digby's description, which he quoted in full.[121]

Charleton declared that 'betwixt this Engine and the Heart, I fancied something of Similitude'. He proceeded to explain the exact bases of their parallel:

> First, As the design or end of the former, was to Coin mony, which is the bloud of all States . . . for the support of the Goverment: so the office and work of the latter is to stamp the character of Vitality upon the mass of bloud, for the maintenance of life in all parts of the body, and regulation of the whole Animal oeconomy.[122]

---

but argued that the 'subtility, acrimony and volatility' of spirits were inconsistent with these characteristics of wine. Charleton, *Enquiries*, pp. 503-4. Charleton cited the lack of evidence over the exact origin of the beneficial effects of wine. It could not be proven whether wine's vivifying effects were the result of its spirituous nature, or 'some other reason joyned to it'. Charleton, *Enquiries*, p. 503. This argument echoed Steno's *Elementorum*, p. 64, in which the latter had expressed doubt as to whether exhaustion cured by imbibing spirit of wine should be ascribed to 'another matter which the fluid spirit provides, or on account of another cause to which the humour, which we call spirit, is closely linked -who will determine'. Steno had argued further in this text that spirit was a misleading and vague term, and that too little was known of what it signified. This same inability to carry through a complete analogic comparison also affected Thomas Willis.

[119] Charleton, *Lectures*, p. 71.

[120] The uses of the metaphor by Hobbes, Digby and Charleton, are discussed by Sawday, 'The mint at Segovia'.

[121] Digby, *Treatise*, chapter 23, page unknown. Digby had compared the multiple functions and components of the Segovian mint with another kind of machine bodies, exemplified by the Toledo engine. Digby, *Treatise*, p. 207.

[122] Charleton, *Lectures*, p. 71. This echoed his 1659 declaration that the preservation of vital heat depended upon 'a perpetual expence of the most pure, i.e. the most volatile spirits of the blood; and consequently necessary, that during life, fresh spirits must be perpetually minted out of the blood, to defray that vast and continual expence.' Charleton, *Natural History*, p. 65.

Charleton compared the heart and mint in terms of their equivalent functions, 'secondly, As the one is moved by a stream of Water, so is the other by a current of bloud'. He echoed Digby's depiction of the mint as 'This Engine, or rather multitude of several Engines, to performe different Operations, all conducing to one work':[123]

> Thirdly, As the Artificial Engine was composed of many less Machines, each of which performed its proper office by a distinct operation; yet all conspired to one common end: So the Natural, being also complex, consisteth of various smaller Machines, viz. the Ears, Valves, Ventricles, Musculose flesh, Fibres of different orders, Chords, Columns, Papillae, &c. all which have their peculiar functions and motions; yet so combined, that they all co-operate to the Vital motion or heat of the bloud, and diffusion of the same.[124]

This was a familiar aspect of the comparison of the body with machines and in the commonwealth metaphor. The whole of the body operates according to the distinct and complementary operation of its component parts, each vital to the overall economy. This was contrasted, by Digby, with the unity of operation of certain other engineers. However Charleton depicted it without comparison:

> Fourthly, By the Segovian Engine Ingots of Silver were distended to a bredth and thinness requisite to make mony: by the heart and its Ears vehemently constringing themselves, and repeting their strokes, the Silver Chyle, or publick revenue of the Animal, is attenuated, its viscid and grumose parts dissolved, the cruder parts concocted, and all by conquassation and compression so perfectly commixt with the bloud, as to be fit to make good and current bloud.[125]

Again he echoed closely Digby's depiction of the transfer of pieces 'into a reserve, in another room, where the Officer, whose charge it is, findeth treasure ready Coined, &c.'[126] This indicates that Charleton's knowledge of the mint was gained entirely through Digby's description, beyond which his own characterisation did not stray:

> Fifthly, From the Mint-engine the new stampt Coin was quickly transferred into a receptacle in another room, thence to be distributed, by orders of the Mint-master: From the Heart is the new Coined bloud instantly transmitted into the Arteries, to be distributed, according to the ordinance of Nature.[127]

Nature's role was equivalent to that of the 'mint-master':

> Sixthly, As the various parts of the greater Engine were so situate, disposed, and connected, as that if any one of them were by chance displaced, broken, or hindred in its

[123] Quoted in Charleton, *Lectures*, p. 71. Digby made a similar declaration, that 'though every part and member be, as it were, a complex thing of itself, yet every one requireth to be putt on its motion by another . . . for the use and service of the whole'. Digby, *Treatise*, p. 208.

[124] Charleton, *Lectures*, p. 72.

[125] Charleton, *Lectures*, p. 72.

[126] Digby, *Treatise*, cited by Charleton, *Lectures*.

[127] Charleton, *Lectures*, pp. 72-3.

> motion and action, presently all the rest must fail to procede in their respective operations, and the work of making coin cease: So in the much more subtil and mysterious Machine of the heart, if any the least part, though but the chord of a Valve, be broken, or arrested in its motions, all the rest will soon be at a stand, and the grand work of making the bloud vital be at an end.[128]

This echoed Digby's characterisation of the Segovian mint.[129] However, the next stage of Charleton's development of the analogy took a different direction. So far, he argued, 'the Parallel held fairly enough, and I was not ill pleased with the ramble of my imagination'. Yet 'when I had attempted to carry on the resemblance a little farther, I soon discovered the disparities to be so many, and so great, that it was impossible to reconcile them into a just Analogy.'[130] He decided not merely that this specific analogy was inadequate to the task of explication, but 'condemning the extravagance of my fancy, I soberly concluded, that the Heart of an Animal is an Engine never to be imitated by human art.'[131]

He applauded Archimedes, for his dignity in 'never attempting to counterfeit the motions of the heart', above his attempts to reproduce the 'order and motions of the Celestial Bodies'.[132] Although he had earlier stated that its motions could be perfectly understood through mechanical reasoning, Charleton here implied that their imitation, or counterfeit, should not be attempted. It seems that while physical, mechanical knowledge was possible, *imitation* was controversial. While understanding, aided by analogy, furthered the glorification of God, imitation threatened the ultimate supremacy of divine creation.

Jonathan Sawday believes Charleton's rejection of the mint comparison constituted a strategic rejection of 'specious tropes and figures', in keeping with the Royal Society's call for plain language.[133] He wished, Sawday claims, to appear before his audience as 'a rigorous scientist who has heeded, and practices, the

[128] Charleton, *Lectures*, p. 73.

[129] Digby described it as a machine in which each element 'considered by it selfe might seeme a distinct complete engine, is but a serving part of the whole; whose office is to make money: and that for this worke, any of them separated from the rest, ceaseth to be the part of a minte, and the whole is maymed and destroyed'. Digby, *Treatise*, p. 207.

[130] Charleton, *Lectures*, p. 73.

[131] Charleton, *Lectures*, p. 73. This analogy had been used by Hobbes: was Charleton therefore careful, in an already dangerously materialist work, to avoid close association with Hobbes, and accusations of atheism? Interestingly, although Borelli did not use this analogy, he made a similar presentation of possible explanations, and subsequent retreat in his opening to this discussion. He proposed the comprehension of unmeasurable actions of the heart through analogy with 'cog-wheels or pendulums', but then retracted these comparisons, claiming 'Not only are such machines not found in the brain but they seem to be opposed to the simplicity in which Nature delights in her operations.' Borelli, *Movement*, Bk II, Prop. 79, p. 283.

[132] Charleton, *Lectures*, p. 73.

[133] Sawday, 'Mint at Segovia', p. 33. Sawday also argues that Charleton recognised the point at which the mint metaphor 'fails to help his argument develop'. Its specific application could not account for the return of the blood to the circulatory system without also implying a nonsense about return of the coin to the mint. 'Mint at Segovia', p. 31.

proscriptions of the Royal Society'.[134] Yet Sawday provides no indication of how such a self-characterisation might relate to the relevant audience for this work, nor does he indicate that the rejection of metaphor was as substantially important in this anatomical context, as he believes it to have been in other areas of natural philosophy. It seems unlikely that rebuttal of the use of analogy and metaphor was behind Charleton's rejection of the mint example. His next discussion, of the motive force of the heart, gave a powerful status to exactly this kind of similitude. Because the heart's shape excluded it from the kind of measurements of 'motive power' performed on other muscles within the body, Charleton argued, citing Borelli verbatim 'we may investigate the greatness of the effects' through 'the Similitude and Analogy, which the Muscule [sic] of the heart seems to hold to other Muscles of the same Animal.'[135]

These examples from *Lectures* illustrate that the role of analogy in demonstration possessed an equal if not greater epistemic status to that of empirical demonstration. While analogic demonstrations were often analysed in great detail for their possible points of convergence with the subject under discussion, experimental instances were often treated with little specificity or contextual detail. The demonstrative power of analogy has a greater status in *Lectures* than in *Natural History*. It contained more extended analogies than did *Enquiries*.

THE QUESTION OF CAUSES

Final causes

The accessibility and relevance of final causes had long been a subject of debate. In Aristotelian and scholastic science the quest for final causes was central, within a teleological understanding of nature. Aristotle insisted that 'All natural things are for the sake of something'.[136] This approach seems to have been taken by Charleton, who viewed the structure of the body as divine artifice, and all of its operations as beneficial. The *Lectures* were concerned with the search for the final and efficient causes of circulation.[137] Following Aristotelian non-intentional teleology, Charleton seems to have used the term 'final cause' to imply the principle by which nature endowed all phenomena with the means for their own preservation and maintenance. The final cause then was 'to what end Nature, all whose counsels and actions are ordained by an infinite wisdom, hath instituted

[134] Sawday, 'Mint at Segovia', p.32.

[135] Charleton, *Lectures*, p. 74. Borelli, *Movement*, Bk II, Proposition 66, p. 271.

[136] Aristotle, *Physics* II.8. In some instances, specification of the form itself could be equivalent to identification of the final cause, 'since the complete substance, as the actualisation by form of certain potentialities in matter, is the end of the process.' S. Nadler, 'Doctrines of Explanation', in *The Cambridge History of Seventeenth Century Philosophy*, Cambridge, Cambridge University Press, 1998, p. 529.

[137] Charleton, *Lectures*, p. 31.

this rapid Circulation of the bloud.' He concluded in this instance that the final cause of the circulation of blood was 'the conservation of its requisite temper and vital constitution'.[138] Borelli had given the same explanation, but without reference to the terminology of final cause.[139] Indeed it seems the English physician was far more concerned to describe his subject in causal terms than was the Italian.

Charleton defended the search for final causes against those who argued that it was presumptuous for man to attempt such knowledge: 'who hath prohibited us to investigate the formal reason and manner of their operations?'[140] Descartes had stated that teleological explanation should be abolished in philosophy, and that therefore only efficient causes, and never final causes, should be sought.[141] It is possible that Charleton was protesting specifically against this Cartesian pronouncement.

The search for final causes had other defenders: Boyle had supported the importance of discovering final causes as a means to glorify God.[142] Therefore, according to Boyle, a proper explanation should incorporate both mechanical and teleological considerations.[143] Knowledge of mechanical causes could be used to account for phenomena, but ultimately the fabric and operation of these causes were the result of higher metaphysical purpose. There were thus several layers to knowledge:

> It is not more certain, that no mortal can know enough of Gods works, than it is, that the more we are able to discover of his wisdom, power, and goodness discernible in the mirrour of his Creatures, the more we find our selves obliged to admire, love and adore him.[144]

The next part of his defence of the search for final causes was cribbed directly from Sprat's *History of the Royal Society*, suggesting a genre convention:

> Equally certain it is also, that no kind of devotion is more acceptable to him, than that which procedes from knowledge of his infinite Perfections: and that the Sacrifice of Praises offer'd up to Heaven from the mouth of one who has well studied what he commends, are more sutable to the Divine Nature, than the blind applauses of the ignorant.[145]

[138] Charleton, *Lectures*, pp. 32-3. Borelli, *Movement*, Bk II, p. 247.

[139] He made the more tentative claim that 'it is likely to these ends that Nature designed the very quick circulation of the blood.' Borelli, *Movement*, Bk II, p. 247.

[140] Charleton, *Lectures*, p. 38.

[141] Descartes, *Principia Philosophiæ*, 1644, I, 28. For further discussion see Nadler, 'Doctrines of Explanation', p. 523.

[142] There are some things in nature so curiously contrived, and so exquisitely fitted for certain operations and uses, that it seems little less than blindness in him, that acknowledges, with the Cartesians, a most wise author of things, not to conclude . . . that they were designed for this use.' R. Boyle, 'A Disquisition about the Final Causes of Natural Things' (1688), in Boyle, *Works*, vol. 4, p. 579.

[143] My discussion is indebted to Nadler, 'Doctrines of Explanation', pp. 529-31.

[144] Charleton, *Lectures*, p. 38.

[145] Charleton, *Lectures*, p. 38. Taken from Sprat, *History*, p. 349.

That this quote was taken verbatim from Sprat illustrates that Charleton's deviations from Borelli were sometimes designed to annex material from other sources, rather than to add a personal note to the text. Thus he combined the content and demonstrations of one (foreign) authority, with the conventions associated with his immediate audience, balancing a contextually relevant identity with new knowledge.

Charleton appears to have reasoned as follows: although we cannot know or imitate the mind of God, we can gain knowledge of the mechanical operations of some elements of his creation. The heart is one of these elements, but only the heart, not the whole body, is mechanical. The heart operates according to mechanical necessity [laws], and therefore we can have true knowledge of the heart's operations. However we cannot claim certainty about the other operations of the body, nor about the higher levels of the heart's action: viz. its divine purpose: 'though the Heart of man be to us inscrutable, as to its . . . thoughts and reserves; it seems not to be inscrutable, as to its Fabric and Conformation.'[146] Through a complex negotiation of knowledges, Charleton was able to claim some certainty for the knowledge he presented, yet also to avoid seeming to claim insight into the divine mind. The latter theological presumption would have earned him stern censure.

Once the efficient cause of the blood's motion was shown to be the heart's pulsation, Charleton turned to consider 'the Mechanism of the heart', the last stage in coming to know 'the true reason and manner of the Pulsation'. Once this was complete, 'we shall so much the more admire and laud the skill of the Divine Engineer, who contrived and made the Machine of the heart of so small a bulk, and yet of so stupendous power and force.'[147] Comprehensive knowledge of the mechanical operations of the heart was to be gained for the glorification of the divine artificer.[148] He proceeded to explain its structure, indicating the possibility of true knowledge, which touched on the understanding of divine intention, while limited in scope to physical and observable evidence. If the mechanism of the heart's operation 'hath been by us rightly explicated (as I am perswaded it hath) in the precedent discourse, no man has reason longer to believe, that the manner of the motion of the heart is a thing to human wit wholly impervestigable.'[149]

Charleton claimed that when Lower recognised the limits of knowledge in *de Corde*, he had been 'out of modesty willing to limit his own curiosity in that

[146] Charleton, *Lectures*, p. 38.
[147] Charleton, *Lectures*, p. 38.
[148] The extraordinary proportion of cause to effect was portrayed as part of the divine creation. This was demonstrated at the end of his work through two extended analogies. The possibility of so large an effect, from so small a cause, showed the presence of divine power. While mechanical operations could be portrayed as purely material and free from divine intelligence, Charleton reasserted that the *nature* of those operations required an insight that only God could endow.
[149] Charleton, *Lectures*, p. 69.

particular', he had not 'set bounds to the future disquisitions of other men.'[150] The searcher ought not, according to Charleton, despair of further discovery. Thus his approach showed respect for the limits of human knowledge of divine intention, but maintained the accessibility of meaningful physical knowledge. This reaffirmed the purpose of anatomy while preserving appropriate theological limits on its aims.

THE EFFICIENT CAUSES OF THE HEART'S PULSATION

Although Charleton followed Borelli in almost all other particulars, the agency of fermentation or chemical reaction was utterly rejected in the *Lectures*. This was the primary dispute between the two men.[151] The efficient causes of the heart's pulsation were the mediate and immediate causes. The muscles of the heart performed pulsation regularly but without the involvement of the will. As outlined in the last lecture of the *Enquiries*, voluntary motion required the intervention of some communication from the brain. The heart's motion, regular but beyond the control of the will, had been identified by Aristotle as distinctive from other varieties of muscular action. The mediate cause then was the means by which the heart's motion was regulated. The immediate cause was the method by which it was moved.

Immediate cause

Like Borelli, Charleton rejected the idea that a faculty was the immediate cause of the heart's movement: the 'Ancients' had argued that the heart moved independently of the will [ie natural rather than animal motion] and 'accordingly constituted and assign'd to the heart a certain blind and unintelligible Pulsifick Faculty'.[152] Charleton invoked Borelli's proposition that 'the immediate Motive cause of the heart, is the very same with that, by which the Muscles of the Limbs are moved Voluntarily.'[153] They agreed that the immediate cause was distention of the pores of the heart's fibres. However, the English physician made one significant alteration. Having rejected the theories of a faculty, spirits, and violent injection of blood,[154] Borelli claimed that 'the direct cause' of the tension of the heart was the swelling of its pores as a result of 'fermentative ebullition of

[150] Charleton, *Lectures*, p. 69. Richard Lower's claim had been that where it was too difficult to understand the reason of the heart's motion, where God alone knew the motion and did not reveal it, then he himself would not waste his effort in further scrutiny. See Lower, *De Corde*, cap. 2.

[151] Nayler is the only historian to explore the dimensions of this disagreement. On Charleton's criticisms of Borelli see Nayler, 'Insoluble Problem', p. 504 and chapter 7 *passim*.

[152] Charleton, *Lectures*, p. 77.

[153] Charleton, *Lectures*, p. 77.

[154] Charleton echoed Borelli's rejection of the ancients' theory of the 'blind pulsifick faculty' almost word for word. See Borelli, *Movement*, Book II, p. 281, Charleton, *Lectures*, p. 77.

elements of tartar of the blood by some spirituous juice instilled from the orifices of the nerves.'[155] Thus the heart's distention arose from chemical reaction.

Charleton also rejected explanation by 'an incorporeal Faculty', or spirits, or by the blood 'violently rushing' into the heart. But he added to these rejected explanations the agency of 'the same bloud to what degree soever rarefied in its Ventricles, or by a Fermentation, conflict and displosions of Acid and Saline juices met together in the heart'.[156] Thus he specifically rejected Borelli's ebullition solution. Charleton's explanation differed from his source only in this final particular. They agreed on the process, and on the actions involved. Charleton acquiesced with Borelli's view that the fibres of the heart were composed of dilatable rhomboid pores. But while Borelli saw fermentation in an active causal role, Charleton argued that it was unnecessary. He dismissed the theory of effervescing reactions between 'tartarous' particles in the blood and 'spirituous nervous juice' along with Willis' explosions.[157] After his grandiose claims for a revision of Borellian theory, Charleton effectively posited a single fluid (the *succus nervosus*) which inflated the muscle fibre pores, in the place of Borelli's several fluids, the reaction of which created expansion.[158] Although he altered certain explanations within the Borellian process, he did not attempt to account for the problems and discrepancies caused by his partial revisions.

Mediate cause

Borelli made no explicit statements about the mediate cause. He seems to have supported the idea that the impulse for the heart derived from the brain (though he did not use the terminology of mediate cause). Charleton, however, confidently asserted: 'Certain it is, that the first and mediate Cause of the Motion of the heart, as well as that of the motion of the other Muscles . . . is derived to it by the Nerves from the Brain'.[159] He provided detailed reasons, drawn from Lower's *de Corde* and Ent's *Antidiatribæ*, to illustrate that the brain was the origin of 'the incitement and continuation' of the heart's motion.[160] Beyond this he argued that it was 'highly probable' the brain was the 'promptuary' as well as the 'laboratory' of the *succus nutritius/nervosus*. Here he cited demonstrations performed by himself with Glisson and Ent.[161]

Charleton claimed, after Borelli, that the release of *succus nervosus* into the heart from the brain was regulated by the distinctive structures of the

155 Borelli, *Movement*, Book II, Proposition 77, pp. 281-2.

156 Charleton, *Lectures*, p. 78.

157 Nayler, 'Insoluble Problem', pp. 98-9.

158 Nayler, 'Insoluble Problem', p. 141.

159 Charleton, *Lectures*, p. 80.

160 See Glisson, *Antidiatribæ*, p. 7.

161 These propositions had been 'formerly with many considerable arguments drawn as well from experiments Anatomical, as from reason, asserted by Sir George Ent, Dr. Glisson, and (if it be lawful for me to put my self into the same period with such excellent men) my self'. Charleton, *Lectures*, p. 82. He then proceeded that if his findings were supposed accurate, it was 'not unreasonable' to reach the conclusions he had.

nerves.[162] This explained the possible transmission of the impulse from the immediate cause (the *succus nervosus* filling the rhomboid pores) via the mediate cause, which was regulation of the flow of *succus nervosus* according to the structure of the nerves. Their thinness caused fluid to move through them only slowly, thereby regulating its release: the *succus nervosus* was expressed 'drop by drop' into the flesh of the heart.[163] Borelli further claimed that the spirit's 'mixing with the humour of the heart results in ebullition and explosion.'[164] Charleton omitted the reference to ebullition, but otherwise repeated Borelli's account.[165] His main deviation was that he made more specific the analogic bases of the *succus nervosus* ('whose consistence is not much thinner than the white of an Egg'), and of the nerves (which were 'like Indian Canes').[166] These analogies were not drawn from Borelli's text, but were those he had invoked in the *Enquiries*, in his discussion of Glisson's *succus nutritius* hypothesis.

While Borelli provided analogies as the *justification* for his theory, Charleton invoked them as its *demonstration*.[167] Borelli had stated the analogic method as a means by which to gain knowledge of primary causes, and in this discussion had run through a range of options for appropriate analogy. The Englishman, by contrast, had outlined the theory, and then given analogic demonstrations to prove his point. His final conclusion was that 'Where we find a parility of Causes, we may rightly expect a similitude of effects.'[168] This illustrates deductive, rather than inductive, reasoning. It shows Charleton's fluid movement between these varieties of epistemology.

Charleton devoted more time to the examination of mediate and immediate causes than had Borelli. He explored the two main problems consequent upon Borelli's hypothesis. The Italian had considered and dismissed these himself, but Charleton gave them far more time. The first problem was that pulsation of the heart continued after it had been removed from the body, therefore threatening the theory that its action was contingent upon influence from the brain.[169] The second was the question of why other muscles did not pulsate, if indeed the configuration of the fibres allowed this to occur in the heart. Having stated confidently the brain's involvement as the mediate cause, Charleton took pains to emphasise the uncertainty surrounding the topic: 'it is not yet certainly known to any mortal man, by what mediate cause the Muscles of the Limbs are moved at the command of the Will'.[170] Charleton here supported the theory that an increased emission of *succus nervosus* was the cause of the action of cardiac

[162] Charleton, *Lectures*, p. 88-9. See Borelli, *Movement*, Bk II, Prop. 79, p. 284.
[163] Borelli, *Movement*, Bk II, Prop 79, p. 283.
[164] Borelli, *Movement*, Bk II, Prop 79, p. 284.
[165] Charleton, *Lectures*, p. 83.
[166] Charleton, *Lectures*, p. 85.
[167] Borelli, *Movement*, Bk II, Prop. 79, p. 283.
[168] Charleton, *Lectures*, p. 85.
[169] Charleton, *Lectures*, p. 85.
[170] Charleton, *Lectures*, p. 87.

fibres.[171] The structure of the cardiac fibres was distinctive in that it allowed pulsation in them (unlike any of the other muscles). Charleton stated that 'Nature hath framed the Originals of the Cardiac Nerves by a different Artifice', such that (requiring no additional impetus) their mechanical structure (the right proportion of the canals and the right conformation of the orifices) ensured that they would consistently supply the necessary liquid to guarantee the regular pulsation of the heart.[172] No anatomical evidence demonstrated this rather neat hypothesis that the nerves' distinctive structure regulated their intake of *succus nervosus*. However, Charleton asserted its veracity on the basis of the presence of distinctive operations throughout the body.[173] He examined the questions and answers associated with this theory in far more detail than had Borelli, and went to great lengths to illustrate the probability of the cardiac nerves' distinctive structure.

The distinctions of the cardiac nerves 'consist in such minute and subtle artifices, as have hitherto eluded our most curiose researches, though assisted by the best sort of Microscopes.'[174] Yet this absence of empirical evidence was no obstacle to assertion of the principle. The existence of comparable specificity in other parts of the body (for example the optic and auditory nerves) regardless of demonstration, indicated the possibility 'that Nature hath given to the Cardiac Nerves also a constitution divers from that of all other nerves'.[175] Although he was not 'so happy, as certainly to know' the exact structures by which nature had been able to distinguish the operations of the nerves, he was 'notwithstanding fully convinced', that a distinctive structure existed.[176] The similitude between the theorised reality and other operations of the body provided sufficient basis for his criteria of possibility. In this instance empiricism was not an important criterion of truth status, or the plausibility, of the hypothesis.

Charleton concluded that the mediate cause of the motion of both the ordinary muscles and the cardiac muscles was the emission of *succus nervosus* from the brain. Yet he acknowledged his own earlier claim that the mediate cause of the heart's motion was different from that of the other muscles. The difference

[171] The 'Mechanism of their Fibres', he claimed, made it 'more probable, that they are moved by immission of some liquor from the Brain . . . by which the rhomboid meshes or pores of their Fibres being all at the same time swell'd and dilated, a contraction of the whole Muscle must in the same moment be effected'. Charleton, *Lectures*, p. 87.

[172] The orifices of the nerves in the brain, which led to the muscles of the limbs, were 'in such a peculiar manner formed, as at no time to take in and convey into the Muscles, more of the roscid liquor than what is sufficient to nourish them, and recruit their vigor'. The exception was when, 'at the command of the Will', the nerve orifices were alerted to dilate and receive/transmit more liquid, to swell them up. Charleton, *Lectures*, p. 88. See Nayler, 'Insoluble Problem', p. 139.

[173] Charleton, *Lectures*, p. 88. 'the Organ of every sense hath its nerves of a peculiar constitution, accommodate to the nature of its proper object'.

[174] Charleton, *Lectures*, pp. 88-9.

[175] Charleton, *Lectures*, p. 89.

[176] Charleton, *Lectures*, p. 89. 'And if so, why may not the difference consist in such an Artifice, as that which I have here supposed and described; since the same is not only possible, but facile also, and sufficient to produce the effect required.'

between their mediate causes was 'in the divers Modes of its effusion from the Brain.' The *succus nervosus* descended through the cardiac nerves to the heart, 'gently, slowly, and by way of instillation, drop after drop'. In the other muscles, he believed it to be 'immitted with great force and velocity, swift as Lightning, at the command of the Will'.[177] As Nayler notes, this seems easily as problematic as (if not directly derivative of) the animal spirit hypothesis that he vehemently rejected.[178] The principle behind this claim was that 'the same cause, used by Nature *diversimodé*, and in the Organs of different conformation, produceth different effects'.[179] This principle seems directly to contradict his own stated method throughout the remainder of the lectures, and illustrates again the fluidity of his epistemology.

These detailed case studies illustrate the integral role of analogic reasoning in Charleton's explanation. The Englishman omitted many of the Italian's physical proofs, but tended to add his own analogic demonstrations. This suggests that he knew his audience would be satisfied with the latter. He pushed certain philosophical aspects of the study further than Borelli.

CHARLETON'S REASONS FOR REJECTING BORELLI'S HYPOTHESIS

Charleton's refutations of Borelli's ebullition hypothesis were founded upon logical reasoning. His argument was preferable because it explained the matter more neatly and simply, by cutting out one part of the alleged process (it removed the need for any fermentive process). He restated the criteria by which his method could be judged. If, he claimed, his arguments could be judged 'consentaneous to right reason, agreeable to the Animal Oeconomy, congruous to the Organical structure of the Heart (to all which I have been careful to adjust them) and in fine consistent among themselves', then he would 'with assurance conclude, *that the Heart is*, as all Automatas are, *moved by* Mechanic necessity.'[180] The epistemological methods by which he justified his rejection of Borelli's 'pretty conceit' (regarding the reaction of saline and acid spirits causing fermentation and inducing muscular motion) were a combination of analogic reasoning and observation [though not manipulation]. He invoked 'the testimony of our sight';[181] a combination of logic and observation;[182] and the

[177] Charleton, *Lectures*, pp. 91-2.

[178] Nayler, 'Insoluble Problem', p. 139.

[179] Charleton, *Lectures*, p. 92.

[180] Charleton, *Lectures*, p. 97.

[181] In that the pulse occurred in *colliquamentum* before any sign of blood. This, he reasoned, was evidence that fermentation of blood was not the cause of the pulse. Charleton, *Lectures*, p. 98.

[182] The other muscles did not have a pulse, which they should if, as Borelli had claimed, pulsation was caused by 'an explosion of mutually hostile spirits' within the fluids within all muscles. Charleton, *Lectures*, p. 99.

logic of the internal hierarchy;[183] and finally, he attacked the analogy invoked by Willis and Borelli.[184]

Charleton affirmed the likelihood of his own claims largely by rebutting the relevance of Borelli's. However, as we have seen throughout the works considered, this was a common method. His theory differed from Borelli's only in its removal of fermentation: therefore attacks upon the necessity or existence of fermentation affirmed his own claims (as his was a negative hypothesis, negative claims were necessary). His aim was to provide a more plausible account than that provided by previous explanations. He could therefore annex the mechanist claim for greater simplicity in explanation. If he had not made his point sufficiently, he claimed, he would 'consolate' himself by recollecting the difficulty of his subject, and that '*Truth* is a tree, whose root is in Heaven, and of which even the wisest of us dim-sighted Mortals here upon earth see nothing but the shadow of its branches.'[185] This was far from his claim that Lower was wrong when he accepted the limits of possible knowledge. It rather falls into a category of generic disclaimer.

In conclusion, methodological differences between this work and *Enquiries* illustrate the profound eclecticism of Charleton's approach. *Enquiries* was philosophical, and broad-ranging in its coverage. *Lectures* was more specific—discussing the operations of a single organ—and more consistent methodologically. While *Enquiries* drew upon a variety of sources, in discussing a wide array of functions, *Lectures* followed one. This is a key reason for the text's relatively uniform methodology. However, both books were within the lecture genre, both were presented to audiences at the College of Physicians, neither within the framework of any established anatomical lecture series. Both were delivered to audiences composed of a mixture of professional peers and students.

The two sets of lectures illustrate continuity in their eclectic and modest self-construction. In both the author deferred to his audience as arbiters of the probability of the hypotheses presented. As in *Enquiries*, there seems to have been a teaching function to these *Lectures*. Textual, rather than experimental, sources provided the greatest understanding in Charleton's presentation. Despite his investigations at the Royal Society, there is no evidence that he wished to be associated with them in

183 The soul could have no dominion if its constituents 'were agitated every moment by Squibbs or Crackers breaking within them'. Charleton, *Lectures*, p. 99.

184 This was the gunpowder analogy. While explosions might be responsible for the diastole [expansion], they could not cause systole [contraction] of the heart. Such an explosion, he argued, could only cause distention, not contraction, of the muscles. Charleton, *Lectures*, pp. 99-100. Charleton's final attack upon Borelli's hypothesis also drew upon analogy. If pulsation resulted from explosions, the pulse should be more frequent when the body contained more of explosive elements: but rapid heart rate was not associated with increase in 'Acid and Saline particles'. 'sucking infants', he claimed, would not have much acid in them, but tended to have a pulse more rapid than that of frequent consumers of acid. [!] Therefore 'tis highly improbable that the Pulse of the heart should be the effect of such explosion.' Charleton, *Lectures*, p. 100. Therefore a negative finding refuted the other's theory.

185 Charleton, *Lectures*, p. 103.

these public presentations on behalf of the College. The epistemology he adopted in this work seems, though different from *Enquiries*, also to support the characterisation that distinguished the College as an authoritative institution, and its learned members from others in the medical marketplace.

The declared aim of the *Lectures* was 'to enquire strictly into the natural necessity or Mechanical reasons of the Motions of the Bloud'.[186] The intention was thus an exclusively physical understanding. Brown uses this to argue for Charleton's complete conversion to mechanistic philosophy. Yet the English physician's reliance upon Borelli complicates such a claim. Charleton was a temporary mechanist, as his eclectic manner would always dictate. Although his agenda was to illustrate the plausibility of a coherently mechanist explanation of cardiac motion, he did not attempt to broaden this approach to provide a mechanical explanation of the whole animal oeconomy.

Charleton devoted a series of lectures, and a publication, to refutation of a published work without adding new experimental knowledge. His work was a compilation and interpretation of Borelli, in the scholastic tradition of commentary, with the addition of new discoveries and refutations, all done within the context of acceptance of the legitimacy of the overall theory and claims. Despite his dramatic claim for the extent of his revision of Borelli, his approach to the subject demonstrated the correct modesty to secure him some authority in a genteel context. He portrayed his own task as uncertain, and did not claim to determine the final answer. However, this publication seems less tentative in its conclusions than *Enquiries*, perhaps partly because it remained very close to Borelli's text. Charleton seems confident in assuming that knowledge of mechanical operations of the heart allowed some degree of certitude, at least in relation to physical form. As this text was concerned with structural agency this allowed him to claim certainty in relation to important causal knowledge (by comparison with his previous works).

Lectures present Charleton as an armchair experimenter, seeking enlightenment through meditation upon texts. His reference to textual authorities, and evident reliance upon them, illustrates that experiment did not necessarily offer the highest order of knowledge. Rather than promoting experiments as personal experiences, Charleton seems content to portray himself as drawing upon natural laws, made evident through body of common knowledge—some derived experimentally, some via analogy. Reliance upon analogy was one of the consequences of the (mechanist) view that nature did not vary her operations, and that all matter would behave therefore according to the same laws. This meant that insights derived from analogous experiments in the visible realm could demonstrate the plausibility of claims about invisible matter. Charleton used analogy as the 'proper philosophic' method by which to gain knowledge. Experiment was not a necessary basis of authoritative knowledge. Available experiments were often omitted. Charleton's demonstrations tended to rely

[186] Charleton, *Lectures*, p. 4.

upon the application of apposite analogies. Many of these were taken directly from Borelli's text, but some were his own invention. It seems he viewed such analogic demonstration as sufficient proof, without inclusion of additional experimental material. Indeed where reports of experiments were available for his incorporation (in Borelli's lemmas), he often omitted them.

The two texts compared within this discussion show substantial agreement in most aspects, including methodological statements. Charleton's most substantial theoretical deviation was the refutation of the agency of fermentation. He thus supported a more absolute mechanistic hypothesis than that of Borelli, excluding the agency of anything but structural configuration. However, no commitment to comprehensive mechanism was expressed in the *Lectures*. My suggestion here is that Charleton's mechanism in this work was largely borrowed from, or derivative of, that of Borelli. This was often the pattern with the English physician's works, as we have seen. *Lectures* exemplify Charleton's eclecticism of methodology—the work gave epistemic primacy to analogy, textual authority, experiment, inductive principles and deductive reasoning in turn.

Charleton's main methodological deviations from Borelli involved the addition of explicit statements about the search for final and efficient causes, the incorporation of extended analogies in demonstration of the theory, and the omission of extensive mechanical and mathematical demonstrations provided by the Italian. He presented experimental evidence as if it were provided by nature, not as if he had experimented himself. When reporting experiments performed by others he often neglected to personify their authors. He tended to depict them in the passive voice, as events which offered insight into natural processes. The actions revealed belonged to no individual but simply illustrated nature's methods. The laws revealed by experiment had little to do with their experimenters, but were unchanging facets of divine creation.

In this, the last of the three works that I consider, Charleton is once again revealed as a man of consistently eclectic methods. His lack of experimental reliance and indeed his heavy use of textual sources indicates that he did not follow the empiricist method so often associated with him. The analyses that I have offered have investigated the ways in which Charleton presented his physiological and anatomical discussions. I have illuminated the fact that his eclectic self-construction played a significant role in the creation of an authoritative identity as a professional physician.

In summary my findings suggest major problems with the Shapinian account of late seventeenth-century epistemology. There are compelling reasons to suspect that this model may not apply with any great accuracy to investigations into the natural world in this period, either in natural philosophy or physic.

CHAPTER VIII

CONCLUSIONS

For decades Walter Charleton has been characterised by historians as an experimental natural philosopher, a portrayal rooted in a view of the scientific revolution that has been increasingly reviled. The significance of this depiction for our understanding of Charleton's identity is one of the themes of this book. He has been categorised as an experimentalist within the mode of 'virtuoso' natural philosophers (by Shapin, Dear *et al*), and as a promoter of mechanist philosophy in medicine (by Brown). I argue that both portrayals neglect crucial aspects in their attempts to understand this complex character.

First, they fail to recognise the importance of Charleton's eclecticism, and the reasons behind it. While both groups of scholars have aimed to demonstrate the presence of a coherent philosophy in Charleton's work, I argue that eclecticism was vital to his authoritative identity as a practising physician. He constructed himself as a modest, solitary, reclusive, melancholy and meditative scholar. His works affirmed the necessity of classical learning, and the coherence of ancient textual and modern empirical knowledge. He presented himself as a syncretist rather than an innovator.

Second, scholarship on Charleton has ignored the discrepancy between his practices and his textual self-presentation. Many historians have ascribed to him an identity defined by his recorded activities: participation in the Royal Society, experimental manipulations and anatomical investigations. Shapin and Dear suggest that the 'virtuosi' (including Charleton) aspired to an experimentalist identity, and therefore referred to experiments wherever possible to increase the authoritative impact of their claims. They posit that the virtuoso asserted his authority by claiming access to a unique set of practices in the production of knowledge, and actively promoted his own role in the collaborative generation of original, empirical 'matters of fact' in the laboratory context. Brown and Frank see Charleton as a key innovator in medical experimentation at the College of Physicians.

Few historians have considered Charleton's medical texts worthy of investigation. Yet these works offer important insight into the construction of his identity as a physician. They exemplify a wide range of methods and styles (with no clear chronological shift from one viewpoint to another) and make no reference to the apparent discrepancies between the views he presented. In this study I present Walter Charleton, for the first time, in the medical context within which he operated, considering the constraints and influences under which he worked and wrote.

Charleton did not pursue the identity of an experimentalist. He did not consistently seek the authority of 'virtual witnessing', nor did he afford epistemic primacy to experiment. The communal establishment of knowledge and the need for consensus, linked to natural philosophical authority by Shapin *et al*, are thus not relevant to Charleton's medical authority. A comparison between his 1659 text and his works of the 1680s illustrates that even after he had achieved considerable experimental expertise, his medical writings did not construct an identity around empiricism. In focusing their interpretations around his experimental practices, most historians miss the more fascinating story revealed by the physician's presentation of himself in relation to these activities.

The model generated by Schaffer, Shapin and Dear neglects important differences between the knowledge sources and the legitimation strategies of physicians and those of natural philosophers. The circumstances and professional demands of physicians presented them with distinctive requirements. Few scholars have considered how the textual strategies of physicians such as Charleton related to their empirical activities. Physic did not follow the same trajectory as natural philosophy in terms of its authorities, knowledges and practices. While some aspects of the so-called 'virtuoso' model can be seen to overlap with Charleton's profile, the authoritative identity to which he aspired was clearly not consistent with that described by Shapin *et al.*

Charleton depicted himself as a scholar, whose motivation was not innovation but compilation. He did not make extensive claims for his work, but followed a classical model of the self as one suited to philosophical contemplation—a most desirable intellectual characteristic within a sceptical philosophy. This identity supported his status as an impartial and reasonable judge of hypotheses, and a true arbiter of knowledge. His reconciliation of ancient with modern knowledge often aimed to illustrate the consonance of recent discoveries with the principles of traditional medical practice. Charleton's medical texts generated a public profile of the physician as a learned and grave professional. Classical tradition circumscribed the relationship between practitioners and their patients, binding physicians to tradition. The Hippocratic ideals of the profession were central to a physician's identity—he was a responsible, educated provider of medical counsel, not a generator of 'matters of fact'. There is no substantial evidence that the identity of an experimentalist would have benefited Charleton. As a physician he operated within a marketplace in which empiricism was linked with the lower, menial status of surgeons. He was financially dependent upon clients and patrons, and it was in his interest to imitate and flatter genteel interests and preoccupations—the ancient authorities he invoked would have echoed the educational background of the aristocratic patrons upon whom he was dependent.

Professional authority depended also upon the preservation of the traditional status of physicians (founded on sober judgement and classical learning). This identity was moulded by the need to gain admission to London's medical regulatory and licensing authority, the College of Physicians. The College placed high value upon traditional learning and was strongly infused with scholastic, rather than practical, emphasis. Charleton's eclecticism allowed him to reconcile an interest in modern developments with a devotion to the ancients. Viewing Charleton in the light of his professional career we gain new insight into his self-construction. In this setting, his emphasis upon scholastic learning above his experimental activities in the Royal Society would have increased his authority, reaffirming the values of genteel clients and the College of Physicians. While Nayler and Lewis have observed his eclectic tendencies, in their accounts of Charleton, the impact of the demands specific to his profession have not previously been recognised.

The existing historiography has neglected the distinctive concerns that characterised the physician's professional career. Of late, Charleton has been squeezed into conformity with the Shapin model, producing an account that gives little insight. Instead I suggest a more nuanced approach—one that is free from restrictive models. This book does not extend its focus and claims further than the case of Walter Charleton. However, its findings could point the way to further research into the relationship between physicians and natural philosophers in the development of early modern natural philosophy and medicine. The works of this individual tell a complex story about seventeenth-century thought. His writings uncover a narrative of continuity rather than change, in response to the circumstances and demands of medical practices.

I have focused upon three of Charleton's works—the most similar of his medical publications—and shown the great experiential divide which separated them. There was a substantial overlap between the subject matter covered by each. Certain topics were touched on by all texts, and the author's negotiation of these subjects in the light of his changing experiences, and in different contexts, offers insight into his epistemology. Composed in the vernacular, these three books reveal how he constructed himself as a learned physician, within an established classical tradition, for an audience within his own country. The texts offer unique insights into his methodological approaches. His syntheses in English, of Continental and local developments in medical knowledge, appear to have been written for a lay audience. His tracts on specific medical issues were published in Latin, for a solely medically trained audience. Charleton bridged several audiences, as well as multiple subjects and approaches. In his vernacular presentation of Continental material we can see how he negotiated the expectations of this audience. Compared with his Latin works, designed for Continental readers and learned English professionals, the vernacular medical texts afford greater insight into local contextual influences upon his professional identity.

In comparing the most similar of Charleton's medical writings, before and after his extensive empirical activities at the Royal Society, I had several aims. First, to illustrate the continuity of his self-construction and epistemology across the broad range of his experimental practices. This continuity highlights the distinction between the activities in which he was involved as an experimentalist and the identity he constructed in his published works. Second, I aimed to illustrate the depth of his eclecticism by looking at how he constructed his arguments.

Natural History reconciled traditional theories of the animal oeconomy with more recent anatomical discoveries, and supported the legitimacy of new ideas through reference to traditional authorities. Endlessly fluid terminologies allowed the author to slip between different theories, sometimes depicting new theories through reference to established nomenclature, sometimes re-figuring traditional ideas in novel language. This work, Charleton's first English medical publication, was composed before his entry into the Royal Society or College of Physicians. His involvement in those institutions indubitably had an impact upon his subsequent publications. An eclectic combination of epistemologies characterised *Natural History*, but a substantial change had occurred between that work and the publication of *Enquiries* in 1680.

By 1680, Charleton's experience incorporated some anatomical experimental research, and the later text included more references to experiment (though notably not his own). Yet his self-presentation in *Enquiries* was dominated by the persistence of classical and textual authority, analogic reasoning, and his devotion to meditation, as much as experiment. His own experimental experience was not proffered as a source of authority for his arguments. *Enquiries* balanced the textual, traditional, doctrinal nature of medical practice with some of the empirical, non-dogmatic values associated by recent scholarship with the virtuosi. These continuities in his self-construction, despite substantial changes in his personal circumstances and scientific experience, suggest that there were significant reasons for the presentation of such an eclectic identity.

Finally, *Enquiries* showed a greater concern with the retreat from dogmatism, evident from the tentative manner in which Charleton presented his conclusions. These differences can be partly explained through the genres within which *Natural History* and *Enquiries* were composed. The latter was designed as a lecture, which required an overview of contemporary thought for an educated audience. As lectures had a teaching purpose, part of their function was to present the appropriate modest and eclectic manner for a practising physician. Charleton's task was to integrate new anatomical discoveries, and to show their correspondence with traditional theory.

Attempts to discover a consistent programme of medical thought in Charleton's writings, particularly in these lectures, are bound to mislead. *Enquiries* offered an array of hypotheses and epistemologies, from which it is easy to assume that one must constitute his 'real' understanding. Theodore Brown sees *Enquiries* as both a declaration of the College's new adherence to iatromechanism, and a statement of Charleton's own mechanist beliefs. Such an assumption obscures the significance of his eclecticism, in order to support a pre-existing hypothesis about the rise of the mechanist world-view. I agree that *Enquiries* represented a declaration of the College's agenda. But I believe that the declaration was one of support for eclecticism, textual authority and the reconciliation of traditional and innovative hypotheses. Brown argues that the point of *Enquiries* was to demonstrate that the College, like the Royal Society, had adopted empiricism and mechanism. I suggest that Charleton's text indicated that the College, like the Royal Society, espoused an eclectic and non-dogmatic approach, while still revering the elements of traditional medicine which underlay therapeutics. After all, it is difficult to see how adherence to mechanism would have attracted new patients. The College of Physicians had been extensively criticised for its intransigence in the decades prior to publication of *Enquiries*. Charleton displayed considerable knowledge of old and new ideas, and demonstrated the congruity between ancient and modern theories. This allowed the College to look well-informed concerning recent innovations, while simultaneously justifying its adherence to traditional thought.

Enquiries does not display the kind of consistent intellectual agenda sought by Brown, nor does it illustrate a determined or sustained attempt to uphold a mechanical explanation of the animal oeconomy. Although it is possible to find apparently mechanist declarations in his writings, these did not indicate the presence of a consistent explanatory framework. The methodological differences between *Enquiries* and *Three Anatomic Lectures* further illustrate the profound eclecticism of Charleton's approach. *Enquiries* was philosophical and broad-ranging, drawing upon a variety of sources and covering a wide array of physiological functions. *Lectures* was more specific—concerned with the operations of a single organ—and far more consistent methodologically. This work followed the views of a single author, Borelli, which explains the text's relatively uniform methodology.

Lectures offered an exclusively materialist understanding of the blood's motion. Brown argues that this shows Charleton's complete conversion to mechanistic philosophy, but I believe that his reliance upon Borelli complicates such a claim. The Englishman's text illustrated the *plausibility* of a coherently mechanistic explanation of cardiac motion. He did not attempt to provide a mechanical explanation of the whole animal oeconomy. He echoed Borelli in almost every aspect, including several methodological declarations. Charleton's main divergences were in his addition of explicit statements about the search for final and efficient causes, the incorporation of extended analogies in support of the theory, and the omission of many of the mechanical and mathematical proofs provided by the Italian. Charleton's demonstrations tended to rely upon the application of apposite analogies. Many of these were taken directly from Borelli's text, but some were his own invention. It seems he viewed such analogic demonstration as sufficient proof, without requiring additional experimental material. Even where experimental reports were available, he often omitted them.

Lectures presented Charleton as an armchair experimenter, seeking enlightenment through meditation upon learned texts, as well as experimental knowledge. Rather than promoting experiment in terms of personal experience, he seemed content to portray himself as drawing upon natural laws, made evident through a body of common knowledge—some derived through empiricism, some through analogy. The laws revealed by experiment were depicted as unchanging facets of divine creation, not facts generated through the agency of experimenters.

Many historians have failed to see the extent to which Charleton was guided by currently fashionable texts. His early works have been depicted as reflecting a shift from hermeticism to atomism, and the later works as examples of mechanistic philosophy in medicine. His statements from Borelli have been treated as representing Charleton's own beliefs, just as his Helmontian and Gassendian statements have each been assumed to be declarations of his adoption of these views at different points in his career. I argue that the most fascinating issue at stake is what Charleton *made* of these texts—how he deviated from them, and where he adhered to them. This is highly instructive about the context in which he wrote, as well as the assumptions underpinning his epistemology, and the issue of his 'progress' toward mechanism.

Throughout his *oeuvre* Charleton invoked a range of methods to substantiate his physiological claims, the status of which varied across his works. These included analogic argument, textually-derived assertions, logical propositions, mathematical and quantitative reasoning and ancient principles of operation, as well as experimental manipulation, anatomical observations, fixed principles, laws of nature and reference to textual authorities and reports of others' demonstrations. It is difficult to distinguish a consistent epistemology even across one work, let alone across the physician's career. The most striking continuity is his profound eclecticism, which is a consistent characteristic of his works.

Charleton's writings were exploratory, rather than prescriptive, and did not attempt to generate a consistent overall framework of explanation. While influenced by explanatory systems (ie Cartesian hydraulics, Glissonian irritability, Borellian muscle-mechanics), he did not himself compose them. Instead he synthesized masses of material, fitting fragments of alternative systems within his own compilation of contemporary and ancient material. His medical works illustrate a determination to avoid resort to single doctrines in explanation.

Charleton's medical publications covered several issues considered by historians to reflect absolute philosophical divisions in seventeenth-century medical (and metaphysical) thought. These included the presence and operations of faculties, the existence of natural perception, seminal essences, the materialistic redefinition of animal spirits and the adoption of corpuscular explanation. Yet he showed no inclination to treat these ideas as exclusive of their theoretical opposites, and apparently saw no contradiction between them. Charleton commonly argued that if two reasonable theories existed, then the most probable solution might be found in a combination of the strengths of both.

His eclecticism has been seen by some as reflecting his uncertainty over fundamental issues. I believe it illustrated his attempt to show the continuity between ancient and contemporary authorities, thereby ratifying the College's adherence to tradition, and supporting the authority of his profession, which relied upon time-honoured modes of treatment. Clearly there were important reasons for the retention of an identity that emphasised tradition and classical learning rather than innovation and experiment. Charleton illustrates that the identity of an eclectic (both in method and subject matter) could be desirable in itself, not merely as a stepping stone on the way to a more 'coherent' philosophy such as empiricism. *Natural History* showed off his extensive classical scholarship and impressive knowledge of contemporary theories, while simultaneously asserting his impartiality and status as an 'equal arbiter' of truth.

I have argued that Charleton's self-presentation did not alter in response to his experimental practices at the Royal Society, largely because the determinants of his self-presentation did not alter with the rise of experimental natural philosophy. The continuity of epistemology in his published works before and after his Royal Society involvement illustrates the increasing divide between his practical actions and textual presentation. His professional identity drew upon an ancient tradition of authority, rather than the tropes described by Shapin as characteristic of the 'new' natural philosopher. I have tried to reveal the influence of Charleton's medical career upon his intellectual interests and activities, and to explore some of the distinctive intellectual and social circumstances of a physician.

Charleton has in the past been studied within the context of a range of overarching theories and models about intellectual change, institutional change and the rise of the new philosophy in seventeenth-century England. This study has disputed two main depictions of Charleton: first, that we can understand him as a representative of a newly-created model of the 'virtuoso' natural philosopher (as Shapin, Dear and others claim); and second, that he pursued an experimentalist agenda as part of a broader authority gambit for the College of Physicians (as Theodore Brown argues). I have aimed to generate an interpretation of Charleton's medical writings which sits outside these systems of explanation, though inevitably informed by their approaches. The significance of the identity he wrote for himself as a physician needs further exploration in relation to other practising physicians of his era. Full of surprises and endlessly complex, Charleton's construction of how a seventeenth-century physician might model himself is one that has not yet been properly investigated. I hope that my research will be suggestive of further areas of study in relation to this extraordinary man, and the era in which he lived.

ANNOTATED BIBLIOGRAPHY OF THE WORKS OF WALTER CHARLETON[1]

Published works

Chorea Gigantum, *or, The Most Famous Antiquity of Great-Britan* [sic], *Vulgarly called Stone-Heng, Standing on Salisbury Plain, Restored to the Danes*, London, Printed for Henry Herringman, 1663.
The Most Notable Antiquity of Great-Britain, Vulgarly called Stone-Heng, on Salisbury Plain, Restored, by Inigo Jones . . . to which are added Chorea Gigantum and Mr Webb's Vindication, London, Printed for D. Browne Junior, and J. Woodman and D. Lyon, 1725.
A facsimile edition of the 1725 edition has been produced, introduced by Stuart Piggot, Farnborough, Gregg, 1971.

Charleton dedicated *Chorea* 'to the King's Most Excellent Majesty', and referred to the monarch's personal interest in the matter. In July 1663, when the physician presented his observations on Stonehenge to the Society, Aubrey was asked to look into the matter.[2] He indicated that the King was quite taken with Charleton's theory about Stonehenge. Charleton and Aubrey attended the King and the Duke and Duchess of York when they visited the area in 1663.[3]
Chorea contributed to contemporary debate about the origins of the monument. It criticised Inigo Jones' *The Most Notable Antiquity of Great Britain, vulgarly called Stone-heng, restored*, which argued for the Roman origin of the stone monuments. Charleton claimed that Stonehenge was in fact the construction of ancient Danes. This initiated considerable controversy, and was ill-received in London. Wood claims Charleton's text was 'exploded by most persons when t'was published'. *Chorea* garnered a severe retaliation from Jones's son-in-law, John Webb.[4]

[1] The works are organised alphabetically, and are designed to provide a quick reference while reading the main body of the book.

[2] Birch, *History*, vol. 1, p. 272.

[3] Aubrey refers to the King 'discoursing with my Lord Brouncker and Dr Charlton conerning Stoneheng'. Charleton and Aubrey attended the king the Duke and Duchess of York in their visit to the area in 1663. A. Powell, *John Aubrey and his Friends*, London, Hogarth, 1988, pp. 106-8.

[4] Wood, *Athenæ Oxonienses*, p. 754. However, it found support with Sir William Dugdale, and some interest from the king, to whom it was dedicated. Charleton in turn recommended Dugdale's *History of imbanking and draining of divers fens and marshes, both in foreign parts and in this kingdom*, London, 1662, at the Royal Society in 1664.

The work was famously prefaced by a poem by John Dryden, addressed 'To my worthy friend, Dr. Charleton', in which the latter was highly praised. Some lines of the poem seem to recognise Charleton's eclectic method:

> Whatever Truths have been, by Art or Chance,
> Redeem'd from Error, or from Ignorance,
> Thin in their Authors, (likc rich veins in Ore)
> Your Works unite, and still discover more.

The Darknes of Atheism *dispelled by the Light of Nature. A physico-theologicall treatise*, London, Printed by J. F. for William Lee . . . , 1652.
Dedicated, in Latin, to Dr Francis Prujean. The 'Epistle' testifies to Charleton's gratitude to Prujean for his support in gaining admission into the College of Physicians, and for his personal assistance when Charleton was struck down by dysentery.
The copy of *Darknes* in the Wellcome Institute Library, London, contains a prefacing letter, to 'Clement Barksdale, Theologist' dated 28 March 1654.[5] Barksdale's preface asserted that authors such as Gassendi, Descartes and Sennert were elevated in status by their inclusion in Charleton's text.
Darknes demonstrated that natural philosophy supported traditional religion, and argued that the immortality and immateriality of the soul, and the existence of God, could be demonstrated irrefutably by reason.[6]
The book exhibited an awareness of French philosophy not present in Charleton's earlier works, and has thus been thought to have constituted something of a turning point in his intellectual trajectory.[7]
The work is treated as pivotal by some historians, including Margaret Osler, who believes that *Darknes* 'established the general framework for Charleton's system of nature'. She argues that the books unifying theme is the 'unhindered exercise of God's will in his dominion over nature', and that this theological conviction underlay Charleton's entire system of natural knowledge.[8]

Deliramenta Catarrhi: *or, The Incongruities, Impossibilities, and Absurdities couched under the Vulgar Opinion of Defluxions, by Joh. Bapt. Van Helmont, &c. The translator and paraphrast Dr. Charleton*, London, printed by E.G. for William Lee at the signe of the Turks-head in Fleet-street, 1650.
The preface, entitled 'The Translator to the Judicious and (therefore) unprejudicate Reader', complained of 'malevolent, severe and uncharitable'

[5] Clement Barksdale (1609-1687) graduated from Merton in the mid-1630s and served as chaplain of Lincoln College. It could have been at around this time that he and Charleton met. Barksdale was interested in the work of Hugo Grotius, whose writings explored the issue of knowledge and the problem of certainty in relation to scepticism.

[6] See Kargon, 'Introduction', p. xix.

[7] This is discussed by Sharp, 'Early Life', p. 323.

[8] Osler, 'Descartes and Charleton', p. 452.

criticisms of his previous work (*Ternary of Paradoxes*). However, Charleton declared his temperament to be 'too Stoical, to feel the weak assaults of that cowardly Pygmie, detraction'.[9]

As its title suggests, *Deliramenta* questioned accepted explanations of catarrhal defluxions, in concurrence with Helmont's belief that traditional explanations were in need of reform. The text outlined 'The errors of physicians', including their failure to free themselves of the mistakes of the past concerning 'the generation of Rheume, its defluxion, manner, way, matter, means, places, and organs; as also of its Revulsion and Remedies.'[10]

Although the work was essentially a paraphrasing of Helmont, *Deliramenta* did not show Charlerton's unqualified support for his source. He depicted Helmont's reasons to be 'stronger at Demolishing the Doctrines of the Antient Pillars our Art then Erecting a more substantial and durable Structure of his own, his Witt more acute and active at Contradiction, then his judgement profound and authentick at Probabtion.'[11]

Deliramenta contained praise for contemporary works, including Thomas Hobbes' *Human Nature*.

***De Scorbuto** Liber Singularis; cui accessit Epiphenomena in Medicatros*, Londini, Typis E. Tyler, & R. Holt, prostant apud Guliel. Wells & Rob. Scot, 1672. [another edition] Ludguni Batavorum, Apud Felicem Lopez, 1672.

De Scorbuto was dedicated to George Ent, then President of the College of Physicians. The author praised Ent for his precision, erudition, honesty and kindness.

In the preface 'To the learned reader', Charleton welcomed the censure of his readers, and declared himself not addicted to any ideas which might be discovered to be false.[12]

The text followed a specific disease through its causes, signs, symptoms, and progress. It encompassed the names by which scurvy was known; the varieties of scurvy; remote and external as well as near and 'contiguous' causes; theories of the origin of scurvy in rancid blood and in fixed salt; the diagnostic indicators and Prognostic Signs to be observed by doctors looking after 'scorbutick' patients; general therapy and the therapy of scurvy caused by rancid blood, fixed salt and acid; cure of severe symptoms. Finally, the text contained a diatribe against quacks.

The work was perhaps composed in response to Thomas Willis's 1667 publication on scurvy in *Pathologia Cerebri & Nervosi Generis Specimina, in quo agitur de morbis convulsivis et de Scorbuto*. Charleton debated the relative significance of various chemical elements, and offered disputations of the

[9] Charleton, 'To the Reader', *Deliramenta Catarrhi*, p. ii.
[10] Charleton, *Deliramenta Catarrhi*, p. 3.
[11] Charleton, 'To the Reader', *Deliramenta*, p. v.
[12] Charleton, *De Scorbuto*, p. vi.

philosophical principles of Willis's text. *De Scorbuto* did not draw from patient reports or experimental procedures.

Its content indicates that *De Scorbuto*'s intended audience was composed of physicians. Its subject matter was circumscribed by an established tradition of how physicians' knowledge should be presented. Considered in the context of contemporary works on the subject of specific diseases, its presentation was formulaic.

De Vita et Rebus Gestis *Nobilissimi Illustrissimique Principis Guilielmi Ducis Novocastrensis commentarii . . . ex Anglico in Latinum conversi. (Appendicula continens paucula Auctoris observata*, Londini, Excudebat T. M., 1668.

The British Library's copy contains manuscript notes in Charleton's own hand.

This was a translation into Latin for the European market of Margaret Cavendish's biography of her husband, the Duke of Newcastle. Her book was entitled *The Life of William Cavendish, Duke, Marquis and Earl of Newcastle, Earl of Ogle, Bothal, and Hepple, &c.*, London, 1667. The English edition was reprinted in 1675.

Margaret Cavendish was both a friend and patron, and Charleton's act of translation would have been a means of further ingratiating himself. Letters between the physician and this intriguing woman exist in Charleton's collected manuscripts.[13] Charleton facilitated the Duchess's visit to the Royal Society in 1667.

[A Brief] ***Discourse Concerning the Different Wits of Men*** *written at the request of a gentleman, eminent in virtue, learning, fortune, in the year 1664*, London, R. W. for William Whitwood, 1669 (published anonymously).

Two discourses I. Concerning the different wits of men: II. Of the mysterie of vintners, second edition enlarged, London, F. L. for William Whitwood, 1675.

Two discourses: The first, concerning the different wits of men: The second, a brief discourse concerning the various sicknesses of wines, and their respective remedies at this day commonly used, delivered to the Royal Society; to which is added, in this third edition, The art and mystery of vintners and wine-coopers, London, Printed for Will. Whitwood, 1692.

This text offered an explanation why 'some have more wit than others'. It covered the names by which the 'wit' had been known, and explored the causes and nature of its variety. The discourse was strongly influenced by Hobbes.[14] Charleton's preface, to an anonymous patron, complained of the great complexity and difficulty of his subject.

[13] Bodleian Library, MS Smith 13 (see my manuscript reference at end of Appendix).

[14] See Thorpe, 'Two disciples of Hobbes', *passim*.

Merrett contributed the final section, *Some Observations Concerning the Ordering of Wines*, to the later editions.

Enquiries into Human Nature, *in VI Anatomic Prælections in the New Theatre of the Royal Colledge of Physicians in London*, London, Printed by M. White, for Robert Boulter, 1680. The Cambridge University Library copy has MS additions by the author.

[*another edition*] London, Printed for J. Conyers, 1697.

A portion of the original manuscript of this work is held at the British Library.[15]

The lectures were delivered at the new Cutlerian anatomy theatre at the College of Physicians (an illustration of which prefaced the work). The theatre had been the gift of Sir John Cutler, and *Enquiries* contained a dedication 'To the Right Worshipfull Sr John Cutler, Knight and Baronet'. Charleton lauded Cutler as having greatly encouraged the 'Art of Dissection'.[16]

The six lectures which constituted the basis of *Enquiries into Human Nature* were delivered in March 1679. The first, second and third lectures were given on a Friday. All discussed the anatomy and actions of the stomach. The audience had an intermission for a meal between the dissection of the stomach and gullet and the explanation of the action and use of the stomach! The fourth was presented the next day, the fifth and sixth the following week.[17]

Enquiries was structured around six sections (the prælections): 'On nutrition'; 'Historia ventriculi'; 'The actions and uses of the ventricle'; 'Of life'; 'Of fevers' and 'Of motion voluntary'. The lectures examined the traditional vital functions in anatomical terms (nutrition, life and voluntary motion), with additional lectures on fevers and anatomy of the stomach.

Though the material overlapped with the 1659 *Natural History*, these 1680 lectures were distinguished by their examination of anatomical knowledge of each subject, and their structural, rather than physiological, emphasis.

Enquiries drew much of its authority from contemporary texts, including those of Francis Glisson, Francis Bacon, William Harvey and Jacob Müller. Charleton did not refer to his own experimental experience and observations.

The Ephesian Matron, London, Printed for Henry Herringman, 1659.

Matrona Ephesia. Sive Lusus serius de amore . . . Anglicè conscriptus, et nunc demum Latinitate donatus à Barth. Harrisio [B. Harris], Londini, 1665.

[15] MS Sloane 1059 (see my manuscript reference at end of Appendix).

[16] Charleton, 'Preface', *Enquiries*, p. ii.

[17] Charleton, *Enquiries*, pp. 384 and 429.

The Ephesian and Cimmerian matrons: Two notable examples of the power of love and wit, London, Printed for Henry Herringman, 1668.
A facsimile of the 1668 edition has been published, introduced by Achsah Guibbory, Los Angeles, William Andrews Clark Memorial Library, 1975.
This text, published anonymously, was evidently popular as it went into a second edition in 1668 and was translated into Latin in 1665.
The book presented the ancient tale of 'The widow of Ephesus', recounted by Eumopolus in the *Satyricon* of Petronius. Charleton claimed to have read the story in Greek, Latin, German and French.
The 1668 edition included a 'Letter Concerning the Ephesian Matron: To a Person of Honour', in which Charleton described himself as having set forth his subject 'according to the fashion of my own phansie, which is most delighted with sad colours, and plain useful garments; so that she may now seem the Mistress rather of a Philosopher than of a courtier'.[18]
The storyline is roughly as follows: a young widow, grieving at her husband's grave, is encountered by a young soldier, whose persuasions encourage her to forget her loss and enter his embrace. While distracted, the body that the soldier had been sent to guard is stolen. The widow offers to replace it with her husband's body, so that the theft will not be discovered.[19]
The exchange requires that her dead husband's body be mutilated to resemble that of the dead soldier.
Charleton added to the tale some emphases of his own, including the Epicurean argument that man must 'rise above the bestial nature of the multitude through the rational control of his passions.'[20]

Epicurus's Morals: *collected partly out of his owne Greek text in Diogenes Laertius and partly out of Marcus Antoninus, Plutarch, Cicero & Seneca, and faithfully Englished*, London, Printed by W. Wilson for Henry Herringman, 1656.
[*another edition*] London, H. Herringman, 1670.
Epicurus's Morals Collected and Faithfully Englished [facsimile of the 1670 edition] with an 'Introduction' by Frederic Manning, London, Peter Davies, 1926.
Epicurus' Morals was dedicated to 'A Person of Honour'. The copy of *Epicurus* in the Bodleian Library contains a letter to Fauconberg, and Sharp suggests that the person to whom the book is dedicated is Fauconberg himself.[21]
Published initially without Charleton's name on the cover, the book was prefaced by an apology for Epicurus, signed by Charleton.

[18] Charleton, *Ephesian Matron*, p. iii.
[19] For a detailed analysis of the work, see Achsah Guibbory's 'Introduction' to *The Ephesian Matron*, Los Angeles, William Andrews Clark Memorial Library, 1975.
[20] Guibbory, 'Introduction', *Ephesian Matron*, p. ii.
[21] Sharp, 'Early Life', p. 332.

Epicurus' Morals was clearly popular as it went into several editions.
Charleton offered 'An Apologie for Epicurus, As to the three Capitall Crimes whereof he is accused.' These were '(1) That the souls of men are mortall . . . (2) That man is not obliged to honour, revere, and worship God, in respect of his beneficence, or out of the hope of any Good or feare of any evill at his hands, but meerly in respect of the transcendent Excellencies of his Nature, Immortality, and Beatitude. (3) That Selfe-homicide is an Act of Heroick Fortitude in case of intollerable or otherwise inevitable Calamity.'[22]
Divided into thirty-one chapters, the book consists of material from various sources, including Epicurus, Diogenes Laertius, Plutarch, Cicero, and Seneca.[23]
The physician defended the piety and worth of Epicurus's ideas, and argued for the continued relevance and value of his doctrines. He excused Epicurus's materialism on the basis of his pagan ignorance.

Exercitationes Pathologicae, *in quibus morborum penè omnium natura, generatio, & caussae, ex novis anatomicorum inventis sedulo inquiruntur a Gualtero Charltono, M.D. & Caroli I. olim, hodie Caroli II. Magnae Britanniae Monarcharum inclytissimorum medico ordinario*, Londini, apud Tho. Newcomb, 28 January 1661. The British Library copy contains manuscript notes in the author's hand.
[*another edition*] Bononiae, Sumptibus Petronii de Ruinettis, 1675.
The full title can be translated as 'Pathological Dissertations, in which the nature, generation, and cause of almost all diseases are most diligently set forth'.
The book was dedicated to Charles II. It preceded closely Charleton's publication of *An Imperfect Pourtraicture of His Sacred Majesty Charls the II* (London, 1661).
Exercitationes pathologicae was presented to the Royal Society on 13 June 1661. It seems to have been composed before Charleton's entry into the Society.
An introduction to the study of pathological 'Physick', the text examined the nature, generation and causes of most known diseases, providing a nosology, a summary of contemporary arguments, and progressing through the origins, signs and causes of diseases.
Exercitationes did not record clinical details. It was a theoretical work, which encompassed theories such as that hatred was the cause of both leprosy and gout. Charleton's notes from his Oxford days, dated 1642, may have constituted the basis for the volume.[24]

[22] Charleton, 'An apologie for Epicurus', *Epicurus' Morals*, pp. v-vi.
[23] Manning identifies Charleton's reliance upon the Tenth Book of *The Lives and Opinions of the Philosophers*, by Diogenes Laertius. Manning, 'Introduction', p. xvii.
[24] British Library, MS Sloane 3412 (see my manuscript reference at end of Appendix).

The Harmony of Natural and Positive Divine Laws, London, Printed for Walter Kettilby, 1682.

Published anonymously. This work was prefaced by a letter from 'The Publisher to the Reader', in which the author's anonymity was depicted as a defence against immoderate detraction.[25]

Harmony argued that man, as a rational agent within the divine design, was destined to rediscover the principles by which his creation had been effected. The precepts of living which could be discovered through the exercise of reason were 'the very same that are promulgated by the Divine Majesty for the laws of the Kingdom of Heaven'.[26]

'By nature all wise men understand the order, method and economy instituted and established by God from the beginning of creation for government and conservation of the world.' All the laws of nature were therefore the laws of God, and 'that which is called "natural" and "moral" is also "divine" law, as well because reason, which is the very law of nature, is given by God to every man for the rule of his actions'.[27]

Charleton outlined 'Right and law in general', and then explored a series of instances in which natural and divine law could be seen to be identical (including theft, rapine and homicide).

The Immortality of the Human Soul, *Demonstrated by the Light of Nature. In two dialogues*, London, Printed by William Wilson for Henry Herringman, 1657.

[*another edition*], London, Printed for Richard Wellington . . . and Edmund Rumbold, 1699.

A facsimile of 1657 edition has been published, edited & introduced by J. M. Armistead, New York, AMS Press, 1985.

The dedication to *Immortality* thanked Charleton's patron, Henry Pierrepont, Marquis of Dorchester, profusely, declaring that 'from you alone I have received more both of Encouragement and Assistance in my studies, than from the whole World besides.'[28]

Immortality consisted of two dialogues between three speakers, in which proper conduct, scientific progress and philosophical method are discussed. The speakers are Lucretius (Evelyn), Athanasius (Charleton) and Isodicastes (Henry Pierrepont).[29] It has been suggested that the format of the work was derived from Digby's 1644 *Two Treatises*.[30]

25 In the preface to the anonymous *Harmony of Natural and Divine Positive Laws*, the publisher (Walter Kettilby) depicted the author writing 'to no other end, but to confirm his Faith by inquiring into the Reasonableness and Purity of it, and to augment his Piety toward God.' His posture was 'worthy a Philosopher and a Christian' as, surely, was his publication.

26 Charleton, *Harmony*, pp. 8-9.

27 Charleton, *Harmony*, pp. 8-9.

28 Charleton, 'Dedication', *Immortality*.

29 He, with Evelyn, Digby and Hobbes, experienced a period of exile in Paris during England's civil strife. J. M. Armistead, 'Introduction', p. viii.

30 Armistead, 'Introduction', p. viii.

Armistead notes that while the most revered thinkers in the body of the text were Epicurus, Bacon, Descartes, and Digby, *Immortality* contained also scattered references to Hermes Trismegistus, Plato, Aristotle, Cicero, and the Church Fathers.[31] As such the text 'exemplifies and endorses' Charleton eclectic vision.

Immortality depicted the College of Physicians as an exemplar of Bacon's 'Solomon's House'.[32]

An Imperfect Pourtraicture of His Sacred Majesty Charls the II. *By the grace of God King of Great Britain, France, and Ireland, defender of the faith, &c. Written by a loyal subject, who most religiously affirms, se non diversas spes, sed incolumitatem Caesaris simpliciter spectare*, London, printed for Henry Herringman, at the sign of the Anchor in the Lower Walk of the New-Exchange, 1661 (published with *Consilium Hygiasticum*).[33]

[reissue] *A character of His Most Sacred Majesty, Charles the Second, King of Great Britain, France, and Ireland, defender of the faith, &c*, London, Printed for Henry Herringman . . . , 1661.

This 'character' of the newly restored monarch depicted religion as one of the greatest concerns of the dissolute Charles II. Charleton's keenness to gain the approval of the new king seems to have been successful, as in July 1660 he was appointed Physician in Ordinary to the restored monarch, with a salary of £100 a year.[34] This represented a substantial improvement in Charleton's fortunes.

Inquisitiones Duæ anatomico-physicae; *Prior de fulmine, Altera de proprietatibus cerebri humani*, Londini, Typis societati Regali infervientibus, impensis vero Octaviani Pulleyn junioris, 1665.

The Royal Society's Imprimatur appeared on the verso page. The work was dated 21 December 1664. It was dedicated to Viscount Brouncker, President of the Royal Society, and patron of Charleton. The two fell out dramatically soon after this publication.

Inquisitiones Duæ consisted of two discourses, the first of which outlined the nature and effects of thunder and lightning. Charleton dismissed vulgar opinions about death being caused by thunder-bolts, and included some additional observations perhaps based on his experiences at the Royal Society. The disquisition was doubtless informed by the talk he gave at the Society on dissection of a boy killed by lightning.[35]

31 Armistead, 'Introduction', p. vii.

32 For this reason the text has been analysed by Charles Webster as an indication of the College's experimental activities. See Webster, 'Solomon's House', *passim*.

33 *Consilium Hygiasticum, pro illustriss. excellentissq. heroë, Dno. Johanne Luca, Marchione Durazzo, extraordinario serenissimæ reip. Genuensis oratore, ad augustissimum Carolum II*, London, 1661.

34 *Calendar of State Papers, Domestic, 1660-61*, pp. 134, 208.

35 A report of the dissection is held in the Royal Society archives (see my manuscript reference at end of this appendix).

The second discourse was a response to Willis's anatomy of the brain (published early 1664). Again this and would have gained its impetus from a presentation at the Royal Society.[36] The publication of *Inquisitiones Duæ* in Latin was perhaps motivated by a desire to present this material, in an impressive and authoritative manner, to roughly the same expert readership who had witnessed the lecture.

Inquisitio Physica de Causis Catameniorum *& Uteri Rheumatismo in qua en Paeodo Probatur Sanguinem in Animali Fermentescere Nunquam*, London, Impensis Gualt. Kettilby, 1685. Approved by the President (Sir Thomas Witherley), Registrar (Dr Samuel Collins) and Censors (Dr Thomas Burwell, Dr Peter Barwick, Sir Thomas Millington and Dr Humphry Brooke) of the College of Physicians. *Inquisitio* was printed with the College Imprimatur.

Inquisitiones Medico-Physicæ, de Causis Catameniorum sive Fluxus Mensui; nec non Uteri Rheumatismo, sive Fluore Albo, Lugduni Batavorum, Apud Petrum vander Aa, 1686.

The title might be translated as 'A physical dissertation on the causes of Certain Feminine Disorders; and of the Rheumatism in the Womb, in which it is proved, that there is no such thing, as fermentation in the blood'. This, Charleton's last published work, examined the names given to the uterus, both Greek and Latin, and 'their etymology and explanations'. It also explored the genealogy of 'uterine rheumatism'. The author treated the uterus as a 'workshop', in this treatise on the physiology and pathology of menstrual flux, and considered its function.

Charleton rejected ancient theories regarding the causes of Catamenia, but also rejected excessive adherence to the more modern theory of the uterine ferment. He attacked Thomas Willis, among others, for over-reliance upon explanation through fermentation. Among these theorists, he complained, 'There was virtually no function or action integral to an animal that they did not deduce from fermentation'.[37] He seems to have wished at times to invoke the use of ferments in physiological explanation, but objected to their use as *factotum*.

He invoked Boyle's writings on the blood,[38] and explored George Ent's nutritive juice theory.[39]

As this text was published only in Latin, its audience was necessarily restricted to readers who were already familiar with other expert writings in the area. Its appeal would have been to those who were medically-trained themselves.

[36] This was preserved as 'Certain Differences' (see my manuscript reference at end of this Appendix).
[37] Charleton, *Inquisitiones Medico-Physicæ*, p. 55.
[38] Charleton, *Inquisitiones Medico-Physicæ*, p. 57.
[39] Charleton, *Inquisitiones Medico-Physicæ*, pp. 72-3.

'The Life of Marcellus, translated from the Greek', published in *Plutarch's Lives*, ed. John Dryden, London, 1684, vol. 2, pp. 401-68.

Another of Charleton's Greek translations, possibly performed during his stay in Cheshire in the 1670s.

Charleton's 'Life of Marcellus' was a straightforward translation, without commentary.

Natural History of Nutrition*, Life, and Voluntary Motion containing all the New Discoveries of Anatomist's and most Probable Opinions of Physicians, concerning the Oeconomie of Human Nature: Methodically delivered in exercitations physico-anatomical*, London, Printed for Henry Herringman, and are to be sold at his shop, 1659.

Oeconomia animalis, Oeconomia superstructa et mechanice explicata (Acc. ejusdem diss. epist. de ortu animæ humanæ, Londini, Typis R. Danielis, & J. Redmanni, 1659.

Exercitationes physico-anatomicæ de oeconomia animali, novis in medicina hypothesibus superstructa, & mechanice explicata, Editio secunda, Amsterdam, Joannem Ravesteynium, 1659.

Gualteri Charletoni Oeconomia animalis novis in medicina hypothesibus superstructa & mechanice explicata accessere ejusdem dissertatio epistolica de ortu animae humanæ; & Consilium hygiasticum, Editio tertia, Londini, Ex officina Rogeri Danielis, 1666.

Gualteri Charletoni Oeconomia animalis novis in medicina hypothesibus superstructa & mechanice explicata . . . , Editio quarta, London, Ex officina Johannis Redmayne, prostant venales apud Johannem Creed: Cantab., 1669.

Exercitationes physico-anatomicæ, de oeconomia animali novis in medicina hypothesibus superstructa, & mechanice explicata, Lugd. Batav, Peter de Graef & Jacob Moukee, 1678.

Exercitationes de oeconomia animalis in medicina hypothesibus superstructa, & mechanice explicata. Quibus accessere Guilielmi Cole . . . De secretione animali cogitata, ad hanc oeconomiam praeipue spectantia. Ed novissima, prioribus emendatior & correctior, Hagae-Comitum, Arnold Leers, 1681.[40]

Both English and Latin editions of *Natural History* were dedicated to Viscount Fauconberg and Sir George Ent. Thomas Belayse, the second Viscount Fauconberg (1627-1700) was one of Charleton's patrons.[41]

[40] This edition incorporated a discussion of William Cole's physiological text *De secretione animali cogitata*, Oxford, 1674.

[41] The dedication portrays Fauconberg as a 'Grand Exemplar' of truth and reason, whose approval constituted verification of fact: 'whatever position hath once received the stamp of your Assent and Approbation' will be accepted as 'Authentique and Current', by even 'the most scrupulous'. *Natural History* p. ii. Fauconberg was a prominent supporter of Cromwell, but converted to royalism at the Restoration, and rose rapidly to privileged status.

There is some evidence to suggest that *Natural History* was composed in the early 1650s, but that printing was delayed.[42] The Epistle to Ent is dated 12 June 1653, and the text was allegedly written a year prior to that. However, the volume was not published until 1659.

Natural History examined the classical triad of faculties (natural, vital, animal) through their key processes: nutrition, vitality and locomotion.[43] It divided the natural faculty into nutrition, generation, digestion and growth. Vital faculties were examined through the subdivisions of the journey of the chyle, the heart's action, depuration of the blood, circulation and respiration. The animal faculties, generation and action of the animal spirits were considered through reference to the issue of locomotion.[44]

The 1659 volume in English corresponded in its contents with the Latin editions of 1659 and 1669, except for the omission from the English edition of a number of acknowledgements of works cited. This made the English edition slightly shorter, and gave it the impression, as Nayler notes, of a more original piece of scholarship. The English version omitted some direct references to authors, along with various quotations.[45] This difference between English and Latin versions might result from a divergence in the expectations of the intended audiences of the two works, or from censure over plagiarism in certain circles.

Natural History of the Passions, London, Printed by T. N. for James Magnes, 1674.

This work was for a long time incorrectly assumed to be a translation from Senault's *De l'usage des passions* (Paris, 1641).[46]

The dedication, to an unnamed person of honour, states that *Passions* was 'the product of my late ten weeks solitude in the Country. Where being remote from my Library, and wanting conversation with learned Men; I knew not how more innocently to shorten the winter evenings, than by

[42] It is difficult to ascertain exactly why the text might have been delayed. The publication dates of the texts which Charleton cited do not illuminate the matter, as relevant material was often circulated in manuscript, and/or presented in the form of lectures, and publication often came after the relevant transfer of ideas. The research basis of Glisson's *Anatomia hepatis* (published in 1654) was delivered in the Gulstonian lectures of 1641. Wharton's research on the glands (published as *Adenographia* in 1656) was presented at the College of Physicians in 1652. Charleton may thus have encountered each of these prior to their publication. Titles and details of the eventual publications could have been added to the completed *Natural History* just prior to its printing.

[43] Luyendijk-Elshout, 'Oeconomia animalis', p. 299.

[44] Luyendijk-Elshout, 'Oeconomia animalis', *passim*.

[45] This difference is noted only by Nayler, 'Insoluble Problem', pp. 99-100. Sabina Fleitmann argues that the 1659 *Oeconomia* was more overtly mechanistic than the English edition. Fleitmann, pp. 203 and 369. Webster notes that the English edition is a 'slightly abbreviated form of the Latin'. Webster, *Great Instauration*, p. 272.

[46] Hunter & Cuttler, 'Mistaken identity'.

spending them in revising some Philosophical papers of my own.'[47] It was motivated 'because my accumulated Misfortunes had . . . reduced me to a necessity of consulting that part of Philosophy, about the most effectual Remedies against Discontent.'[48] It appears he composed the book in his retreat at Cheshire.

The texts he had with him appear to have included the works of 'those three excellent men, Gassendus, Des Cartes, and our Mr Hobbes', as well as Digby's *Two Treatises*. Charleton also had some notes he had previously made 'out of the best Authors'.

The book was composed by 'reading and meditation'. Reading first, then 'Meditation, that I might examin [sic] the weight of what I read, by comparing it with what I had daily observed within the theatre of my own breast'.[49]

Natural History of the Passions delineated the interaction of body and soul, focusing on the conflict between reason and emotion.

Charleton posited that 'all the Good and Evil of this life depends upon the various Passions incident to the Mind of man', the conclusiveness of which he claimed to know from 'my own dearly bought experience'. Like afflictions of the body, those of the mind might be more easily cured through understanding of their 'nature, causes, motions [effects] &c.' Charleton's treatise, in explicating these features of the passions, aimed to discover possible 'Remedies against their Excesses.'

Passions exemplified the practising physician's concern to provide learned counsel. The principle of self-knowledge as a key to health, prominent in this text, referred to an essentially classical definition of self-discipline toward achievement of well-being and the prolongation of life.

The volume concluded with a recommendation for 'the *Ethicks* of *Epicurus*' as second only to the Holy writ as a source of cures for 'all distempers, incident to the mind of Man.'[50] While the *Passions* made reference to the physiological and material aspects of the passions, innovations deriving from the rise of the corpuscular philosophy, its emphasis was on the importance and nature of the passions in accordance with traditional physic.

Onomasticon Zoicon *plerorumque animalium differentias & nomina propria pluribus linguis exponens: cui accedunt mantissa anatomica: et quaedam de variis fossilivm generibus*, Londini, Apud Jacobum Allestry Regalis Societatis typographum, 1668.

[47] Charleton, *Natural History of the Passions*, p. i-ii.
[48] Charleton, *Natural History of the Passions*, p. ii.
[49] Charleton, *Natural History of the Passions*, p. vi.
[50] The second edition of *Epicurus' Morals* had been printed in 1670, and Hunter & Cuttler suggest that this was almost certainly an added promotion for it, 'Mistaken Identity', p. 91.

Onomastikon zoikon continens plerorumque animalium quadrupedum, serpentium, insectorum, avium & piscium differentias, eorumque nomina propria diversis linguis exposita, cui accedunt mantissa anatomica, et nonnulla de variis fossilium generibus, cum figuris, Londini, Apud Jacobum Allestry, 1671.
Onomastikon zoikon, Gualteri Charletoni exercitationes de differentiis & nominibus animalium quibus accedunt mantissa anatomica, et quaedam de variis fossilium generibus, deque differentiis & nominibus colorum, Editio secunda, duplo fere auctior priori, novisque iconibus ornata, Oxoniae, e Theatro Sheldoniano, 1677.
The British Library's copy of the 1668 edition contains manuscript notes by the author.
An anatomical compendium, which listed the names of all known animals in several languages, and offered descriptions of their habits and habitats where known.
Onomasticon also contained an account of Charles II's menagerie at St James's Park, an anatomical appendix, and some observations on the varieties of fossils.[51]
The keen interest in categorisation illustrated by the *Onomasticon* is evident in various of Charleton's projects throughout his years of Royal Society attendance.[52]
This work contained some record of the author's own observations.

***Oratio Anniversaria** habita in theatro inclyti Collegii Medicorum Londinensium 5 Augusti, Anno Domini 1680*, Londini, Sumptibus Joannis Baker, 1680.
This oration was, as its title suggests, delivered in the Theatre of the Royal College of Physicians at London. The Harveian Oration was an annual feast, endowed by William Harvey in 1656. At this occasion benefactors were praised and fellows of the College of Physicians were exhorted 'to search and studdy [sic] out the secrett [sic] of Nature by way of Experiment.'[53]
The Harveian Oration was thrice delivered by Charleton: on 5 August 1680, in 1702 and on 16 August 1705.[54] The standard presentation, at which Charleton excelled, was the dedication of praise to his friend Harvey, and laudatory remarks concerning the College's activities. Though there is no record of his 1702 'Oration', his 1705 presentation was published as: *Oratio*

[51] See also W. H. Mullens, 'Walter Charleton and his *Onomasticon Zoicon*', *British Birds*, vol. 5, 1911, pp. 64-71.
[52] The preservation and classification of flora were also of interest to Charleton. In March 1666 he was elected to a committee with Wilkins, Cornwall, Hill, Evelyn, Goddard, Cock, Hooke, Harrington and Graunt, to assess the preservation and taxonomy of the Society's inventory of plant specimens. Birch, *History*, vol. 2, p. 73.
[53] Frank, *Oxford Physiologists*, p. 25
[54] Though Munk notes that the 1704 oration was published, I have not been able to find a copy in print. Munk, *Roll*, p. 391.

anniversaria, recitata in Theatro anatomico inclyti Medicorum Lond. Collegii, in commemorationem beneficorum a Dno Dre Harveo, aliisq; munificis vivis . . . eidem Collegio praestitorum . . . 16 die Augusti, An. Dom. 1705, London, 1705.

Physiologia *Epicuro-Gassendo-Charltoniana, or, A Fabrick of Science Natural, upon the hypothesis of atoms founded by Epicurus, repaired [by] Petrus Gassendus, augmented [by] Walter Charleton*, London, Printed by Tho. Newcomb for Thomas Heath, 1654.

A facsimile edition has been printed, introduced by Robert Kargon, London, Johnson Reprint Co., 1966.

Dedicated to Mrs Elizabeth Villiers, wife to Sir Robert Villiers, and Charleton's host at the time of the text's composition.[55]

This work was effectively a translation into English and commentary upon Pierre Gassendi's *Animadversiones in decimum librum Diogenes Laertii* (1649). It was the first systematic presentation of Gassendian views in the vernacular.

Physiologia is seen by many to signal Charleton's complete rejection of Helmontianism. However, while his recantation of Helmont is interesting, it is certainly not evidence of a comprehensive shift in his philosophy.[56]

Physiologia claimed the primacy of non-purposive, physical factors in guiding the action of the internal body. Actions were explicable through reference to function, rather than sentience. Invisible processes, he argued, were not immaterial, but were rather effected by 'Corporeal, though both impalpable and invisible Organs.'[57]

The operations of these organs were deduced, through analogy, from actions in the corporeal sphere. Man could explore the uncertainties of the invisible through his reasoned knowledge of the visible.[58] This echo of Cartesian notions of the similarity between invisible and visible entities illustrates Descartes' influence on Charleton's thought in these years.

[55] Elizabeth Villiers was the daughter of a prominent parliamentarian. Sharp suggests that Charleton's choice of her as a patron might have been politically astute, as protection from a prestigious person of quiescent status would have advanced his career. Sharp, 'Early Life', pp. 331-3.

[56] This perspective on Charleton's apparent reversal is shared by Lewis, 'Early modern eclecticism', p. 664.

[57] Charleton, *Physiologia*, p. 346.

[58] 'The means used in every common and sensible Attraction and Complection of one Bodie by another, every man observes to be Hooks, Lines, or some such intermediate Instrument contained from the Attrahent to the Attracted; and in every Repulsion or Disjunction of one Bodie from another, there is used some Pole, lever, or other Organ intercedent, or somewhat exploded or discharged from the Impellent to the Impulsed.' Charleton, *Physiologia*, p. 344.

***Plato, His apology of Socrates**, and Phædo, or Dialogue concerning the Immortality of Mans Soul, and Manner of Socrates his Death; carefully translated out of the Greek, and illustrated by Reflections Upon both the Athenian Laws, and Ancient Rites and Traditions concerning the Soul, therein mentioned*, London, Published anonymously, in 1675.

This work was dedicated to Arthur Annesley, the Earl of Anglesey.[59]

These and his other translations signal Charleton's continued interest in widening the audience for ancient scholarly works, and his thorough knowledge of Greek.

Plato, His Apology was published after Charleton's involvement at the Royal Society, and the translation was probably completed during his stay at Crewe Hall in Cheshire.

***Spiritus Gorgonicus**, vi sua saxipara exutus; sive De causis, signis &sanatione lithiaseως, diatriba*, Ludg. Batav. Ex Officina Elseviriorum, 1650.

The title might be translated as 'The Gorgonic Spirit deprived of its stone-producing power, or a discourse of the cause, symptoms and cure of the stone.'

The text discussed 'the universal stone-forming spirit', depicted as the origin of urinary concretions or 'stones' in man, and of rock deposits in the material world.[60]

Spiritus clearly illustrated a strong interest in, and knowledge of, the hermetic tradition. The text covered the traditional subject matter: material and remote causes, symptoms, prognosis and diagnosis, varieties of the affliction, and therapeutic recommendations.

Containing cabbalistic and neo-Platonic material, the text drew reference to hermetic authors, such as Hartmann, Severinus and especially Paracelsus. Jean Baptista van Helmont's 1644 *De Lithiasis*, on the treatise on urinary calculus, was perhaps the single greatest influence.

Charleton's book was referred to by the shortened title *De Lithiasi Diatriba*, among peers.[61]

***A Ternary of Paradoxes**. The Magnetick Cure of Wounds. Nativity of Tartar in Tine. Image of God in Man. Written originally by Joh. Bapt. Van Helmont, and translated, illustrated, and ampliated by Walter Charleton*, London, printed by James Flesher for William Lee, 1650.

[59] Three letters to Annesley, written in October and November 1672, exist in Charleton's miscellaneous papers, Bodleian MS Smith 13, ff. 5-7. It seems he was courting Anglesey as a patron.

[60] The British Library's edition of *Spiritus Gorgonicus* has been bound, in a volume entitled *Medical Tracts 1591-1683,* with the following texts on this and related subjects: two editions of Henricus à Bra, *Medicamentorum simplicium* (1589 and 1591); Christophe Cachet's *Apologia dogmatica in hermetici cuiusdam . . . calculi* (1617); Frischman de Ehrencron, Mattheus Johannes, N. Franchimont a Franckenfeld . . . *Lithotomia Medica* (1683); Martin Pansa, *Consilium Antiphreneticum* (1615).

[61] This was noted by Wood, *Athenæ Oxonienses*, p. 752.

Ternary was dedicated to Viscount Brouncker. The dedicatory epistle of *Ternary* provides biographical material on the relationship between Brouncker and Charleton, which suggests that the two men were close.
This was the first English translation from Latin of Flemish physician and author Johannes (Jean) Van Helmont, whose works were gaining audiences on the Continent.
Charleton depicted his role in *Ternary* as having been 'to clear the prospect, by the necessary remove of such *Doubts*, as seemed very much to obscure the resplendent lustre of *Magnetisme*, and render the Excellencies of *Sympatheticall Remedies* imperceptible'.[62]
He did not explicitly argue against the Helmontian spiritualist perspective, but claimed to leave the illumination of spiritual matters to those writers best equipped to illustrate them.[63] He cited the preceding works of Baptista della Porta, Severinus, Hortmann, Kircher, Cabeus, Rob. de Fluctibus, and Digby as a point of reference for those interested to know more about 'how *Sympathetically magnetick* Agents transmit their *Spirituall Energy*, unto determinate Patients, at vast and intermediate distance'.[64]
Although supportive of the Helmont's discoveries, and of a great many of his cures, his acceptance of the idea of magnetic cure at a distance did not rely upon the existence of an *anima mundi*. Instead Charleton hinted at a mechanical explanation, in which atoms were pivotal.[65] He redescribed the operation of magnetism through an account which excluded the spiritual aspects so crucial to Helmont.

***Three Anatomic Lectures**: Concerning 1. The Motion of the Bloud through the veins and arteries, 2. The organic structure of the heart, 3. The efficient causes of the hearts pulsation : read on the 19, 20 and 21 days of March 1682, in the Anatomic Theatre of His Majesties Royal College of Physicians in London*, London, Printed for Walter Kettilby, 1683.
Thomas Coxe, President of the College of Physicians, ordered the lectures to be printed. Published by Kettilby, they appeared for purchase in the same year.[66]
Three Anatomic Lectures constituted a response to *De Motu Animalium* by Neapolitan anatomist Giovanni Alfonso Borelli (1608-1679).[67] Borelli, a key member of the Accademia del Cimento, had studied muscle physiology, respiration, the nervous system and cardiac motion. *De motu animalium* was published (posthumously) in two parts. The first, published in 1680, dealt with the locomotion of man and animals. The second, printed in

62 Charleton, 'Prolegomena', *Ternary*, p. XVIII.
63 Charleton, 'Prolegomena', *Ternary*, pp. VIII-XIX.
64 Charleton, 'Prolegomena', *Ternary*, pp. XVIII-XIX.
65 Mulligan, 'Right reason', p. 381.
66 Charleton's *Three Anatomic Lectures* was prefaced merely by a Latin 'Præloquium'.
67 Giovanni Alfonso Borelli, *De Motu Animalium*, Rome, 2 vols, 1680 and 1681.

1681, directly influenced Charleton's *Lectures*. It examined the internal motions of animals and their causes. Borelli's text had been published only in Latin. Charleton was thus the first to bring this author's theories into English.

Charleton's lectures covered, respectively, the circulation of the blood, and the effects of circulation, the heart's pulsation, and lastly, the efficient causes of the heart's pulsation.

Manuscript Sources

British Library

MS Sloane 53	'Adversaria Medicinalia' (1680).
MS Sloane 698	ff. 74-87, 'Certain differences observable betwixt the brains of men and of all other animals'. Read before the Royal Society 8 June 1664.
MS Sloane 1059	manuscript copy, in English, of prælections IV-VI, delivered in 1679 at the College of Physicians and subsequently published as *Enquiries into Human Nature* (1680).
MS Sloane 1532	ff. 41-61, 'Epitome libri ejus de scorbuto' (circ 1672). Latin notes on the topic of scurvy. Made prior to the publication of *De Scorbuto*.
MS Sloane 1828	ff. 96b-115b, 'Receptæ variæ' Undated, these notes in Latin cover a various of subjects, including recipes for numerous medical syrups, conversion tables of the times of the day and of the year, in both Latin and English. This set of documents contains a rare record of the actual medical compositions that Charleton used in his medical practice.
MS Sloane 1833	ff. 155-59, 'Oratio Harveiana', a manuscript edition of the Harveian Oration that Charleton delivered at the College of Physicians on 5 August 1680. The Oration was subsequently published (1680).
MS Sloane 2082	ff. 1-73, 'De omnibus symptomatibus'; ff. 74-81, 'Tables of materia medica'. Dated 1642-3, these notes in Latin cover a range of medical subjects, and examine the virtues of various medicaments. They seem to have been made during Charleton's education at Oxford.
MS Sloane 3412	ff. 2-102, 'Methodus febrium cuandarum'; ff. 103-28, 'De morbis'. Both dated around 1643. These Latin notes also seem to have been notes made during Charleton's time at

Oxford. The notes on fevers cover the names and methods by which fevers have been understood and cured across a range of Greek, Latin and Arabic scholarship. Charleton cites a large number of ancient and recent scholars on the subject of fevers.

MS Sloane 3413 — 'Loci communes' (Charleton's commonplace book). Composed in a variety of Greek, Latin English and French, it is extraordinarily hard to read. The volume contains excerpts from the writings of various authors, including Tacitus, Lucian, Demnocritus, Palladius, Possidonius, etc, along with quotes from Hippocrates; excerpts from the work of Reinier de Graaf on reproduction, and from Bernard Swabe on the pancreas; a catalogue of the contents of Thomas Browne's personal museum and gallery; notes on the differences between colours (on which experiments had been performed at the Royal Society during Charleton's involvement); Andrew Marvell's poem on Colonel Blood, in Latin; and 'de arborum natura', a taxonomic article on trees.

Bodleian Library

MS Aubrey 11 — 'Letter to John Aubrey', in which Charleton complained to Aubrey about his falling out with Brouncker. A copy is held in the British Library, MS Egerton 2231, f. 166.

MS Rawl D 257 — Abridgement of Charleton's *Physiologia* (1654).

MS Smith 13 — miscellaneous papers of Walter Charleton. The bound volume contains a total of eighty-nine documents, in both English and Latin. There are too many documents to list individually, but some notable inclusions are the letters by Charleton, including epistles to George Ent, Margaret Cavendish, Richard Lower, Daniel Elzevir, Paulus Sarotti (at the University of Padua) and a number of unidentified individuals. The volume contains short tracts by Charleton on subjects including the origin of Aristotelian teaching in European universities, 'human felicity and infelicity', reflections on the Talmud, 'love refined', and some notes on Saturn, Jupiter and Pluto. There are several epigraphs, written by Charleton in the 1670s regarding friends and mentors. This is a significant collection of notes and observations by Charleton.

MS Wood F 40 — ff. 360-1, 'Letter to a friend of Wood', dated 20 January 1671. In this Charleton outlined his living arrangements in Cheshire at Crewe Hall.

Library of the Royal Society

Not all of the investigations listed in Birch's *History* have been filed. Below is a record of those that I was able to find. Most were contained in 'Early Letters and Classified Papers, 1660-1740' (microfilm).

13 August 1662	'Account of a boy killed by lightning'. Classified Papers, IV (1) 3.
17 September 1661	'Apparatus Phonocampticus, or what enquiries are principally to be made by such, who would attain to the certain knowledge of the nature of echos'. Filed in the Royal Society Register Book, vol. 1, p. 197 (a more legible copy exists in Charleton's papers, Bodleian MS Smith 13).
10 September 1662	'Echoes'. Classified Papers, II 35.
8 July 1663	'Plan and description of the circles of Avebury, near Marlburgh, Wiltshire'. Classified Papers, XVI 18.
13 January 1664	'Experiments concerning freezing and snow'. Classified Papers, IV (1) 10.
8 June 1664	'Observations on the differences between the brains of men and brutes'. Register Book.
15 February 1665	'Merganser versus'. Classified Papers, XX (1) 2.

Library of the Royal College of Physicians of London

Manuscript collection: Medical recipe book, circa. 1667. This volume is believed to have been Walter Charleton's. It contains recipes for syrups and medical potions, and reveals the author's methods of clinical treatment to have been determinedly Galenic.

PRIMARY SOURCES

Aubrey, John, *Brief Lives, Chiefly of Contemporaries, set down by John Aubrey between the years 1669 and 1696*. ed. Andrew Clark, Oxford, Clarendon Press, 2 vols, 1898.

Bacon, Francis, *The Works of Francis Bacon*, eds J. Spedding, R. Leslie Ellis and D. Denon Heath, 7 vols, London, Longmans, 1857-61.

Barbati, Hieronimi, *Dissertatio Elegantissima de Sanguine Et Eivs Sero*, Parisiis, Apud Robertum de Ninville, sun signo Scuti Francicæ & Navarre, in Quadrivio poutis S. Michaelis, 1667 (the text was edited by Walter Charleton, and contained a section on his and Wharton's exposition of the muscles).

Barbette, Paul, *The chirurgical and anatomical works of Paul Barbette, M.D. . . . composed according to the circulation of the blood and other new inventions of the moderns: together with a treatise of the plague, illustrated with observations*, London, J. Darby for Moses Pitt, 1672.

Bartholin, Thomas, *Bartholinus Anatomy Made from the precepts of his father, And from the Observations of all Modern Anatomists, together with his own*, trans. Nicholas Culpeper, London, John Streater, 1668.

Bartholin, Thomas, *Vasa lymphatica, nuper Hafniæ in animantibus inventa, et hepatis exsequiæ*, Parisiis, Apud viduam M. du Puis, 1653.

Beddevole, Dominique, *Essays of Anatomy: in which the construction of the organs and their mechanical operations are clearly explained*, Edinburgh, George Mosman, 1691.

Biggs, Noah, *Mataeotechnia medicinae praxeus. The vanity of the craft of physick. Or, a new dispensatory. Wherein is dissected the errors, ignorance, impostures and supinities of the schools in their main pillars of purges, blood-letting, fontanels or issues, and diet, &c*, London, 1650.

Birch, Thomas, *The History of the Royal Society of London for the Improving of Natural Knowledge, from its First Rise*, 4 vols, London, 1756-57, New York, Johnson Reprint Co., 1968.

Blottesandaeus, Benedict [i.e. Olaus Borrichius.], *Deusingius Heautontimorumenos, sive epistolæ selectæ eruditorum, quæ immaturis Antonii Deusingii . . . scriptis larvam strictim, sed sincerè detrahunt, et clarissimi nominis viros . . . à supercilio, & censurâ ejusdem* . . ., Hamburgi, Typis N. Molybdii, 1661 (discusses the dispute between Walter Charleton and Antony Deusing).

Bohn, Johannes, *Circulus anatomico-physiologicus, seu Œconomia corporis animalis, etc*, Lipsiæ, sumtibus Joh. Friedrich Gleditsch, typis Christophori Fleischeri, 1686.

Bontekoe, Cornelis, *Metaphysica, et liber singularis de motu, nec non ejusdem Oeconomia Animalis*, Lugduni Batavorum, J. de Vivié & F. Haaring, 1688.

Borelli, Giovanni Alfonso, *De motu animalium (1680-81): On the Movement of Animals*, ed. and transl. Paul Maquet, Berlin, Springer-Verlag, 1989.

Boyle, Robert, 'The Christian Virtuoso: Appendix to the First Part, and the Second Part' (1744), in *The Works of the Honourable Robert Boyle*, ed. T. Birch, London, J. & F. Rivington, 1772, vol. VI, pp. 673-796.

Boyle, Robert, 'A disquisition about the final causes of natural things: wherein it is inquir'd, whether, and (if at all) with what cautions a naturalist should admit them?' (1688), *Works*, vol. 5, pp. 392-444.

Boyle, Robert, *Certain physiological essays and other tracts; wherein some of the tracts are enlarged by experiments and the work is increased by the addition of a discourse about the absolute rest in bodies*, Second edition, London, Henry Herringman, 1669.

Boyle, Robert, 'Memoirs for the natural history of human blood' (1684), *Works*, vol. 4, pp. 595-759.

Boyle, Robert, *New experiments and observations touching cold, or, An experimental history of cold begun: to which are added an examen of antiperistasis and an examen of Mr. Hobs's doctrine about cold / by the Honorable Robert Boyle . . .; whereunto is annexed An account of freezing, brought in to the Royal Society by the learned Dr. C. Merret* . . ., London, Printed for John Crook, 1665.

Boyle, Robert, 'Of the reconcileableness of specifick medicines to the corpuscular philosophy: to which is annexed a discourse about the advantages of the use of simple medicines' (1685), *Works*, vol. 5, pp. 74-129.

Boyle, Robert, 'Some considerations touching the Usefulnesse of experimental naturall philosophy' (1663), *Works*, vol. 2, pp. 1-246.

Boyle, Robert, 'The Sceptical chymist, or, Chymico-physical doubts & paradoxes: touching the spagyrist's principles commonly call'd hypostatical, [a]s they are wont to be propos'd and defended by the generality of alchymists' (1661), *Works*, vol. 1, pp. 458-586.

Broeckhuysen, Benjamin van, *Oeconomia corporis animalis sive cogitationes succinctae de Mente, Corpore et utriusque conjuncione iuxta Methodum Philosophiae Cartesianae deductae,* Nijmegen, 1672.

Browne, John, *A compleat treatise of the muscles as they appear in humane body, and arise in dissection: with diverse anatomical observations not yet discover'd*, London, Tho. Newcombe for the author, 1681.

Browne, John, *Myographia nova, or, A graphical description of all the muscles in humane body, as they arise in dissection: distributed into six lectures . . . together with an accurate and concise discourse of the heart, and its use, as also of the circulation of the blood, and the parts of which the sanguinary mass is made and framed, written by the late learned Dr. Lower*, London, Tho. Milbourn for the author, 1697.

Brown, Thomas, *Nature's cabinet unlock'd. Wherein is discovered the natural causes of metals, stones, precious earths, juyces, humors, and spirits, the nature of plants in general; their affections, parts, and kinds in particular. Together with a description of the individual parts and species of all animate bodies, similar and dissimilar, median and organical, perfect and imperfect*, London, Edw. Farnham, 1657.

Browne, Thomas, *Pseudodoxia epidemica, or, Enquires into very many received tenents and commonly presumed truths; together with some marginall observations and a table alphabeticall at the end*, Third edition, London, R. W. for Nath. Ekins, 1658.

Browne, Thomas, *Religio Medici and Christian Morals*, ed. Geoffrey Keynes, London, Thomas Nelson and Sons, 1940.

Burton, Robert, *The Anatomy of Melancholy* (1628), ed. F. Dell and P. Jordan-Smith, New York, Tudor Publishing Co., 1927.

Butler, Samuel, 'An occasional reflection on Dr Charleton's feeling a dog's pulse at Gresham College by R[obert] B[oyle]', in *The Genuine Remains in Verse and Prose of Mr Samuel Butler*, 1749, vol. 1, pp. 404-10.

Calendar of the Proceedings of the Committee for Advance of Money [1642-1656], ed. M. A. Green, London, 1888, part 2.

Calendar of State Papers, Domestic, Addenda 1625-1649, ed. W. D. Hamilton, London, 1897.

Castle, George, *The Chymical Galenist: Wherein the Practise of the Ancients is Reconcil'd to the New Discoveries in the Theory of Physick*, London, 1667.

Cavendish, Margaret, *The Life of . . . William Cavendish*, London, A. Maxwell, 1667.

Cavendish, Margaret, *Letters and Poems in Honour of the Incomparable Princess Margaret, Duchess of Newcastle*, London, 1676.

Cockburn, William, *Œconomia corporis animalis . . . Juxta editionem Londinensem, de' anno 1695, Recusa*, Impensis Kronigeri & hæredum Gœbelij, 1696.

Collins, Samuel, *A Systeme of Anatomy Treating of the Body of Man, Beasts, Birds, Fish, Insects and Plants*, London, Thomas Newcombe, 1685.

Craanen, Theodor, *Oeconomia animalis in duas partes distributa. Nec non Oeconomiæ animalis ad circulationem sanguinis & chyli dispositæ, brevis delineatio*, Goudæ, Ex officina Guilhelmi vander Hoeve, 1685.

Craanen, Theodor, *Tractatus physico-medicus de homine, in quô status ejus tam naturalis, quam præternaturalis, quoad theoriam rationalem mechanicè demonstratur*, Lugduni Batavorum, Apud Petrum Vander Aa, 1689.

Crook, Helkiah, *Mikrokosmographia: a description of the body of man : together with the controversies and figures thereto belonging / collected and translated out of all the best authors of anatomy, especially out of Gaspar Bauhinus and Andréas Laurentius*, London, R. C., 1651.

Croone, William, *De ratione motus musculorum. Editio secunda. priori emendatior*, Amstelodami, Apud Casparum Commelinum, 1667.

Descartes, René, *De homine*, transl. Florentius Schuyl, Leiden, 1662.

Descartes, René, *Discourse on the Method of Properly Conducting One's Reason and of Seeking the Truth in the Sciences*, Penguin, Middlesex, 1968.

Descartes, René, 'Principia', part II, chap. xxxvi, in *Descartes: Philosophical writings*, ed. and trans. Elizabeth Anscombe and Peter Thomas Geach, London, Nelson, 1964.

Deusing Antony, *Oeconomia Corporis animalis: ac speciatium de ortu humanæ dissertatio. In qua demonstratur, non esse homini simpliciter impossibile, per naturale intellectus lumen seipsum nosse. Opposita conceptibus . . . D. Gualt. Charletonis*, Groningæ, Typis Johannis Cöllenii, 1661.

Diemerbroek, Isbrand, *The Anatomy of Human Bodies, Comprehending the most Modern Discoveries and Curiosities in that Art*, trans. William Salmon, London, Edward Brewster, 1689.

Digby, Kenelm, *Two Treatises, In the one of which the nature of bodies, in the other, the nature of mans soule is looked into: in way of discovery of the immortality of reasonable soules*, Paris, 1644.

Dugdale, William, *History of imbanking and draining of divers fens and marshes, both in foreign parts and in this kingdom*, London, Alice Warren, 1662.

Ent, George, *Animadversiones in Malachiæ Thrustoni, M.D., Diatribam de respirationis usu primario*, London, Insigne Bibliorum, 1685.

Ent, George, *Apologia pro circulatione sanguinis: qua respondetur Aemilio Parisano medico Veneto*, London, Robert Young & William Hope, 1641.

Evelyn, John, *Diary and Correspondence*, ed. W. Bray, London, Routledge, 1952.

Fernel, Joannes, *Universa medicina, tribus et viginti libris absoluta. Ab ipso authore . . . recognita et quatuor libris . . . aucta. Postea autem studio . . . G. Plantii . . . elimata et in librum Therapeutices septimum scholiis illustrata*, Genevae, 1619.

Galen, *On the Usefulness of the Parts of the Body*, ed. M. T. May, 2 vols, Ithaca, Cornell University Press, 1968.

Gibson, Thomas, *The Anatomy of Humane Bodies Epitomized*, London, M. Flesher for T. Flesher, 1682.

Glanvill, Joseph, 'Of Scepticism and Certainty: In a short Reply To the Learned Mr. Thomas White', *Essays on Several Important Subjects in Philosophy and Religion*, London, J. D. for Baker and H. Mortlock, 1676.

Glanvill, Joseph, *Plus ultra, or, The progress and advancement of knowledge since the days of Aristotle: in an account of some of the most remarkable late improvements of practical, useful learning, to encourage philosophical endeavours*, London, James Collins, 1668.

Glanvill, Joseph, *Scepsis scientifica The vanity of dogmatizing, or, Confidence in opinions: manifested in a discourse of the shortness and uncertainty of our knowledge, and its causes: with some reflexions on peripateticism, and an apology for philosophy*, London, E. C. for Henry Eversden, 1661.

Glisson, Francis, *Anatomia hepatis, cui praemittuntur quaedam ad rem anatomicam universe spectantia et, ad calcem operis, subjiciuntur nonnulla de lymphae ductibus nuper repertis*, London, Du-Gardianis, 1654.

Glisson, Francis, *English manuscripts*, ed. Andrew Cunningham, (i) From *Anatomia Hepatis* (1654), Cambridge, Wellcome Texts and Documents, 1993.

Glisson, Francis, *De Rachitide sive morbo puerili qui vulgo The Rickets didictur*, London, Laurence Sadler & Robert Beaumont, 1650.

Glisson, Francis, *Tractatus de natura substantiæ energetica, seu, De vita naturæ: ejúsque tribus primis facultatibus, I. perceptiva, II. appetitiva, III motiva, & [brace] naturalibus, &c.*, London, E. Flesher for H. Brome & N. Hooke , 1672.

Glisson, Francis, *Tractatus de ventriculo et intestinis: cui præmittitur alius, de partibus continentibus in genre: & in specie, de ïis abdominis*, London, Henricum Brome, 1677.

Glisson, Francis, 'Philosophical papers, materials related to *De natura substantiae energetica* (On the energetic nature of substance) 1672', ed. Guido Giglioni, Cambridge, Wellcome Unit for the History of Medicine, 1996.

Goodall, Charles, *The Colledge of Physicians vindicated, and the true state of Physick in this Nation, faithfully represented in answer to a . . . pamphlet* [by A. Huyberts] *entituled, The Corner Stone, etc.*, London, 1676.

Goodall, Charles, *The Royal College of Physicians of London; Founded and Established by Law*, London, The College, 1684.

Graaf, Reinier de, *De succo pancreatico, or, A physical and anatomical treatise of the nature and office of the pancreatick juice*, trans. Christopher Pack, London, N. Brook, 1676.

Harvey, Gideon, *Morbus Anglicus: or The anatomy of consumptions, containing the nature, causes, subject, progress, change, signs, prognosticks, preservatives and several methods of curing all consumptions*, London, Thomas Johnson for Nathanael Brook, 1672.

Harvey, William, *De generatione animalium* (1651). *Disputations touching the generation of animals*, trans. Gweneth Whitteridge, Oxford, Boston, Blackwell Scientific, 1981.

Harvey, William, *De motu cordis. An anatomical disputation concerning the movement of the heart and blood in living creatures*, ed. and trans. Gweneth Whitteridge, Oxford, Blackwell Scientific, 1976.

Harvey, William, *Exercitatio anatomica de motu cordis et sanguinis in animalibus. Movement of the heart and blood in animals: an anatomical essay*, trans. K. J. Franklin, Oxford, Blackwell, 1957.

Harvey, William, *De Motu Locu Animalium* (1627), ed. and trans. Gweneth Whitteridge, Cambridge, Cambridge University Press, 1959.

Harvey, William, *Prelectiones anatomiae universalis. The anatomical lectures of William Harvey*, ed. and trans. Gweneth Whitteridge, Edinburgh, E. & S. Livingston for the Royal College of Physicians, London, 1964.

Helmont, Jan Baptista van, *Oriatrike or, Physick refined. The common errors therein refuted, and the whole art reformed & rectified: being a new rise and progress of phylosophy and medicine*, trans. John Chandler, London, printed for Lodowick Loyd, 1662.

Heseler, Baldasar, *Andreas Vesalius' First Public Anatomy at Bologna, 1540. An eyewitness report by B. Heseler . . . together with his notes on Matthæus Curtius' lectures on Anatomia Mundini*, ed. and trans. Ruben Eriksson, Uppsala & Stockholm, Almqvist & Wiksells Boktrykkeri, 1959.

Highmore, Nathaniel, *The history of generation. Examining the several opinions of divers authors, especially that of Sir Kenelm Digby, in his Discourse of bodies . . . To which is joyned a Discourse of the cure of wounds by sympathy, or without any real applycation of medicines*, London, printed by R. N. for John Martin, 1651.

Hobbes, Thomas, *The life of Mr Thomas Hobbes of Malmesbury/Thomae Hobbesii Malmesburiensis vita* (1679), Exeter, University of Exeter, 1979.

Hobbes, Thomas, *Leviathan*, ed. Richard Tuck, New York, Cambridge University Press, 1991.

Hobbes, Thomas, *The Correspondence*, ed. Noel Malcolm, 2 vols, Oxford, Clarendon Press, 1994.

Hobbes, Thomas, *The Elements of Law, Natural and Politic*, ed. J. C. A. Gaskin, New York, Oxford University Press, 1994.

Hogehlande, Cornelius, *Cogitationes: Quibus de existentia; item animae spiritalitas, et possibilis cum corpore unio, Demonstratur: Nec non, brevis historia oeconomiae corporis animalis proponitur, atque mechanice explicatur*, Leiden, 1646.

Hooke, Robert, 'Method of Improving Natural Philosophy', in R. Waller (ed.), *The Posthumous Works of Robert Hooke, MD, SRS, Geom. Prof. Gresh &c.*, New York, Johnson Reprint Co., 1969.

Hooke, Robert, *The Diary of Robert Hooke, 1672-1680*, ed. H. W. Robinson and W. Adams, London, Wykenham Publications, 1968.

Jewel, Edward, *A Brief Discourse of the Stomach and Parts subservient Unto it*, London, s.n., 1678.

Keill, John *The Anatomy of the Humane Body Abridged; or a short and full view of all the parts of the body. Together with their several Uses drawn from their Compositions and Structures*, London, William Keblewhite, 1698.

Laurentius, Andreas, *Historia anatomica*, Frankfurt, 1602.

Layton, Henry, 'Observations upon Dr. Charltons treatise intituled, The immortality of the humane soul' in *Observations upon Mr. Wadsworth's book of the souls immortality*, London, [s.n.], 1670.

Lower, Richard, *De catarrhis* (1672) trans. and ed. Richard Hunter and Ida Macalpine. London, Dawsons, 1963.

Lower, Richard, *Diatribæ Thomæ Willisii Doct. Med. & Prof. Oxon. de febribus vindicatio adversus Edmundum de Meara Ormoniensem Hibernum M.D.*, London, Jo. Martyn & Ja. Allestry, 1665.

Lower, Richard, *Richard Lower's Vindicatio: A defence of the experimental method*, transl. Leofranc Holdford-Strevens, ed. Kenneth Dewhurst, Oxford, Sanford, 1983.

Lower, Richard, *Tractatus de corde: item de motu & colore sanguinis et chyli in cum transitu*, Londini, Typis Jo. Redmayne impensis Jacobi Allestry, 1669.

Mayerne, Théodore Turquet de, *Praxis Medica, Ad Exemplar Londinensis 1690, Impressum novissimè recusa, De Cura Gravidarum, Accessit, Tractatus De Arthritide*, Genevæ, Samuelem de Tournes, 1690.

Maynwaring, Everard, *The Frequent But Unsuspected Progress of Pains, Inflammations, Tumors, Apostems, Ulcers, Cancers, Gangrenes, and Mortifications, Internal*, London, 1679.

Maynwaring, Everard, *The Method and Means of Enjoying Health, Vigour and Long Life*, London, 1683.

Mayow, John, *Medico-Physical works, being a translation of the Tractatus Quinque medico-physicae* (1674), Edinburgh, Alembick Club Reprints, 1907.

Merrett, Christopher, *Pinax Rerum Naturalium Britannicarum, continens Vegetabilia, Animalia et Fossilia in hac insula reperta*, London, 1665/6, 1667.

Molins, William, *Myotomia: Or the Anatomical Administration of all the Muscles of the Body*, London, John Field for Edward Husband, 1648.

Müller, Jacobus, *Resp. De natura motus animalis et voluntarii, exercitatio. Præs. G. Horstio*, Giesæ, 1617.

Needham, Marchamount, *Medela Medicinæ, A Plea for the Free Profession and a Renovation of the Art of Physick*, London, Richard Lownds, 1665.

Oldenburg, Henry, *The Correspondence Of Henry Oldenburg*, ed. A. Rupert Hall and Marie Boas Hall, 13 vols, Madison, University of Wisconsin Press, 1965.

Paré, Ambroise, *The Works of that famous chirurgeon Ambrose Parey*, trans. Thomas Johnson, London, Mary Clark, 1678.

Pecquet, Jean, *Dissertatio anatomica de chyli motu*, 1654.

Pecquet, Jean, *New Anatomical Experiments of J. Pecquet, by which the hitherto unknown receptacle of the chyle, and the transmission from thence to the subclavial veins by the now discovered lacteal chanes of the thorax, is plainly made appear in brutes*, 2 vols, London, T. W. for Octavian Pulleyn, 1653.

Pepys, Samuel, *The Diary of Samuel Pepys,* a new and complete transcript, ed. R. Latham and W. Matthews, London, Bell, 1970-1983.

Power, Henry, *Experimental philosophy, in three books: containing new experiments microscopical, mercurial, magnetical: with some deductions, and probable hypotheses, raised from them, in avouchment and illustration of the now famous atomical hypothesis*, London, T. Roycroft, for John Martin and James Allestry, 1664.

Read, Alexander, *The Manuall of the Anatomy . . . Methodically Digested into Six Books*, London, 1642.

Read, Alexander, *The Workes of that Famous Physitian Dr Alexander Read, Containing Chirurgical Lectures of Tumors and Ulcers...Delivered in several Lectures at Barbar-Chirurgians-Hall*, 3rd Edition, Printed by E. T. for Richard Thrale, London, 1659.

Regius, Henricius, *Fundamenta physices*, Amsterdam, L. Elzevir, 1646.

Remmelin, Johann, *A Survey of the Microcosme, or The Anatomy of the Bodies of Man and Woman*, London, James Moxon, 1695.

Riolan, Jean, *A sure guide, or, The best and nearest way to physick and chyrurgery: that is to say, the arts of healing by medicine and manual operation: being an anatomical description of the whole body of man and its parts: with their respective diseases demonstrated from the fabrick and use of the said parts: in six books*, trans. Nicholas Culpeper, London, Printed by Peter Cole, 1657.

Riolan, Jean, *Anthropographia et Osteologia omnia recognita*, Parisiis, 1626.

Riolan, Jean, *Opuscula anatomica nova quæ nunc primum in lucem prodeunt. Instauratio magna physicæ et medicinæ, per novam doctrinam de motu circulatorio sanguinis in corde*, Londini, typis Milonis Flesher, 1649.

Rudbeck, Olof, *Disputatio anatomica de circulatione sangvinis, nova excercitatio anatomica exhibens ductus hepaticos aquosos et vasa glandularum serosa*, ed. Wolfram Kock, Stockholm, Rediviva, 1977.

Securis, John, *Detection and querimonie of the daily enormities and abuses Comitted in physick*, Amsterdam, Theatrum Orbis Terrarum, W. J. Johnson, 1976.

Seneca, *Naturales Quaestiones*, trans. T. H. Corcoran, 2 vols, London, Heinemann, 1971.

Sprat, Thomas, *History of the Royal Society of London, for the Improving of Natural Knowledge*, London, 1667.

Steno, Niels, 'New Structure of the Muscles and Heart' (1663) and 'Specimen of Elements of Myology' (1667), transl., M. Emmanuel Collins with Paul Maquet and Troels Kardel, in *Steno on Muscles*, Philadelphia, American Philosophical Society, 1994.

Steno, Nicolaus, *Lecture on the Anatomy of the Brain*, ed. Gustav Scherz, Copenhagen, Arnold Busck, 1965.

Sydenham, Thomas, 'Anatomie' (1668), in K. Dewhurst (ed.), *Dr Thomas Sydenham (1624-1689): His Life and Original Writings*, Berkeley, University of California Press, 1966.

Sydenham, Thomas, *The Works of Thomas Sydenham MD, translated from the Latin edition of Mr Greenhill, with a life of the author by R. G Latham*, Birmingham, Classics of Medicine Library, 1979.

Tyson, Edward, *Carigueya, seu marsupiale Americanum, or, The anatomy of an opossum dissected at Gresham College*, London, Sam. Smith and Benj. Walford, 1698.

Vesalius, Andreas, *De humani corporis fabrica. Vesalius on the Human Brain*, ed. Charles Singer, London, Oxford University Press, 1952.

Vesling, Johann, *The Anatomy of the Body of Man: wherein is exactly described every part thereof, in the same manner as it is commonly shewed in publick anatomies*, transl. Nicholas Culpeper, London, Peter Cole, 1653.

Walaeus, Johannes, 'Two Epistles Concerning the Motion of the Chyle and the Blood', in *Bartholinus Anatomy*, London, John Streater, 1668.

Webster, John, *Academiarum Examen: or, the Examination of Academies: Wherein is discussed and examined the matter, method, and customes of academick and scholastick learning, and the insufficiency thereof discovered and laid open*, London, 1654.

Wharton, Thomas, *Adenographia sive Glandularum totius corporis descriptio* (1656), trans. Stephen Freer, Oxford, Clarendon Press, 1996.

Wilkins, John, *Of the Principles and Duties of Natural Religion*, 4th edition, London, 1699.

Willis, Thomas, *Diatribæ duæ medico-philosophicæ, quarum prior agit de fermentatione . . . altera de febribus*, Editio secunda, ab authore . . . locupletata. Londini, 1660. 12o.

Willis, Thomas, *Oxford Lectures*, ed. and introduced K. Dewhurst, Sandford, Oxford, 1980.

Willis, Thomas, *A medical-philosophical discourse of fermentation, or, Of the intestine motion of particles in every body*, trans. S. P., London, T. Dring, C. Harper, J. Leigh & S. Martin, 1681.

Willis, Thomas, *De anima brutorum quæ hominis vitalis ac sensitiva est, exercitationes duæ: prior physiologica djusdem naturam, partes, potentias & affectiones tradit: altera pathologica morbos quiipsam, & sedem ejus primariam, nempe cerebrum & nervosum genus afficiunt, explicat, corunque therapeias instituit*, Oxonii, E Theatro Sheldoniano, Ric. Davis, 1672.

Willis, Thomas, *Dr. Willis's practice of physick: being the whole works of that renowned and famous physician wherein most of the diseases belonging to the body of man are treated of*, London, T. Dring, C. Harper, and J. Leigh, 1684.

Willis, Thomas, *Pathologiæ cerebri, et nervosi generis specimen: in quo agitur de morbis convulsivis, et de scorbuto . . .*, Oxonii, Guil. Hall, Ja. Allestry, 1667.

Willis, Thomas, *Pharmaceutice rationalis, or, An exercitation of the operations of medicines in human bodies: shewing the signs, causes, and cures of most distempers incident thereunto: in two parts: as also a treatise of the scurvy*, London, T. Dring [etc.], 1679.

Willis, Thomas, *The Anatomy of the Brain and Nerves*, trans. 1681 by S. Pordage, ed. W. Feindel, Montreal, McGill University Press, 1965.

Willis, Thomas, *The London practice of physick, or, The whole practical part of physick contained in the works of Dr. Willis*, London, T. Basset, 1689.

Willis, Thomas, *The Remaining Medical Works of that Famous and Renowned Physician Dr Thomas Willis*, trans. S. Pordage, London, Printed for T. Dring, C. Harper, J. Leigh and S. Martyn, 1681.

Willis, Thomas, *Two discourses concerning the soul of brutes, which is that of the vital and sensitive of man. The first is physiological, shewing the nature, parts, powers, and affections of the same. The other is pathological, which unfolds the diseases which affect it and its primary seat*, trans. S. Pordage, London, Thomas Dring, 1683.

Winston, Thomas, *Anatomy Lectures at Gresham College*, London, R. Daniel for Thomas Eglesfield, 1659.

Wood, Anthony a, *Atheniæ Oxonienses . . . to which are added the fasti or annals of the said university etc.*, London, Johnson Reprint Co., 1967.

Manuscripts

Glisson, Francis, lectures, British Museum MS Sloane 3306.

Glisson, Francis, original draft of *Anatomia*, British Museum MS Sloane 3315, ff 165-9.

Annals of the Royal College of Physicians of London. Typescript translation into English from the original Latin. Library of the Royal College of Physicians of London. Volumes III-VII cover the period from 1608 to 1710. My thanks to the Royal College of Physicians for permission to cite their records.

SECONDARY SOURCES

Ackerknecht, Erwin H., *A Short History of Medicine*, New York, Ronald Press Co, 1968.

Adamson, Ian, 'The Royal Society and Gresham College, 1660-1711', *Notes and Records*, vol. 33, no. 1, 1978, pp. 1-21.

Adelman, H. B., *Marcello Malpighi and the Evolution of Biology*, 5 vols, Ithaca, Cornell University Press, 1966.

Albrecht, Michael, *Eklektik. Eine Begriffsgeschichte mit Hinweisen auf die Philosophie-und Wissenschaftsgeschichte*, Stuttgart, Frommann-Holzboog, 1994.

Albury, W. Randall, 'Confessional anatomy in the Counter-Reformation period: "Self-dissecting" anatomical illustrations in the 17th century', in 'Individuals and Institutions in the History of Medicine, Proceedings from the 6th Biennial Conference of the Australian Society of the History of Medicine', 1999.

Allen, Phyllis, 'Medical Education in Seventeenth Century England', *Journal of the History of Medicine*, vol. 1, 1946, pp. 115-43.

Alpers, Svetlana, *The Art of Describing: Dutch art in the seventeenth century*, London, John Murray, 1983.

Anselment, Raymond A., *The Realms of Apollo: Literature and healing in seventeenth century England*, London, Assoc. University Press, 1995.

Arakelian, Paul G., 'The myth of a Restoration style shift', *The Eighteenth Century: Theory and Interpretation*, vol. 209, 1979, pp. 227-45.

Arbib, M. A. and Hesse, M. B., 'Language, Metaphor and a New Epistemology', in *The Construction of Reality*, Cambridge, Cambridge University Press, 1986, pp. 147-170.

Ariew, Roger, 'Damned if you do: Cartesians and censorship, 1663-1706', *Perspectives on Science*, vol. 2, no. 3, 1994, pp. 255-74.

Armistead, J. M., 'Introduction', to Walter Charleton, *The Immortality of the Human Soul*, New York, A.M.S. Press, 1985.

Armstrong, D., 'Bodies of Knowledge: Foucault and the Problem of Human Anatomy', in G. Scrambler (ed.), *Sociological Theory and Medical Sociology*, London, Tavistock, 1987, pp. 59-76.

Ashworth, William B., 'Light of reason, light of nature: Catholic and Protestant metaphors of scientific knowledge', *Science in Context*, vol. 3, 1989, pp. 89-107.

Aughterson, Kate, 'The Waking Vision: Reference in the *New Atlantis*', *Renaissance Quarterly*, vol. 45, no. 1, 1992, pp. 119-39.

Axtell, James L., 'Education and status in Stuart England: The London physician', *History of Education Quarterly*, vol. 10, 1970, pp. 141-59.

Babb, Lawrence, *The Elizabethan Malady: A Study of Melancholia In English Literature from 1580-1642*, East Lansing, Michigan State College Press, 1951.

Baker, G. P. and Morris, K. J., 'Descartes Unlocked', *British Journal for the History of Philosophy*, vol. 1, no. 1, 1993, pp. 5-27.

Barkan, Leonard, *Nature's Work of Art: the Human Body as Image of the World*, London, Yale University Press, 1977.

Barker, Francis, *The Tremulous Private Body: Essays on Subjection*, London, Methuen, 1984.

Barnes, Barry and Shapin, Steven, *Natural Order: Historical Studies of Scientific Culture*, California, Sage, 1979.

Barnes, Barry, Bloor, David and Henry, John (eds), *Scientific knowledge: a sociological analysis*, London, Athlone, 1996.

Barnouw, Jeffrey, 'Passion as "Confused" Perception or Thought in Descartes, Malebranche and Hutcheson', *Journal of the History of Ideas*, vol. 53, no. 3, 1992, pp. 397-424.

Bastholm, E., *The history of muscle physiology from the natural philosophers to Albrecht von Haller*, Copenhagen, Munksgaard, 1950.

Bates, Donald G., 'Thomas Willis and the Epidemic Fever of 1661; A Commentary', *Bulletin of the History of Medicine*, vol. 39, 1965, pp. 393-414.

Bates, Donald G., 'Sydenham and the medical meaning of "Method"', *Bulletin of the History of Medicine*, vol. 51, 1977, pp. 324-38.

Bates, Donald G., 'Thomas Willis and the fevers literature of the seventeenth century', *Medical History*, suppl. no. 1, 1981, pp. 45-70.

Bates, Donald G., 'Harvey's account of his discovery', *Medical History*, vol. 36, no. 4, 1992, pp. 361-78.

Bates, Donald G., 'Closing the circle: How Harvey and his contemporaries played the game of truth', *History of Science*, vol. 36, 1998, pp. 213-32 and 245-67.

Bazerman, Charles, *Shaping Written Knowledge: The Genre and Activity of the Experimental Article in Science*, Madison, University of Wisconsin Press, 1988.

Beaune, Jean-Claude, 'The classical age of automata: an impressionistic survey from the sixteenth to the nineteenth century', in Feher, Nadaff & Tazi (eds), *Fragments for a History of the Human Body; Part I*, Massachusetts, MIT Press, 1989, pp. 431-80.

Bechler, Zev, 'The essence and soul of seventeenth-century scientific revolution', *Science in Context*, vol. 1, no. 1, 1987, pp. 87-101.

Beck, Daniel A., 'Miracle and the mechanical philosophy: The theology of Robert Boyle', PhD dissertation, University of Notre Dame, 1986.

Beier, Lucinda M., *Sufferers and Healers: The Experience of Illness in Seventeenth Century England*, New York, Routledge and Kegan Paul, 1987.

Benjamin, Andrew E., Cantor, Geoffrey N. and Christie, John R. R. (eds), *The Figural and the Literal: Problems of Language in the History of Science and Philosophy, 1630-1800*, Manchester, Manchester University Press, 1987.

Bennett, J. A., 'A note on theories of respiration and muscular action in England c. 1660', *Medical History*, vol. 20, 1976, pp. 59-69.

Bennett, J. A., 'Robert Hooke as mechanic and natural philosopher', *Notes and Records of the Royal Society of London*, vol. 35, 1980, pp. 33-48.

Bennett, J. A., 'The mechanic's philosophy and the mechanical philosophy', *History of Science*, vol. 24, 1986, pp. 1-28.

Bennett, Joan, 'An aspect of the evolution of seventeenth century prose style', *Review of English Studies*, vol. 17, 1941, pp. 284-88.

Berger, Peter and Luckman, Thomas, *The Social Construction of Reality: A treatise in the sociology of knowledge*, London, Penguin, 1971.

Berman, Morris, *The Re-enchantment of the World*, London, Cornell University Press, 1981.

Biagioli, Mario, 'The scientific revolution is undead', *Configurations*, vol. 6, no. 2, 1998, pp. 141-8.

Biographia Britannica: Or the lives of the most eminent persons who have flourished in Great Britain and Ireland from the earliest ages down to the present times, 1747-1766, Hildesheim, Georg Olms Verlag, 1969-1973.

Birken, William Joseph, 'The Puritan connexions of Sir Edward Alston, President of the Royal College of Physicians, 1655-1666', *Medical History*, vol. 18, 1974, pp. 370-74.

Birken, William Joseph, 'Dr John King (1614-1681) and Dr Asseurus Regemorter (1615-1650)', *Medical History*, vol. 20, no. 3, 1976, pp. 276-95.

Birken, William Joseph, 'The social problem of the English physician in the early seventeenth century', *Medical History*, vol. 31, 1987, pp. 201-16.

Birken, William Joseph, 'The Royal College of Physicians of London and its support of the Parliamentary cause in the English Civil War', *Journal of British Studies*, vol. 23, no. 1, 1983, pp. 47-62.

Birken, William Joseph, 'The dissenting tradition in English medicine of the seventeenth and eighteenth centuries', *Medical History*, vol. 39, no. 2, 1995, pp. 197-218.

Black, Max, 'More about metaphor', in Ortony, A. (ed.), *Metaphor and Thought*, New York, Cambridge University Press, 1993, pp. 19-41.

Blackwell, Constance, 'Sturm, Morhof and Brucker vs. Aristotle: Three Eclectic Philosophers View the Aristotelian Method', in *Method and order in Renaissance philosophy of nature: The Aristotle commentary tradition* Daniel A. Di Liscia, Eckhard Kessler and Charlotte Methuen (eds), Aldershot, Ashgate, 1998.

Bliss, Robert M., *Restoration England, 1660-1688*, London, Methuen, 1985.

Bloch, H., 'Francis Glisson, MD (1597-1677): The Glissonian irritability phenomenon, its roots, links and confirmation', *Southern Medical Journal*, vol. 81, no. 11, 1988, pp. 1433-6.

Bono, James J., 'Reform and the languages of Renaissance theoretical medicine: Harvey versus Fernel', *Journal of the History of Biology*, vol. 23, no. 3, 1990, pp. 341-87.

Bono, James J., 'Science, Discourse and Literature: the Role/Rule of Metaphor in Science', in S. Peterfreund (ed.), *Literature and Science: Theory and Practice*, Boston, Northeastern University Press 1990, pp. 59-90.

Bono, James J., *'The Word of God' and the 'Languages of Man', Interpreting Nature and Texts in Early Modern Science and Medicine*, Madison, University of Wisconsin Press, 1995.

Bordieu, Pierre, *Outline of a Theory of Practice*, trans. Richard Nice, New York, Cambridge University Press, 1977.

Bordo, Susan, *The Flight to Objectivity: Essays on Cartesianism and Culture*, New York, State University of New York Press, 1987.

Boss, Jeffrey, '*Doctrina de circulatione sanguinis haud immutat antiquam medendi methodum*: an unpublished manuscript (1662) by Francis Glisson (1597-1677) on the implications of Harvey's physiology', *Physis*, vol. 20, no. 1-4, 1978, pp. 309-35.

Boss, Jeffrey, 'Helmont, Glisson and the doctrine of the common reservoir in the seventeenth century revolution in physiology', *British Journal for the History of Science*, vol. 16, no. 3, 1983, pp. 261-72.

Boss, Jeffrey, 'The "Methodus Medendi" as an Index of Change in the Philosophy of Medical Science in the Sixteenth and Seventeenth Centuries', *History and Philosophy of the Life Sciences*, vol. 1, 1979, pp. 13-42.

Bracegirdle, Brian, 'The Microscopical Tradition', in R. Porter and W. Bynum (eds), *Companion Encyclopedia of the History of Medicine* Vol. I, London, Routledge, 1993, pp. 102-19.

Brandt, Alan M., 'Emerging Themes in the History of Medicine', *Milbank Memorial Quarterly*, vol. 69, no. 2, 1991, pp. 199-214.

Brazier, Mary A. B., A *History of Neurophysiology in the Seventeenth and Eighteenth Centuries: From Concept to Experiment*, New York, Raven Press, 1984.

Brockliss, Laurence, 'Seeing and believing: Contrasting attitudes towards observational autonomy among French Galenists in the first half of the seventeenth century', in W. F. Bynum and Roy Porter (eds), *Medicine and The Five Senses*, Cambridge, Cambridge University Press, 1993, pp. 69-84.

Bronowski, Jacob and Mazlish, Bruce, 'The Scientific Revolution', in *The Western Intellectual Tradition: From Leonardo to Hegel*, London, Hutchison, 1960, pp. 133-53.

Brown, Terence, 'Toward a scientific ethic: Man and nature in John Dryden's "To Dr Charleton" ', *Artes Liberales*, vol. 4, 1977, pp. 13-22.

Brown, Theodore M., 'The College of Physicians and the acceptance of iatromechanism in England, 1665-1695', *Bulletin of the History of Medicine*, vol. 44, 1970, pp. 12-30.

Brown, Theodore M., 'Physiology and the mechanical philosophy in mid-seventeenth century England', *Bulletin of the History of Medicine*, vol. 51, 1977, pp. 25-54.

Brown, Theodore M., *The Mechanical Philosophy and the 'Animal Oeconomy': A Study in the Development of English Physiology In the Seventeenth and Early Eighteenth Century*, New York, Arno, 1981.

Brown, Theodore M., 'Descartes, dualism and psychosomatic medicine', in William F. Bynum, M. Shepherd and Roy Porter (eds), *The Anatomy of Madness*, Vol. I, 'People and Ideas', London, Tavistock, 1985, pp. 40-62.

Brown, Theodore M., 'Medicine in the shadow of the *Principia*', *Journal of the History of Ideas*, vol. 48, no. 4, 1987, pp. 629-49.

Buckland-Wright J., 'Readers of anatomy at the Barber-Surgeons Company on the Tudor period', *Journal of the Royal Society of Medicine*, vol. 78, 1985, pp. 802-11.

Burchell, Howard, 'Mechanical and hydraulic analogies in Harvey's discovery of the circulation', *Journal of the History of Medicine and the Allied Sciences*, vol. 36, no. 3, 1981, pp. 260-77.

Burnham, John C., 'Garrison Lecture: How the Concept of profession evolved in the work of historians of medicine', *Bulletin of the History of Medicine*, vol. 70, no. 1, 1996, pp. 1-24.

Bury, M. R., 'Social constructionism and the development of medical sociology', *Sociology of Health and Illness*, vol. 8, no. 2, 1986, pp. 135-69.

Butterfield, Herbert, *The Origins of Modern Science, 1300-1800*, London, G. Bell, 1957.

Bylebyl, Jerome J., 'Nutrition, quantification and circulation', *Bulletin of the History of Medicine*, vol. 51, no. 3, 1977, pp. 369-85.

Bylebyl, Jerome J., *William Harvey and His Age: The Professional and Social Context of the Discovery of Circulation*, Baltimore, Johns Hopkins University Press, 1979.

Bynum, William F., 'The anatomical method, natural theology, and the functions of the Brain', *Isis*, vol. 64, no. 22, 1973, pp. 445-68.

Bynum, William F. and Nutton, V. (eds), *Theories of Fever from Antiquity to the Enlightenment*, London, Wellcome Institute for the History of Medicine, 1981.

Bynum, William F., '"C'est Une Malade": Animal models and concepts of human diseases', *Journal of the History of Medicine and the Allied Sciences*, vol. 45, no. 3, 1990, pp. 397-413.

Bynum, William F. and Porter, Roy (eds), *Medicine and the Five Senses*, Cambridge, Cambridge University Press, 1993.

Canguilhem, Georges, 'The role of analogies and models in biological discovery', in A. Crombie (ed.), *Scientific Change*, New York, Basic Books, 1963, pp. 507-20.

Canguilhem, Georges, *The Normal and the Pathological*, trans. C. R. Fawcett, New York, Zone, 1989.

Caplan, A. L., McCartney, J. J. and Engelhardt, H. T. (eds), *Concepts of Health and Disease: Interdisciplinary Perspectives*, Massachusetts, Addison Wesley, 1981.

Carter, Richard B., *Descartes' Medical Philosophy: The organic solution to the mind-body problem*, Baltimore, Johns Hopkins University Press, 1985.

Cassirer, Ernst, *The Individual and the Cosmos in Renaissance Philosophy*, Oxford, Clarendon Press, 1963.

Chalmers, Alan, 'The lack of excellency of Boyle's mechanical philosophy', *Studies in the History and Philosophy of Science*, vol. 24, no. 4, 1993, pp. 541-64.

Chalmers, Gordon, 'Effluvia: The history of a metaphor', *Modern Language Association of America Publications*, vol. 52, 1937, pp. 1031-50.

Chambers, Douglas, 'John Evelyn and the Construction of the scientific self', in W. G. Marshall (ed.), *The Restoration Mind*, Newark, University of Delaware Press, 1997.

Christie, J. R. R., 'Influencing people', *Annals of Science*, vol. 33, 1976, pp. 311-18.

Christie, J. R. R., 'Bad news for the body', *Art History*, vol. 9, 1986, pp. 263-70.

Clagett, Marshall, *The Science of Mechanics in the Middle Ages*, Madison, Wisconsin University Press, 1961.

Clark, George, *A History of the Royal College of Physicians of London*, Oxford, Clarendon Press, 2 volumes, 1964-6.

Clarke, Bruce and Aycock, Wendell (eds), *The Body and the Text: Comparative Essays in Literature and Medicine*, Lubbock, Texas, Texas Tech University Press, 1990.

Clarke, Edwin, 'The doctrine of the hollow nerve in the seventeenth and eighteenth centuries', in Lloyd G. Stevenson & Robert P. Multhauf (eds), *Medicine, Science and Culture, Essays in Honour of Owsei Temkin*, Baltimore, Johns Hopkins University Press, 1968, pp. 123-41.

Clarke, Edwin, 'The Neural Circulation: The use of analogy in medicine', *Medical History*, 1978, vol. 22, pp. 291-307.

Clericuzio, Antonio, 'A redefinition of Boyle's chemistry and corpuscular philosophy', *Annals of Science*, vol. 47, 1990, pp. 561-89.

Clericuzio, Antonio, 'From Van Helmont to Boyle: A study of the transmission of Helmontian chemical and medical theories in seventeenth century England', *British Journal of the History of Science*, vol. 26, 1993, pp. 303-34.

Clericuzio, Antonio, 'The internal laboratory: The chemical reinterpretation of medical spirits in England (1650-1680)', in A. Clericuzio and P. Rattansi (eds), *Alchemy and Chemistry in the 16th and 17th Centuries*: Proceedings Warburg colloquium 1989, Boston, Kluwer Academic Publishers, 1994.

Cloie, R. L., 'Some Paradoxes in the Language of Things', in J. A. Mazzeo (ed.), *Reason and the Imagination: Studies in the History of Ideas 1600-1800*, New York, Columbia University Press, 1962, pp. 93-129.

Clucas, S., 'The atomism of the Cavendish circle: A reappraisal', *The Seventeenth Century*, vol. 9, 1994, pp. 247-73.

Cohen, Henry, 'The evolution of the concept of disease' in Caplan, Engelhardt & McCartney (eds), *Concepts of Health and Disease: Interdisciplinary Perspectives*, Massachusetts, Addison Wesley, 1981, pp. 29-219.

Cohen, I. Bernard, 'Harrington and Harvey: a theory of the state based on the new physiology', *Journal of the History of Ideas*, vol. 55, no. 2, 1994, pp. 187-210.

Coleman, William, 'Mechanical philosophy and hypothetical physiology', *Texas Quarterly*, vol. 10, 1967, pp. 259-69.

Collingwood, R. G., *The Idea of Nature*, Oxford, Clarendon Press, 1945.

Condren, Conal, *George Lawson's Politica and the English Revolution*, New York, Cambridge University Press, 1989.

Condren, Conal, 'Foucault's Cave: the reification of power and the metaphorical legacy of early modern political discourse', *Common Knowledge*, vol. 3, no. 3, 1994, pp. 20-38.

Condren, Conal, *The Language of Politics in Seventeenth-century England*, Basingstoke, Macmillan, 1994.

Condren, Conal, 'Fragmented continuities: Reflections on metaphor, narrative construction and the early modern historian', *Parergon*, vol. 1, 1998.

Conger, George Perrigo, *Theories of Macrocosms and Microcosms in the History of Philosophy*, New York, Russell & Russell, 1967.

Cook, Harold J., *The Decline of the Old Medical Regime in Stuart London*, Ithaca, Cornell University Press, 1986.

Cook, Harold J., 'Physicians and the New Philosophy: Henry Stubbe and the Virtuosi-Physicians', in A. Wear and R. French (eds), *The Medical Revolution of the Seventeenth Century*, Cambridge, Cambridge University Press, 1989, pp. 246-71.

Cook, Harold J., 'The new philosophy and medicine in seventeenth century England', in D. C. Lindberg and R. S. Westman (eds), *Reappraisals Of The Scientific Revolution*, Cambridge, Cambridge University Press, 1990, pp. 397-436.

Cook, Harold J., 'Good Advice and Little Medicine: The professional authority of early modern English physicians', *Journal of British Studies*, vol. 33, no. 1, 1994, pp. 1-31.

Cook, Harold J., *Trials of an Ordinary Doctor: Joannes Groenevelt in Seventeenth-century London*, Baltimore, Johns Hopkins University Press, 1994.

Cook, Harold J., 'Institutional Structures and personal belief in the London College of Physicians', in O. P. Grell and A. Cunningham (eds), *Religio Medici: Medicine and Religion in Seventeenth-Century England*, Brookfield, Scolar Press, 1996.

Cook, Harold J., 'The Society of Chemical Physicians, The new philosophy and the Restoration court', *Bulletin of the History of Medicine*, vol. 61, no. 1, 1987, pp. 61-77.

Cope, Jackson I., 'Evelyn, Boyle and Dr Wilkinson's "Mathematico-chymico-mechanical school" ', *Isis*, vol. 50, 1959, pp. 30-32.

Crosland, Maurice, 'Changes in chemical concepts and language in the seventeenth century', *Science in Context*, vol. 9, no. 3, 1996, pp. 225-40.

Costello, William T., *The Scholastic Curriculum at Early Seventeenth Century Cambridge*, Massachusetts, Harvard University Press, 1958.

Cunningham, Andrew, 'The kinds of anatomy', *Medical History*, vol. 19, 1975.

Cunningham, Andrew, 'Thomas Sydenham: Epidemics, experiment and the "Good Old Cause" ', in A. Wear and R. French (eds), *The Medical Revolution of the Seventeenth Century*, Cambridge, Cambridge University Press, 1989, pp. 164-90.

Cunningham, Andrew & Williams, Perry, 'De-centring the "big picture" ', *British Journal of the History of Science*, 26, 1993.

Cunningham, Andrew, 'Historical context of Wharton's work on the glands', in *Thomas Wharton's Adenographia*, Oxford, Clarendon Press, 1996.

Cunningham, Andrew, *The Anatomical Renaissance: The resurrection of the Anatomical Projects of the Ancients*, Ashgate, Scolar Press, 1997.

Daston, Lorraine J., 'Galilean Analogies: Imagination at the bounds of sense', *Isis*, vol. 75, 1984, pp. 302-10.

Daston, Lorraine J., 'Baconian facts, academic civility and the pre-history of objectivity', *Annals of Scholarship*, vol. 8, 1991, pp. 337-63.

Daston, Lorraine J., 'Probabilism' in D. Garber & M. Ayers (eds), *The Cambridge History of Seventeenth Century Philosophy*, Cambridge, Cambridge University Press, 1998.

Davidson, Donald, 'What Metaphors Mean', in *Inquiries into Truth and Interpretation*, Oxford, Oxford University Press, 1984.

Davis, Audrey B., 'Some Implications of the circulation theory for disease theory and treatment in the seventeenth century', *Journal of the History of Medicine*, vol. 26, 1971.

Davis, Audrey B., 'The circulation of blood and chemical anatomy', in A. G. Debus (ed.), *Science, Medicine and Society in the Renaissance: Essays in Honour of Walter Pagel*, New York, Science History Publications, 1972, vol. 2, pp. 25-37.

Davis, Audrey B., *Circulation Physiology and Medical Chemistry in England 1650-1680*, Lawrence, Conrado, 1973.

Dawbarn, Frances, 'Learned physicians, medical heterodoxy and court patronage: Conflict in early modern England', *Endeavour*, vol. 22, 1998, pp. 98-100.

Debru, Claude (ed.), *Essays in the History of the Physiological Sciences: Proceedings of network symposium of the European Association for the History of Medicine and Health, University Louis Pasteur, Strasbourg, March 1993*, Amsterdam, Rodopi, 1995.

De Beer, E. S., 'Earliest Fellows of the Royal Society', *Notes and Queries*, vol. 7, 1949, pp. 172-92.

De Folter, R. J., 'A newly discovered *Oeconomia Animalis*, by Pieter Muis of Rotterdam (1645-1721)', *Janus*, vol. 65, 1978.

De Moulin, D., 'Paul Barbette, MD: A seventeenth century Amsterdam author of best-selling textbooks', *Bulletin of the History of Medicine*, vol. 59, no. 4, 1985, pp. 509-14.

De Quehen, A. H., 'A parenthetical allusion in Dryden's "To Dr Charleton" ', *Notes and Queries*, vol. 222, 1977, pp. 544-5.

De Solla Price, D. J., 'Automata and the origins of mechanism and mechanistic philosophy', *Technology and Culture*, vol. 5, 1964, pp. 9-23.

Dear, Peter, 'Totius in Verba: Rhetoric and authority in the early Royal Society', *Isis*, vol. 76, 1985, pp. 145-61.

Dear, Peter, 'Miracles, experiments, and the ordinary course of nature', *Isis*, vol. 81, 1990, pp. 663-83.

Dear, Peter, 'Narratives, anecdotes and experiments: Turning experience into science in the seventeenth century', in P. Dear (ed.), *The Literary Structure of Scientific Argument: Historical Studies*, Philadelphia, University of Pennsylvania Press, 1991, pp. 135-63.

Dear, Peter, 'From Truth to Disinterestedness in the Seventeenth Century', *Social Studies of Science*, vol. 22, 1992, pp. 619-31.

Dear, Peter, *Discipline and Experience: The Mathematical Way and the scientific revolution*, Chicago, Chicago University Press, 1996.

Dear, Peter (ed.), *The Scientific Enterprise in Early Modern Europe: Readings from Isis*, Chicago, Chicago University Press, 1997.

Deason, Gary B., 'Reformation theology and the mechanistic conception of nature', in Ron L. Numbers and David C. Lindberg (eds), *God and Nature: Essays on the encounter between Christianity and science*, Berkeley, University of California Press, 1986.

Debus, Allen G., *The English Paracelsians*, New York, Franklin Watts Inc, 1966.

Debus, Allen G. (ed.), *Science and Education in the Seventeenth Century: The Webster-Ward Debate*, London, Macdonald, 1970.

Debus, Allen G. (ed.), *Medicine in Seventeenth Century England*, Berkeley, University of California Press, 1974.

Debus, Allen G., 'The chemical debates of the seventeenth century: The reaction to Robert Fludd and Jean Baptiste van Helmont', pp. 19-48 in M. L. Righini Bonelli, & W. R. Shea (eds), *Reason, Experiment and Mysticism in the Scientific Revolution*, New York, Macmillan, 1975.

Debus, Allen G., 'The medico-chemical world of the Paracelsians', in Robert Young and Mikulas Teich (eds), *Changing perspective in the history of science: essays in honour of Joseph Needham*, Dordrecht, Reidel, 1973.

Debus, Allen G., *The Chemical Philosophy: Paracelsian science and Medicine in the Sixteenth and Seventeenth Centuries*, New York, Science History Publications, 1977.

Delaporte, Francois, 'The history of medicine according to Foucault', in J. Goldstein (ed.), *Foucault and the Writing of History*, New York, Blackwell, 1994, pp. 137-49.

Dew, Nicholas, 'The Politics of the Body in Restoration England: Anatomy and Theology in the Work of Walter Charleton', MSc Economic and Social History Dissertation, Oxford, 1995.

Dewhurst, Kenneth, 'Locke and Sydenham on the Teaching of Anatomy', *Medical History*, vol. 2, 1958, pp. 1-12.

Dewhurst, Kenneth, *John Locke (1632-1704) Physician and Philosopher: A Medical Biography with an Edition of the Medical Notes in his Journals*, London, Wellcome Historical Medical Library, 1963.

Dewhurst, Kenneth, *Thomas Sydenham (1624-1689) His Life and Writings*, Berkeley, University of California Press, 1966.

Dewhurst, Kenneth, 'Some Letters of Dr Thomas Willis (1621-1675)', *Medical History*, vol. 16, 1972, pp. 63-77.

Dewhurst, Kenneth, 'Introduction', *Thomas Willis's Oxford Lectures*, Sandford, Oxford, 1980.

Dictionary of National Biography, From the earliest times to 1900, ed. Leslie Stephen and Sidney Lee, London, Oxford University Press, 1917.

DiGiacomo, S. M., 'Metaphor as Illness; Postmodern dilemmas in the representation of body, mind and disorder', *Medical Anthropology*, vol. 14, 1992, pp. 109-37.

Dijksterhuis, Eduard J., *The mechanization of the world picture: Pythagoras to Newton*, New Jersey, Princeton University Press, 1986.

Dingwall, Helen, *Physicians, Surgeons and Apothecaries: Medical Practice in Seventeenth Century Edinburgh*, East Linton, Tuckwell Press, 1995.

Dobson, Jessie and Walker, R. M., *Barbers and Barber-Surgeons of London: A History of the Barbers' and Barber-Surgeons' Companies*, Oxford, Blackwell Science Publications, 1979.

Duchesneau, Francois, 'Malpighi, Descartes and the epistemological problems of iatromechanism', in M. L. Righini Bonelli & W. R. Shea (eds), *Reason, Experiment and Mysticism in the Scientific Revolution*, New York, Macmillan, 1975.

Duden, Barbara, *The Woman Beneath the Skin*, trans. T. Dunlap, London, Harvard University Press, 1991.

Dunn, John, 'The identity of the history of ideas', *Philosophy*, vol. 43, 1968, pp. 85-104.

Eales, Nellie B., 'The history of the lymphatic system, with special reference to the Hunter-Monro controversy', *Journal of the History of Medicine and the Allied Sciences*, vol. 29, no. 3, 1974, pp. 280-94.

Eamon, William, *Science and the Secrets of Nature: Books of secrets in medieval and early modern culture*, New Jersey, Princeton University Press, 1994.

Easlea, Brian, *Witch-hunting, Magic and the New Philosophy*, New Jersey, Harvester Press, 1980.

Eastwood, B. S., 'Descartes on refraction: Scientific versus rhetorical method', *Isis*, vol. 75, 1984, pp. 481-502.

Edelstein, Ludwig, *Ancient Medicine*, ed. O. Temkin and C. Lilian Temkin, trans. C. L. Temkin, Baltimore, Johns Hopkins University Press, 1967.

Elmer, Peter, 'Medicine, science and the Quakers: The Puritanism-Science debate reconsidered', *Journal of the Friends' Historical Society*, vol. 54, no. 6, 1981, pp. 265-86.

Elmer, Peter, 'Medicine, Religion and the Puritan Revolution', in R. French and A. Wear (eds), *The Medical Revolution of the Seventeenth Century*, Cambridge, Cambridge University Press, 1989, pp. 10-45.

Featherstone, M., Hepworth, M. and Turner, B. S., *The Body: Social Process and Cultural Theory*, London, Sage, 1991.

Feingold, Mordechai, 'The Universities and the Scientific Revolution: the case of England', in R. P. W. Visser, H. J. M. Bos, L. C. Palm and H. A. M. Snelders (eds), *New Trends in the History of Science - proceedings from a conference held at the University of Utrecht*, Amsterdam, Rodopi, 1989.

Feingold, Mordechai, 'When facts matter', *Isis*, vol. 87, 1996, pp. 131-9.

Feingold, Mordechai, 'Mathematical Sciences and New Philosophy', in N. Tyacke (ed.), *The History of The University of Oxford*, vol. IV, 'Seventeenth Century', Oxford, Clarendon Press, 1997.

Feldhay, Rivka and Heyd, Michael, 'The discourse of pious science', *Science in Context*, vol. 3, no. 1, 1989, pp. 109-42.

Feldhay, Rivka, 'Narrative constraints on historical writing: the case of the scientific revolution', *Science in Context*, vol. 7, 1994, pp. 7-24.

Ferrari, G., 'Public anatomy lessons and the carnival: The anatomy theatre of Bologna', *Past and Present*, vol. 117, 1987, pp. 50-106.

Findlen, Paula, 'Empty signs? Reading the book of nature in Renaissance science', *Studies in the History and Philosophy of Science*, vol. 21, 1990, pp. 511-18.

Findlen, Paula, 'Jokes of nature and jokes of knowledge: The playfulness of scientific discourse in early modern Europe', *Renaissance Quarterly*, vol. 43, 1990, pp. 292-331.

Findlen, Paula, 'Between Carnival and Lent: The Scientific revolution at the margins of culture', *Configurations*, vol. 6, no. 2, 1998, pp. 243-67.

Fish, Stanley E., *Self-consuming artifacts: The experience of seventeenth-century literature*, Berkeley, University of California Press, 1972.

Fleitmann, Sabina, *Walter Charleton (1620-1707), "Viruoso": Leben und Werke*, Frankfurt, Peter Lang, 1986.

Force, James E., 'Secularization, the language of God and the Royal Society at the turn of the seventeenth century', *History of European Ideas*, vol. 2, 1981, pp. 221-35.

Force, James E. and Popkin, Richard H., (eds), *The books of nature and Scripture: recent essays on natural philosophy, theology, and biblical criticism in the Netherlands of Spinoza's time and the British Isles of Newton's time*, Boston, Kluwer Academic, 1994.

Folter, R. J., 'A newly discovered Oeconomia Animalis by Peter Muis of Rotterdam (c. 1645-1721)', *Janus*, vol. 65, 1978, pp. 183-204.

Foucault, Michel, *The Order of Things: An archaeology of the human sciences*, New York, Pantheon Books, 1970.

Fournier, Marian, *The Fabric of Life: Microscopy in the Seventeenth Century*, Baltimore, Johns Hopkins University Press, 1996.

Fox, Christopher, 'Crawford, Willis, and Anthropologie Extracted: Some Early English Uses of Psychology', *Journal of the History of the Behavioural Sciences*, vol. 24, no. 4, 1988, pp. 378-80.

Frank, A. W., 'Bringing Bodies Back in: A Decade Review', *Theory, Culture and Society*, vol. 7, 1990, pp. 131-62.

Frank, Robert G., 'Science, Medicine and the Universities of early Modern England: Background and Sources', *History of Science*, vol. 1, 1973, pp. 194-216, 239-69.

Frank, Robert G., 'The John Ward Diaries: Mirror of seventeenth century science and Medicine', *Journal of the History of Medicine*, vol. 29, 1974, pp. 147-79.

Frank, Robert G., 'The Physician as Virtuoso in seventeenth-century England', in R. G. Frank and B. Shapiro (eds), *English Scientific Virtuosi in the Sixteenth and Seventeenth Centuries*, Los Angeles, University of California Press, 1979, pp. 57-114.

Frank, Robert G., *Harvey & the Oxford Physiologists: Scientific Ideas and Social Interaction*, Los Angeles, University of California Press, 1980.

Frank, Robert G., 'Viewing the Body: Reframing man and disease in Commonwealth and Restoration England', in W. G. Marshall (ed.), *The Restoration Mind*, Newark, University of Delaware Press, 1997.

Frank, Robert G., 'Medicine', in N. Tyacke (ed.), *The History of The University of Oxford*, vol. IV, 'Seventeenth Century', Oxford, Clarendon Press, 1997.

Franklin, Kenneth James, 'Some textual changes in successive editions of Richard Lower's *Tractatus De Corde*', *Annals of Science*, vol. 4, 1939, pp. 283-94.

French, Roger, *The History of the Heart Thoracic Physiology from ancient to modern times*, Aberdeen, Equipress, 1979.

French R. K. and Wear, A. (eds), *The Medical Renaissance of the Sixteenth Century*, Cambridge, Cambridge University Press, 1985.

French, Roger, 'Berengario da Carpi and the use of commentary in anatomical teaching', in I. M. Lonie, A. Wear and R. K. French (eds), *The Medical Renaissance of the Sixteenth Century*, Cambridge, Cambridge University Press, 1985, pp. 42-74.

French, R. and Wear, A. (eds), *The Medical Revolution of the Seventeenth Century*, Cambridge, Cambridge University Press, 1989.

French, Roger, 'Harvey in Holland: Circulation and the Calvinists', in A. Wear and R. French (eds), *The Medical Revolution of the Seventeenth Century*, Cambridge, Cambridge University Press, 1989, pp. 46-86.

French, Roger, 'The Anatomical Tradition', in W. Bynum & R. Porter (eds), *The Companion Encyclopedia of the History of Medicine*, London, Routledge, 1994.

French, Roger, 'The language of William Harvey's natural philosophy', *Journal of the History of Medicine and the Allied Sciences*, vol. 49, no. 1, 1994, pp. 24-51.

French, Roger, *William Harvey's Natural Philosophy*, New York, Cambridge University Press, 1994.

Fulton, J. F., 'A note on the origin of the term "physiology" ', *Yale Journal of Biology and Medicine*, vol. 3, 1930, pp. 59-62.

Funkenstein, Amos, *Theology and the Scientific Imagination from the Middle Ages to the Seventeenth Century*, New Jersey, Princeton University Press, 1984.

Gadow, Sally, 'Body and self: A dialectic', *Journal of Medicine and Philosophy*, vol. 5, no. 3, 1980, pp. 172-85.

Galison, Peter, 'Descartes' Comparisons: from the invisible to the visible', *Isis*, vol. 75, 1984, pp. 311-26.

Garber, Daniel and Ayers, Michael (eds), *The Cambridge History of Seventeenth Century Philosophy*, New York, Cambridge University Press, 1998.

Gariepy, Thomas P. 'Mechanism without metaphysics: Henricus Regius and the establishment of Cartesian medicine', PhD dissertation, Yale University, 1990.

Gascoigne, John, 'A reappraisal of the role of universities in the scientific revolution', in D. C. Lindberg and R. S. Westman (eds), *Reappraisals of the*

Scientific Revolution, Cambridge, Cambridge University Press, 1990, pp. 207-60.

Gaukroger, Stephen (ed.), *The Uses of Antiquity: The scientific revolution and the classical tradition*, Boston, Kluwer Academic Publishers, 1991.

Gaukroger, Stephen, *Descartes: an intellectual biography*, Oxford, Clarendon Press, 1997.

Gaukroger, Stephen (ed.), *The soft underbelly of reason: The passions in the seventeenth century*, London, Routledge, 1998.

Gaukroger, Stephen, *Francis Bacon and the Transformation of Early-Modern Philosophy*, New York, Cambridge University Press, 2001.

Gelbart, Nina R., 'The Intellectual Development of Walter Charleton', *Ambix,* 1971, vol. 18, no. 3, pp. 149-68.

Gerhart, Mary and Russell, Allan, *Metaphoric Process; the Creation of Scientific and Religious Understanding*, Fort Worth, Texas Christian University Press, 1984.

Geyer-Kordesch, Johanna, 'Passions and the ghost in the machine: or, what not to ask about science in seventeenth- and eighteenth-century Germany', in A. Wear and R. French (eds), *The Medical Revolution of the Seventeenth Century*, Cambridge, Cambridge University Press, 1989, pp. 145-63.

Ghiselin, Michael T., 'William Harvey's Methodology in *De Motu Cordis* from the standpoint of comparative anatomy', *Bulletin of the History of Medicine*, vol. 40, no. 4, 1966.

Gibson, W. C., 'The biomedical pursuits of Christopher Wren', *Medical History*, vol. 14, 1970, pp. 331-41.

Gieryn, Thomas, 'Distancing science from religion in seventeenth century England', *Isis*, vol. 79, 1988, pp. 582-93.

Giglioni, Guido, 'Anatomist Atheist? The "Hylozoistic" foundations of Francis Glisson's anatomical research', in O. P. Grell and A. Cunningham (eds), *Religio medici: medicine and religion in seventeenth century England*, Brookfield, Scolar Press, 1996.

Gilbert, G. Nigel and Mulkay, Michael, 'Experiments are the key: participants' histories and historians' histories of science', *Isis*, vol. 75, 1984, pp. 105-25.

Gillespie, Charles, 'Physick and Philosophy: A study of the influence of the College of Physicians of London upon the foundation of the Royal Society', *Journal of Modern History*, vol. 19, 1947, pp. 210-25.

Gillespie, Neal C., 'Natural history, natural theology and social order: John Ray and Newtonian ideology', *Journal of the History of Biology*, vol. 201, 1987, pp. 1-49.

Glidden, David K., 'The Hellenistic Background for Gassendi's Theory of Ideas', *Journal of the History of Ideas*, vol. 49, no. 3, 1988, pp. 405-25.

Gogel, Edward L. and Terry, James S., 'Medicine as Interpretation: The Uses of Literary Metaphors and Methods', *Journal of Medicine and Philosophy*, vol. 12, 1987, pp. 205-17.

Golden, S. A., 'Dryden's praise of Dr Charleton', *Hermathena*, vol. 103, 1966, pp. 59-65.

Golinski, Jan V., 'Robert Boyle: Scepticism and authority in seventeenth-century chemical discourse', in A. E. Benjamin, G. N. Cantor and J. R. R. Christie (eds), *The Figural and the Literal: Problems of language in the history of science and philosophy, 1630-1800*, Manchester, Manchester University Press, 1987.

Golinski, Jan V., 'Chemistry in the scientific revolution: Problems of language and communication', in D. C. Lindberg and R. S. Westman (eds), *Reappraisals of the Scientific Revolution*, New York, Cambridge University Press, 1990.

Good, Byron, *Medicine, rationality and experience: an anthropological perspective*, New York, Cambridge University Press, 1994.

Goodfield, Gwyneth June, *The Growth of Scientific Physiology: Physiological Method and the Mechanist-Vitalist Controversy, Illustrated by the Problems of Respiration and Animal Heat,* London, Hutchinson, 1960.

Gordon, D. R., 'Tenacious assumptions in western medicine', in M. Lock and D. Gordon (eds), *Biomedicine Examined*, Dordrecht, Kluwer, 1988, pp. 19-56.

Gorham, Geoffrey, 'Mind-body dualism and the Harvey-Descartes controversy', *Journal of the History of Ideas*, vol. 55, no. 2, 1994, pp. 211-34.

Gouk, Penelope, 'Performance practice: Music, medicine and natural philosophy in Interregnum Oxford', *British Journal of the History of Science*, vol. 29, 1996, pp. 257-88.

Graham, Peter W., 'Harvey's *De Motu Cordis*: the rhetoric of science and the science of rhetoric', *Journal of the History of Medicine and the Allied Sciences*, vol. 33, no. 4, 1978, pp. 469-76.

Grant, Douglas, *Margaret the First: A biography of Margaret Cavendish, Duchess of Newcastle, 1623-1673*, London, Rupert Hart-Davis, 1957.

Greenblatt, Stephen, 'Toward a universal language of motion: Reflections on a seventeenth century muscle man', in Susan Leigh Foster (ed.), *Choreographing History*, Bloomington, Indiana University Press, 1995.

Grell, Ole Peter, 'Conflicting duties: Plague and the obligations of early modern physicians toward patients and commonwealth in England and the Netherlands', in A. Wear, J. Geyer-Kordesch and R. French (eds), *Doctors and Ethics: The Earlier Historical Setting of Professional Ethics*, Amsterdam, Rodopi, 1993.

Gross, Alan G., *The Rhetoric of Science,* London, Harvard University Press, 1990, pp. 21-32.

Guerrini, Anita, 'Newtonian Matter Theory, Chemistry and Medicine, 1690-1713', PhD Dissertation, Indiana University, 1983.

Guerrini, Anita, 'Archibald Pitcairn and Newtonian Medicine', *Medical History*, vol. 31, 1987, pp. 70-83.

Hacking, Ian, *The Emergence of Probability*, Cambridge, Cambridge University Press, 1975.

Hall, A. Rupert, *The Scientific Revolution, 1500-1800: The Formation of the Modern Scientific Attitude*. 2nd ed., Boston, Beacon Press, 1966.

Hall, A. Rupert, 'Medicine and the Royal Society', in A. G. Debus (ed.), *Medicine in Seventeenth Century England*, London, University of California Press, 1974.

Hall, A. Rupert, 'Magic, metaphysics and mysticism in the scientific revolution', in M. L. Righini Bonelli & W. R. Shea (eds), *Reason, Experiment and Mysticism in the Scientific Revolution*, New York, Macmillan, 1975, pp. 275-82.

[Hall] Marie Boas, 'The establishment of the mechanical philosophy', *Osiris*, vol. 10, 1952, pp. 412-541.

[Hall], Marie Boas, *Robert Boyle and Seventeenth Century Chemistry*, Cambridge, Cambridge University Press, 1958.

Hall, Marie Boas, *Promoting Experimental Learning: Experiment and the Royal Society, 1660-1727*, New York, Cambridge University Press, 1991.

Hall, Marie Boas, 'Frederic Slare, F.R.S. (1648-1727)', *Notes and Records of the Royal Society of London*, vol. 46, no. 1, 1992, pp. 23-41.

Hall, Roland, 'Some seventeenth century antedatings of OED from Charleton', *Notes and Queries*, vol. 32, 1985, p. 192.

Hall, Thomas S., *Ideas of Life & Matter, Studies in the History of general Physiology, 600BC-1900AD*, Chicago, University of Chicago Press, 1969.

Hall, Thomas S., 'Euripus: or, the ebb and flow of the blood', *Journal of the History of Biology*, vol. 8, no. 2, 1975, pp. 321-50.

Haraway, Donna, 'The Biopolitics of Postmodern Bodies; Constitutions of Self in immune systems discourses', in *Simians, Cyborgs and Women: The Reinvention of Nature*, New York, Routledge, 1991, pp. 203-30.

Harley, David, 'Medical metaphors in English moral theology', *Journal of the History of Medicine and the Allied Sciences*, vol. 48, no. 4, 1993, pp. 396-435.

Harley, David, 'Political postmortems and English morbid anatomy in the seventeenth century', *Social History of Medicine*, vol. 7, 1994, pp. 1-28.

Harrison, Peter, *The Bible, Protestantism and Theories of Natural Science*, New York, Cambridge University Press, 1998.

Hearne, Thomas, *Remarks and Collections*, vol. IV, ed. D. W. Rannie, Oxford, Clarendon Press, 1898.

Henry, John, 'Occult qualities and the experimental philosophy: active principles in pre-Newtonian matter theory', *History of Science*, vol. 24, no. 4, 1986, pp. 335-81.

Henry, John, 'Medicine and Pneumatology; Henry More, Richard Baxter, and Francis Glisson's *Treatise on the Energetic Nature of Substance*', *Medical History*, vol. 31, no. 1, 1987, pp. 15-40.

Henry, John, 'The matter of souls: medical thought and theology in seventeenth century England', in A. Wear and R. French (eds), *The Medical Revolution of the Seventeenth Century*, Cambridge, Cambridge University Press, 1989, pp. 87-113.

Henry, John, 'Magic and science in the seventeenth and eighteenth centuries', in R. C. Olby, G. N. Cantor, J. R. R. Christie and M. J. S. Hodge (eds), *Companion to the History of Modern Science*, London, Routledge, 1990, pp. 583-96.

Henry, John, and Hutton, Sarah (eds), *New perspectives on Renaissance thought: essays in the history of science, education and philosophy: in memory of Charles B. Schmitt*, London, Duckworth, 1990.

Henry, John, 'Doctors and healers: popular culture and the medical profession', in S. Pumfrey, P. L. Rossi and M. Slawinski (eds), *Science, Culture and Popular Belief in Renaissance Europe*, New York, Manchester University Press, 1991, pp. 191-221.

Henry, John, 'The Scientific Revolution in England', in Roy Porter and Mikulas Teich (eds), *Scientific Revolution in National Context*, New York, Cambridge University Press, 1992, pp. 178-210.

Henry, John, *The Scientific Revolution and the Origins of Modern Science*, New York, St. Martin's Press, 1997.

Hesse, Mary B., *Revolutions and Reconstructions in the Philosophy of Science*, Bloomington, Indiana University Press, 1980.

Hesse, Mary, 'Reason and evaluation in the history of science', in Robert Young and Mikulas Teich (eds), *Changing Perspectives in the History of Science: Essays in honour of Joseph Needham*, Dordrecht, D. Reidel Pub. Co., 1973.

Heyd, Michael, 'The new experimental philosophy: A manifestation of "enthusiasm" or an antidote to it?', *Minerva*, vol. 25, 1987, pp. 423-40.

Hill, Christopher, 'Science in Seventeenth-Century London', *The Listener*, vol. 67, 1962, p. 943.

Hill, Christopher, 'William Harvey and the idea of monarchy', *Past and Present*, vol. 27, 1964, pp. 54-72.

Hill, Christopher, *Intellectual Origins of the English Revolution*, Oxford, Clarendon Press, 1965, pp. 28-9, 74-84.

Hill, Christopher, 'The medical profession and its radical critics', in C. Hill (ed.), *Change and Continuity in Seventeenth Century England*, London, Weidenfeld and Nicolson, 1974, pp. 157-78.

Hirschauer, Stefan, 'The manufacture of bodies in surgery', *Social Studies of Science*, vol. 21, 1991, pp. 279-319.

Hochstrasse, T. J., *Natural Law Theories in the Early Enlightenment*, Cambridge, Cambridge University Press, 2000.

Hookyaas, R., *Religion and the Rise of Modern Science*, Edinburgh, Scottish Academic Press, 1973.

Hoopes, Robert, *Right Reason in the English Renaissance*, Massachusetts, Harvard University Press, 1962.

Houghton, W. E., 'The history of trades: its relation to seventeenth century thought', *Journal of the History of Ideas*, vol. 2, 1941, pp. 33-60.

Howell, W. S., *Logic and Rhetoric in England, 1500-1700*, New York, Russell and Russell, 1956.

Hudson-Jones, Anne, 'Literature and Medicine: Traditions and Innovations', in B. Clarke and W. Aycock (eds), *The Body and the Text: Comparative Essays in Literature and Medicine*, Lubbock, Texas, Texas Tech University Press, 1990, pp. 11-24.

Hughes, J. Trevor, *Thomas Willis, 1621-1675: His life and work*, London, Royal Society of Medicine, 1991.

Hunter, Michael, *Science and Society in Restoration England*, New York, Cambridge University Press, 1981.

Hunter, Michael, *The Royal Society and its Fellows 1660-1700: The morphology of an early scientific institution*, Chalfont St. Giles, British Society for the History of Science, 1982.

Hunter, Michael and Schaffer, Simon (eds), *Robert Hooke, New Studies*, Woodbridge, Boydell Press, 1989.

Hunter, Michael, *Establishing the New Science: The experience of the early Royal Society*, Woodbridge, Suffolk, Boydell Press, 1989.

Hunter, Michael, 'Alchemy, Magic and Moralism in the Thought of Robert Boyle', *British Journal of the History of Science*, vol. 23, 1990, pp. 387-410.

Hunter, Michael, *Science and the Shape of Orthodoxy: Intellectual change in late seventeenth-century Britain*, Woodbridge, Suffolk, Boydell Press, 1995.

Hunter, Michael, 'How Boyle became a scientist', *History of Science*, vol. 33, 1995, p. 59-103.

Hunter, Michael, 'Robert Boyle and the dilemma of biography in the scientific revolution', in M. Shortland and R. Yeo (eds), *Telling Lives in Science, Essays on Scientific Biography*, Cambridge, Cambridge University Press, 1996, pp. 115-37.

Hunter, Michael (ed.), *Archives of the Scientific Revolution: The formation and exchange of ideas in seventeenth-century Europe*, Rochester, NY, Boydell Press, 1998.

Hunter, Richard A. and Cuttler, Emily, 'Walter Charleton's *Natural History of the Passions* (1674) and J. F. Senault's *The Use of the Passions* (1649): A case of mistaken identity', *Journal of the History of Medicine and the Allied Sciences*, vol. 13, 1958, pp. 87-92.

Huntley, Frank L., 'Sir Thomas Browne MD, William Harvey, and the metaphor of the circle', *Bulletin of Medical History*, vol. 25, 1951, pp. 236-47.

Hutchison, Keith, 'What happened to Occult Qualities in the scientific revolution?', *Isis*, vol. 73, 1982, pp. 233-53.

Hutchison, Keith, 'Supernaturalism and the Mechanical Philosophy', *History of Science*, vol. 21, no. 53, 1983, pp. 297-333.

Hutchison, Keith, 'Dormitive Virtues, Scholastic Qualities and the New Philosophies', *History of Science*, vol. 29, no. 85, 1991, pp. 245-78.

Isaacs, L. N., 'The effecting of all things possible: Molecular biology and Bacon's vision', *Perspectives in Biology and Medicine*, vol. 30, no. 3, 1987, pp. 402-32.

Isler, Hansreudi, *Thomas Willis*, New York, Hafner, 1968.

Jackson, Stanley W., *Melancholia and Depression: From Hippocratic Times to Modern Times*, New Haven, Yale University Press, 1986, pp. 104-15.

Jacob, James R., *Robert Boyle and the English Revolution: a study in social and intellectual change*, New York, B. Franklin, 1977.

Jacob, Margaret, 'Science and social passion: the case of seventeenth century England', *Journal of the History of Ideas*, vol. 43, no. 2, 1982, pp. 331-9.

Jacob, Margaret, 'Constructing, deconstructing and reconstructing the history of science', *Journal of British Studies*, vol. 36, no. 4, 1997, pp. 459-67.

Jardine, Lisa, *Francis Bacon: Discovery and the Arts of Discourse*, London, Cambridge University Press, 1974.

Jardine, Nicholas, 'Writing off the scientific revolution', *Journal of the History of Astronomy*, vol. 22, 1991, pp. 311-18.

Jaynes, Julian, 'The problem of animate motion of the seventeenth century', *Journal of the History of Ideas*, vol. 31, 1970, pp. 219-34.

Jensen, H. James, 'Controversy in seventeenth-century art and criticism: Sir Walter Charleton's redefinition of nature, reason and imagination', in J. P. Strelka (ed.), *Literary Theory and Criticism; Festschrift presented to Rene Wellek in honor of his 80th birthday*, Frankfurt, Lang, 1984, pp. 315-33.

Jewson, Nicholas, 'Medical knowledge & the patronage system in eighteenth century England', *Sociology*, vol. 8, 1974, pp. 369-85.

Jobe, T. H., 'Medical theories of melancholia in the seventeenth and early eighteenth centuries', *Clio Medica*, vol. 11, no. 4, 1976, pp. 217-31.

Johns, Adrian, *The Nature of the Book: Print and knowledge in the making*, Chicago, University of Chicago Press, 1998.

Jones, Kathleen, *A Glorious Fame: The life of Margaret Cavendish, Duchess of Newcastle, 1623-1673*, London, Bloomsbury, 1988.

Jones, Richard Foster, *Ancients and Moderns, a study of the rise of the scientific movement in seventeenth-century England*, St Louis, Washington University Press, 1961.

Jones, Richard Foster, 'The Humanistic defence of learning in the mid-seventeenth century', in J. A. Mazzeo (ed.), *Reason and the Imagination: Studies in the History of Ideas 1600-1800*, New York, Columbia University Press, 1962, pp. 71-93.

Jones, Richard Foster, *The Rhetoric of Science in the Mid-Seventeenth Century*, Chicago, University of Chicago Press, 1963, pp. 5-24.

Jordanova, Ludmilla, 'Has the social history of medicine come of age?', *Historical Journal*, vol. 36, no. 2, 1993, pp. 437-49.

Jucovy, Peter M., 'Circle and circulation: The language and imagery of William Harvey's discovery', *Perspectives in Biology and Medicine,* vol. 20, 1976, pp. 92-107.

Kahn, Victoria, *Rhetoric, Prudence and Skepticism in the Renaissance*, Ithaca, Cornell University Press, 1985.

Kantorowicz, Ernst H., *The King's Two Bodies: A Study in Mediaeval Political Sociology*, New Jersey, Princeton University Press, 1957.

Kaplan, Barbara, *"Divulging of Useful Truths in Physick"; The Medical Agenda of Robert Boyle*, Baltimore, Johns Hopkins University Press, 1993.

Kardel, Troels, 'Stensen's myology in historical perspective', in *Steno on Muscles*, Philadelphia, American Philosophical Society, 1994.

Kargon, Robert H., 'Walter Charleton, Robert Boyle, and the Acceptance of Epicurean Atomism in England', *Isis*, vol. 55, no. 2, 1964. pp. 184-92.

Kargon, Robert H., 'Atomism in England: Walter Charleton and the *Physiologia*', Introduction to Charleton, *Physiologia Epicuro-Gassendo Charletoniana*, New York and London, Johnson Reprint Corporation, 1966.

Kargon, Robert H., *Atomism in England From Harriot to Newton*, Oxford, Clarendon Press, 1966.

Kassler, Jamie, *Inner Music: Hobbes, Hooke and North on Internal Character*, London, Athlone, 1995.

Kearney, Hugh, *Science and Change: 1500–1700*, London, Weidenfeld and Nicolson, 1971.

Kearns, M. S., *Metaphors of Mind in Fiction and Psychology*, Lexington, University of Kentucky Press, 1987.

Keele, Kenneth D., 'Thomas Willis on the Brain', *Medical History*, vol. 11, 1967, pp. 194-200.

Keele, Kenneth D., 'The Sydenham-Boyle theory of morbific particles', *Medical History*, vol. 18, 1974, pp. 240-48.

Keevil, J. J., 'Sir Charles Scarburgh', *Annals of Science*, vol. 8, 1952, pp. 113-21.

Keller, Evelyn Fox, *Secrets of Life, Secrets of Death: Essay on Language, Gender and Science*, New York, Routledge, 1992.

Kelley, Donald R., 'Eclecticism and the History of Ideas', *Journal of the History of Ideas*, vol. 62, no. 4, 2001, pp. 577-92.

Kenyon, John P., *The Popish Plot*, London, Heinemann, 1972.

Kim, Yung Sik, 'Another look at Robert Boyle's acceptance of the mechanical philosophy', *Ambix*, vol. 38, no. 1, 1991, pp. 1-10.

King, Lester S., *The Growth of Medical Thought*, Chicago, University of Chicago Press, 1963.

King, Lester S., 'The road to scientific therapy: Signatures, "sympathy" and controlled experiment', *Journal of the American Medical Association*, vol. 197, 1966, pp. 250-6.

King, Lester S., *The Road to Medical Enlightenment 1650-1695*, London, Macdonald and Co., 1970.

King, Lester S., 'Empiricism and rationalism in the works of Thomas Sydenham', *Bulletin of the History of Medicine*, vol. 44, 1970, pp. 1-11.

King, Lester S., 'The Transformation of Galenism', in A. Debus (ed.), *Medicine in Seventeenth Century England*, California, University of California Press, 1974, pp. 7-31.

King, Lester S., 'Some basic explanations of disease: An historian's viewpoint', in A. L. Caplan, H. T. Engelhardt and J. J. McCartney (eds), *Concepts of Health and Disease: Interdisciplinary Perspectives*, Massachusetts, Addison-Wesley, 1981, pp. 231-45.

Kirmayer, L. J., 'Mind and Body as Metaphors; Hidden Values in Biomedicine', in M. Lock and D. Gordon (eds), *Biomedicine Examined*, Dordrecht, Kluwer, 1988, pp. 57-94.

Knowlson, James, *Universal Language Schemes in England and France, 1600-1800*, Toronto, University of Toronto Press, 1975.

Kress, Gunther and Aers, David, *Literature, Language and Society in England, 1580-1680*, Dublin, Gill and Macmillan, 1981.

Kristeller, Paul Oskar, *Renaissance Thought II, Papers on Humanism and the Arts*, New York, Harper & Row, 1965.

Krivatsy, P., *Seventeenth Century Printed Books*, Maryland, National Library of Medicine, 1989.

Kroll, Richard F., *The Material Word: Literate Culture in the Restoration and Early Eighteenth Century*, Baltimore, Johns Hopkins University Press, 1991.

Kudlien, Fridolf, 'Medicine as a liberal art and the question of the physician's income', *Journal of the History of Medicine*, vol. 31, 1976, pp. 448-59.

Kuhn, Thomas S., *The Structure of Scientific Revolutions*, Chicago, University of Chicago Press, 1964.

Kutzer, Michael, 'Tradition, metaphors, anatomy of the brain: The physiology of insanity in the late XVIth and XVIIth centuries', in C. Debru (ed.), *Essays in the History of the Physiological Sciences*, Amsterdam, Rodopi, 1995.

LaCapra, Dominick, *History and Criticism*, Ithaca & London, Cornell University Press, 1985.

Lakoff, George and Johnson, Mark, *Metaphors We Live By*, Chicago, University of Chicago Press, 1980.

Laqueur, Thomas, *Making Sex: Body and Gender from the Greeks to Freud*, Massachusetts, Harvard University Press, 1990.

Larmore, Charles, 'Scepticism', in D. Garber & M. Ayers (eds), *The Cambridge History of Seventeenth Century Philosophy*, Cambridge, Cambridge University Press, 1998, vol. 2, pp. 1145-92.

Laudan, Laurens, 'The clock metaphor and probabilism: The impact of Descartes on English methodological thought, 1660-1665', *Annals of Science*, vol. 22, 1966, pp. 73-104.

Lawrence, C., 'Alexander Monro Primus, and the Edinburgh manner of anatomy', *Bulletin of the History of Medicine*, vol. 62, 1988, pp. 193-214.

Lawrence, Susan C., 'Educating the senses: Students, teachers and medical rhetoric in eighteenth-century London', in W. Bynum and R. Porter (eds), *Medicine and the Five Senses*, Cambridge, Cambridge University Press, 1993.

Lawrence, Susan C., 'Anatomy and address: Creating medical gentlemen in eighteenth century London, in V. Nutton and R. Porter (eds), *The History of Medical Education in Britain*, Amsterdam, Rodopi, 1995.

Leary, David E. (ed.), *Metaphors in the History of Psychology*, New York, Cambridge University Press, 1990.

Leatherdale, W. H., *The Role of Analogy, Model and Metaphor in Science*, Amsterdam & Oxford, North Holland Publishing Co., 1974.

Leder, Drew, 'Medicine and paradigms of embodiment', *Journal of Medicine and Philosophy*, vol. 9, 1984.

Leder, Drew, 'A tale of two bodies: The Cartesian corpse and the lived body', in Drew Leder (ed.), *The Body in Medical Thought and Practice*, Netherlands, Kluwer, 1992, pp. 17-35.

Leslie, M. and Raylor, T. (eds), *Culture and cultivation in early modern England*, Leicester, Leicester University Press, 1994.

Lett, Hugh, 'Anatomy at the Barber-Surgeons Hall', *British Journal of Surgery*, vol. 31, 1943, pp. 101-11.

Levin, David M. and Solomon, George F., 'The Discursive Formation of the Body in the History of Medicine', *Journal of Medicine and Philosophy*, vol. 15, no. 5, 1990, pp. 515-37.

Lewis, Gillian, 'The faculty of medicine', in J. K. McConica (ed), *History of the University of Oxford*, vol. 3, 'The Collegiate University', Oxford, Clarendon Press, 1984.

Lewis, Eric, 'Walter Charleton and early modern eclecticism', *Journal of the History of Ideas*, vol. 62, no. 4, 2001.

Leyden, W. van, 'Antiquity and authority: A paradox in the Renaissance theory of history', *Journal of the History of Ideas*, vol. 19, 1958, pp. 473-92.

Lindberg, D. C. and Westman, R. S. (eds), *Reappraisals of the Scientific Revolution*, Cambridge, Cambridge University Press, 1990.

Lindberg, David C., 'Conceptions of the scientific revolution from Bacon to Butterfield: A preliminary sketch', in D. C. Lindberg and R. S. Westman (eds), *Reappraisals of the Scientific Revolution*, Cambridge, Cambridge University Press, 1990, pp. 1-26.

Lindeboom, Gerrit, 'Dog and Frog: Physiological experiments at Leiden during the seventeenth century', in T. H. Lunsingh Scheurleer & G. H. M. Posthumus Meyjes (eds), *Leiden University in the Seventeenth Century: An Exchange of Ideas*, Leiden, Brill, 1975.

Lindeboom, Gerrit, *Descartes and Medicine*, Amsterdam, Rodopi, 1979.

Linden, Stanton J., 'Walter Charleton and Henry Vaughan's cock crowing', *Notes and Queries*, vol. 36, no. 1, 1989, p. 38.

Lonie, Iain M., 'Hippocrates the Iatromechanist', *Medical History*, vol. 25, 1981, pp. 113-50.

Lovejoy, Arthur, *The Great Chain of Being*, Massachussetts, Harvard University Press, 1936.

Lupton, Deborah, *Medicine as Culture: Illness, Disease and the Body in Western Societies*, London, Sage, 1994.

Lux, David S. and Cook, Harold J., 'Closed circles or open networks? Communicating at a distance during the scientific revolution', *History of Science*, vol. 36, 1998, pp. 179-211.

Luyendijk-Elshout, Antonie M., 'Oeconomia animalis, pores and particles: The rise and fall of the mechanical philosophy of Theodoor Craanen', in T. H. Lunsingh Scheurleer & G. H. M. Posthumus Meyjes (eds), *Leiden University in the Seventeenth Century: An Exchange of Ideas*, Leiden, Brill, 1975.

Lynch, William T., 'Politics in Hobbes' mechanics: The social as enabling', *Studies in the History and Philosophy of Science*, vol. 22, no. 2, 1991, pp. 295-320.

MacCormac, Earl R., *Metaphor and Myth In Science and Religion*, Durham, Duke University Press, 1976.

Macdonald, Michael, 'Insanity and the Realities of History in Early Modern England', *Psychological Medicine*, vol. 11, 1981, pp. 11-25.

Macrae, D. G., 'The Body and Social Metaphor', in J. Benthall and T. Polhemus (eds), *The Body as a Medium of Expression: An Anthology*, New York, Dutton, 1975, pp. 59-73.

Maia Neto, José R., 'Academic skepticism in early modern philosophy', *Journal of the History of Ideas*, vol. 58, no. 2, 1997, pp. 199-220.

Malet, Antoni, 'Isaac Barrow on the Mathematization of Nature: Theological voluntarism and the rise of geometrical optics', *Journal of the History of Ideas*, vol. 58, no. 2, 1997, pp. 265-87.

Mani, Nikolaus, 'Glisson's hepatology and Wepfer's work on apoplexy', in Lloyd G. Stevenson (ed.), *Celebration of Medical History*, Baltimore & London, Johns Hopkins University Press, 1982, pp. 37-63.

Marcovich, Anne, 'Concerning the continuity between the Image of Society and the Image of the Human Body: An Examination of the Work of the English Physician J. C. Lettsom (1746-1815)', in Peter Wright & Andrew Treacher (eds), *The Problem of Medical Knowledge: Examining the Social Construction of Medicine*, Edinburgh, Edinburgh University Press, 1982, pp. 69-86.

Markley, Robert, 'Robert Boyle on language: Some considerations touching the style of the Holy Scriptures', *Studies in Eighteenth Century Culture*, vol. 14, 1985, pp. 159-71.

Markley, Robert, *Fallen Languages: Crises of Representation in Newtonian England, 1600-1740*, Ithaca, Cornell University Press, 1993.

Martensen, Robert, ' "Habit of Reason": Anatomy and Anglicanism in Restoration England', *Bulletin of the History of Medicine*, vol. 66, no. 4, 1992, pp. 511-35.

Martensen, Robert, 'The Circles of Willis: Physiology, Culture and the Formation of the Neurocentric Body in England, 1640–1690', PhD Dissertation, University of California, 1993.

Martensen, Robert, 'The transformation of Eve: Women's bodies, medicine and culture in early modern England', in R., Porter and M. Teich (eds), *Sexual Knowledge, Sexual Science: The History of Attitudes to Sexuality*, Cambridge, Cambridge University Press, 1994.

Martin, Emily, 'The End of the Body?', *American Ethnologist*, vol. 19, 1992, pp. 121-40.

Mason, S. F., 'The scientific revolution and the Protestant reformation', in R. Olson (ed.), *Science as Metaphor*, Wadsworth, California, 1971.

Matsuo, Y., 'A Digression of the Animal Spirits; the changes of the concept of 'spirit' in the seventeenth century science', *Historia Scientiarum*, vol. 25, 1983, pp. 1-15.

Maulitz, Russell C., 'The Pathological Tradition', in R. Porter and W. Bynum (eds), *The Companion Encyclopedia of the History of Medicine*, London, Routledge, 1993, pp. 169-91.

Mazzeo, Joseph A., 'Universal Analogy and the Culture of the Renaissance', *Journal of the History of Ideas*, vol. 15, 1954, pp. 299-303.

Mazzeo, Joseph A. (ed.), *Reason and the Imagination: Studies in the History of Ideas 1600-1800*, New York, Columbia University Press, 1962.

McKenna, M., 'William Harvey and the Royal College of Physicians versus the empirics', *Canadian Journal of Surgery*, vol. 20, no. 3, 1987, pp. 215-7.

McKie, Douglas, 'English writers on atomism before Dalton', *Endeavour*, vol. 25, 1966, pp. 13-15.

McMahon, C. E., 'Harvey on the soul', *Journal of the History of Behavioural Science*, vol. 11, no. 3, 1975, pp. 276-83.

McMahon, C. E., 'The role of imagination in the disease process: Pre-Cartesian history', *Psychological Medicine*, vol. 6, 1976, pp. 179-84.

Meier, Richard Y., ' "Sympathy" in the neurophysiology of Thomas Willis', *Clio Medica*, vol. 17, no. 2-3, 1982, pp. 95-112.

Meinel, Christoph, 'Early seventeenth century atomism: Theory, epistemology and the insufficiency of experiment', *Isis*, vol. 79, 1988, pp. 68-103.

Meinel, Christoph, 'Empirical support for the corpuscular theory in the seventeenth century', in D. Batens and J. P. van Bendegem (eds), *Theory and Experiment: Recent Insights and New Perspectives on their Relation*, Boston, Reidel, 1988.

Mendelsohn, J. A., 'Alchemy and politics in England, 1949-1665', *Past and Present*, vol. 135, 1992, pp. 30-78.

Merchant, Caroline, *The Death of Nature: Women, Ecology and the Scientific Revolution*, San Francisco, Harper & Row, 1980.

Merton, Robert K., *Science, technology & society in seventeenth century England*, New York, H. Fertig, 1970.

Meyer, Alfred, 'Willis' place in the history of muscle physiology', *Proceedings of the Royal Society of Medicine*, vol. 57, 1964, pp. 687-92.

Meyer, Alfred and Hierons, Raymond, 'On Thomas Willis' concepts of neurophysiology', *Medical History*, vol. 9, 1965, pp. 1-15, 142-55.

Miller, Jonathan, *The Body In Question*, New York, Random House, 1978.

Mintz, Samuel, 'The Duchess of Newcastle's visit to the Royal Society', *Journal of English and German Philology*, vol. 51, 1952, pp. 168-76.

Mintz, Samuel, *The Hunting of Leviathan: Seventeenth century reactions to the materialism and moral philosophy of Thomas Hobbes*, Cambridge, Cambridge University Press, 1962.

Montgomery, S. L., 'Codes and Combat in Biomedical Discourse', *Science as Culture*, vol. 2, no. 3, 1991, pp. 341-91.

Morna, Bruce T., *Patronage and Institutions: Science, Technology and Medicine at the European Court, 1500-1700*, Boydell, Woodbridge, 1991.

Morse, Louis Trenchard, 'Boyle as Alchemist', *Journal of the History of Ideas*, vol. 11, 1941, pp. 61-76.

Mowry, Brian, 'From Galen's theory to William Harvey's theory: a case study in the rationality of scientific theory change', *Studies in the History and Philosophy of Science*, vol. 16, no. 1, 1985, pp. 49-82.

Muhlhausler, Peter, 'Metaphors others live by', *Language and Communication*, vol. 15, no. 3, 1995, pp. 281-8.

Mullens, W. H., 'Walter Charleton and his *Onomasticon Zoicon*', *British Birds*, vol. 5, 1911, pp. 64-71.

Mulligan, Lotte, 'Anglicanism, Latitudinarianism and science in seventeenth century England', *Annals of Science*, vol. 30, 1973, pp. 213-19.

Mulligan, Lotte and Mulligan, Glen, 'Reconstructing Restoration science', *Social Studies of Science*, vol. 11, 1981, pp. 327-64.

Mulligan, Lotte, '"Reason", "right reason" and "revelation" in mid-seventeenth-century England', in B. Vickers (ed.), *Occult and Scientific Mentalities in the Renaissance*, New York, Cambridge University Press, 1984.

Mulligan, Lotte, 'Robert Hooke's "Memoranda": Memory and Natural History', *Annals of Science*, vol. 49, 1992, pp. 47-61.

Mulligan, Lotte, 'Robert Boyle, "Right Reason" and the Meaning of Metaphor', *Journal of the History of Ideas*, vol. 55, no. 2, 1994, pp. 235-58.

Mulligan, Lotte, 'Self scrutiny and the study of nature: Robert Hooke's diary as natural history', *Journal of British Studies*, vol. 35, 1996, pp. 311-42.

Mulsow, Martin, 'Eclecticism or skepticism? A problem of the early enlightenment', *Journal of the History of Ideas*, vol. 58, no. 3, 1997, pp. 465-77.

Munk, William, *The Roll of the Royal College of Physicians of London*, 2nd ed., 3 vols, London, 1878, vol. 1.

Munz, Peter and Vickers, Brian, 'The Rhetoric of Rhetoric', *Journal of the History of Ideas*, vol. 51, 1990, pp. 121-59.

Nadler, S., 'Doctrines of Explanation', in Garber, Daniel and Ayers, Michael (eds), *The Cambridge History of Seventeenth Century Philosophy*, Cambridge, Cambridge University Press, 1998.

Nagy, Doreen, *Popular Medicine in Seventeenth Century England*, Bowling Green, Ohio/Bowling Green University Popular Press, 1988.

Nakamura, Teiri, 'How and why did Descartes support the theory of blood circulation?', *Japanese Studies in the History of Science*, vol. 13, 1974, pp. 75-80.

Nakamura, Teiri, 'Thomas Willlis' theory of the heart Movement', *XIVth International Congress of the History of Science, Proceedings* No. 3, Science Council of Japan, 1975, pp. 55-7.

Nakamura, Teiri, 'Thomas Willis' and Lower's physiology, with special reference to the theory of heart movement', *Japanese Studies in the History of Science*, vol. 16, 1977, pp. 23-41.

Nayler, Margaret A., 'The Insoluble Problem: Muscle in the Mid- to Late-Seventeenth Century', PhD Dissertation, University of Melbourne, 1993.

Nicolson, Marjorie H. (ed.), *The Conway Letters, The Correspondence of Ann, Viscount Conway, Henry More and their friends, 1642-1684*, New Haven, Yale University Press, 1930.

Nicolson, Marjorie Hope, *Pepys' Diary and the New Science*, Charlottesville, University Press of Virginia, 1965.

Niebyl, Peter H., 'Old Age, Fever and the Lamp Metaphor', *Journal of the History of Medicine*, vol. 26, no. 4, 1971, pp. 351-68.

Niebyl, Peter H., 'Sennert, Van Helmont and medical ontology', *Bulletin of the History of Medicine*, vol., 45, 1971, pp. 115-37.

Niebyl, Peter H., 'Science and metaphor in the medicine of Restoration England', *Bulletin of the History of Medicine*, vol. 47, 1973, pp. 356-74.

Nutton, Vivian, *Medicine at the Courts of Europe, 1500-1837*, London, Routledge, 1990.

Nutton, Vivian, 'From medical certainty to medical amulets: Three aspects of ancient therapeutics', *Clio Medica*, vol. 22, 1991, pp. 13-23.

O'Malley, Charles D., 'John Evelyn and medicine', *Medical History*, vol. 12, 1968, pp. 219-31.

O'Malley, Charles Donald, 'Helkiah Crooke, MD, FRCP, 1576-1648', *Bulletin of the History of Medicine*, vol. 42, 1968, pp. 1-18.

O'Neill, J., *Five Bodies: The Human Shape of Modern Society*, Ithaca and London, Cornell University Press, 1985.

O'Toole, F., 'Qualities and Powers in the Corpuscular Philosophy of Robert Boyle', *Journal of the History of Philosophy*, vol. 22, 1974, pp. 295-315.

Oakley, Francis, 'The absolute and ordained power of God in sixteenth- and seventeenth-century theology', *Journal of the History of Ideas*, vol. 59, no. 3, 1998, pp. 437-61.

Oller í Adam, A. M., 'Walter Charleton: Filosofia Natural, Teologia Natural i Etica', PhD dissertation, University of Barcelona, 1995.

Olson, Richard (ed.), *Science as metaphor: The historic role of scientific theories in forming western culture*, Belmont, Wadsworth, 1971

Ormsby-Lennon, Hugh, 'Rosicrucian linguistics: Twilight of a Renaissance Tradition', in I. Merkel and A. G. Debus (eds), *Hermeticism and the Renaissance*, Washington, Folger, 1988, pp. 311-41.

Ortony, Andrew (ed.), *Metaphor and Thought*, New York, Cambridge University Press, 2nd edition, 1993.

Osborn, James M., *John Dryden: Some biographical facts and problems*, Gainesville, University of Florida Press, 1965.

Osler, Margaret J., 'Descartes and Charleton on Nature and God', *Journal of the History of Ideas*, vol. 40, 1979, pp. 445-56.

Osler, Margaret J., 'Providence and divine will in Gassendi's views on scientific knowledge', *Journal of the History of Ideas*, vol. 44, 1983, pp. 549-60.

Osler, Margaret J., 'Eternal truths and the laws of nature: The theological foundations of Descartes' philosophy of nature', *Journal of the History of Ideas*, vol. 46, 1985, pp. 349-62.

Osler, Margaret J. and Farber, Paul L. (eds), *Religion, Science and World View: Essays in honour of Richard S. Westfall*, Cambridge, Cambridge University Press, 1985.

Osler, Margaret J. (ed.), *Atoms, Pneuma and Tranquillity: Epicurean and Stoic themes in European thought*, Cambridge, Cambridge University Press, 1991.

Osler, Margaret J., 'The intellectual sources of Robert Boyle's philosophy of nature: Gassendi's voluntarism and Boyle's physico-theological project', in

R. Kroll, R. Ashcroft and P. Zagorin (eds), *Philosophy, Science and Religion in England*, 1640-1700, New York, Cambridge University Press, 1992.

Osler, Margaret J., *Divine Will and the Mechanical Philosophy: Gassendi and Descartes on contingency and necessity in the created world*, New York, Cambridge University Press, 1994.

Osler, Margaret J., 'Mixing metaphors: Science and religion or natural philosophy and theology in early modern Europe', *History of Science*, vol. 36, 1998, pp. 91-113.

Osler, Margaret J., 'Revolution or resurrection?', *Configurations*, vol. 7, no. 1, 1999, pp. 91-100.

Oster, Malcolm, 'The scholar and the craftsman revisited: Robert Boyle as aristocrat and artisan', *Annals of Science*, vol. 49, no. 3, 1992, pp. 255-76.

Outram, Dorinda, *The Body and the French Revolution: Sex, Class and Political Culture*, New Haven and London, Yale University Press, 1989.

Pagel, Walter, 'Harvey and Glisson on irritability, with a note on Van Helmont', *Bulletin of the History of Medicine*, vol. 41, 1967, pp. 497-514.

Pagel, Walter, 'Religious motives in the medical biology of the seventeenth century', *Bulletin of the Institute of the History of Medicine*, vol. 3, 1935, pp. 97-128; 213-31; 265-312.

Pagel, Walter, 'Giordano Bruno: The Philosophy of Circles and the Circular Movement of the Blood', *Journal of the History of Medicine*, vol. 6, 1951, pp. 116-24.

Pagel, Walter, 'William Harvey and the purpose of the circulation', *Isis*, vol. 42, 1951, pp. 22-38.

Pagel, Walter, 'The reaction to Aristotle in seventeenth century biological thought: Campanella, van Helmont, Glanvill, Charleton, Harvey, Glisson, Descartes', in E. A. Underwood (ed.), *Science, Medicine and History: Essays in Honour of Charles Singer*, London, Oxford University Press, 1953, vol. 1, pp. 489-509.

Pagel, Walter, *William Harvey's Biological Ideas: selected aspects and historical background*, Basel, Karger, 1967.

Pagel, Walter, 'The spectre of Van Helmont and the idea of continuity in the history of chemistry', in Robert Young and Mikulas Teich (eds), *Changing perspective in the history of science: essays in honour of Joseph Needham*, Dordrecht, Reidel, 1973.

Pagel, Walter, *New Light on William Harvey*, New York, Karger, 1976.

Park, Katherine, 'The criminal and saintly body: Autopsy and dissection in Renaissance Italy', *Renaissance Quarterly*, vol. 476, 1994, pp. 1-33.

Park, Katherine, Daston, Lorraine J. and Galison, P. L., 'Bacon, Galileo, Descartes on imagination and analogy', *Isis*, vol. 75, 1984, pp. 287-326.

Paster, Gail Kern, *The Body Embarrassed: Drama and the disciplines of shame in early modern England*, Ithaca, Cornell University Press, 1993.

Payne, L. M., 'Sir Charles Scarburgh's Harveian Oration, 1662', *Journal of the History of Medicine*, vol. 12, 1975, pp. 158-64.

Pelling, Margaret, 'Appearance and reality: Barber Surgeons, the body and disease', in A. L. Beier and R. Finlay (eds), *London, 1500-1700: The Making of the Metropolis, London*, Longman, 1985.

Pepper, Stephen, *World Hypotheses: A Study in Evidence*, Berkeley, University of California Press, 1948.

Pocock, J. G. A., 'The reconstruction of discourse: towards the historiography of political thought', *Modern Language Notes*, vol. 96, 1981, pp. 959-80.

Pooley, Roger, 'Language and loyalty: Plain style at the Restoration', *Literature and History*, vol. 6, 1980, pp. 2-18.

Porter, Roy and Rousseau, G. S., *The Ferment of Knowledge: Studies in the Historiography of Eighteenth-century Science*, New York, Cambridge University Press, 1980.

Porter, Roy, 'Laymen, doctors and medical knowledge in the eighteenth century: the evidence of the Gentleman's Magazine', in R. Porter (ed.), *Patients and Practitioners: Lay Perceptions of Medicine in pre-Industrial society*, Cambridge, Cambridge University Press, 1985.

Porter, Roy (ed.), *Patients and Practitioners: Lay Perceptions of Medicine in Pre-Industrial Society*, Cambridge, Cambridge University Press, 1985.

Porter, Roy, 'The patient's view: Doing medical history from below', *Theory & Society*, vol. 14, 1985, 175-98.

Porter, Roy, *Disease, Medicine and Society in England 1550-1860,* Basingstoke, Macmillan Education, 1987.

Porter, Roy, 'The early Royal Society and the spread of medical knowledge', in A. Wear and R. French (eds), *The Medical Revolution of the Seventeenth Century*, Cambridge, Cambridge University Press, 1989.

Porter, Dorothy and Porter, Roy, *Patient's progress: Doctors and doctoring in eighteenth-century England*, Oxford, Polity Press with Blackwell, 1989.

Porter, Roy, 'History of the Body', in Peter Burke (ed.), *New perspectives on historical writing*, Pennsylvania, Pennsylvania State University Press, 1991, pp. 206-33.

Porter, Roy and Burke, Peter (eds), *Language, self, and society: A social history of language*, Cambridge, Polity Press, 1991.

Porter, Roy and Bynum, W. F. (eds), *The Companion Encyclopedia of the History of Medicine*, London, Routledge, 2 Vols, 1993.

Porter, Roy and Jones, Colin (eds), *Reassessing Foucault: Power, Medicine and The Body*, London and New York, Routledge, 1994.

Porter, Roy, *The Greatest Benefit To Mankind: A Medical History of Humanity from Antiquity to the Present*, London, HarperCollins, 1997.

Pouchelle, Marie-Christine, *The Body and Surgery in the Middle Ages*, trans. R. Morris, Cambridge, Polity Press, 1990.

Powell, Anthony, *John Aubrey and his Friends*, London, Hogarth, 1988.

Poynter, F. N. L., *The Evolution of Medical Practice in Britain*, London, Pitman, 1961.

Pumfrey, Stephen, 'Mechanizing magnetism in restoration England: The decline of magnetic philosophy', *Annals of Science*, vol. 44, 1987, pp. 1-22.

Pumfrey, Stephen, 'Ideas above his station: A social study of Hooke's curatorship of experiments', *History of Science*, vol. 29, no. 83, 1991.

Punt, Hendrick, *Bernard Siegfried Albinus (1697-1770) on 'Human Nature':, Anatomical and Physiological Ideas in Eighteenth Century Leiden*, Amsterdam, B. M. Israel, 1983.

Pyle, Andrew J., 'Animal generation and the mechanical philosophy: Some light on the role of biology in the scientific revolution', *History and Philosophy of the Life Sciences*, vol. 9, 1987, pp. 225-54.

Rather, L. J., 'Old and new view of the emotions and bodily changes: Wright and Harvey versus Descartes, James and Cannon', *Clio Medica*, vol. 1, 1965, pp. 1-25.

Rather, L. J., 'On the use of military metaphor in western medical literature', *Clio Medica*, vol. 7, no. 3, 1972, pp. 201-8.

Rather, L. J., 'On the source and development of metaphorical language in the history of western medicine', in L. G. Stevenson (ed.), *A Celebration of Medical History*, Baltimore, Johns Hopkins University Press, 1982, pp. 135-56.

Rattansi, Pyarali, 'Paracelsus and the Puritan Revolution', *Ambix*, vol. 11, 1963, pp. 24-32.

Rattansi, Pyarali, 'The Helmontian-Galenist Controversy in Restoration England', *Ambix*, vol. 12, 1964, pp. 1-23.

Rattansi, Pyarali, 'Some evaluations of reason in sixteenth and seventeenth century natural philosophy', in Robert Young and Mikulas Teich (eds),

Changing Perspectives in the History of Science, Dordrecht, Reidel, 1973, pp. 148-66.

Reif, Patricia, 'The textbook tradition in natural philosophy, 1600-1650', *Journal of the History of Ideas*, vol. 30, 1969, pp. 17-32.

Reiser, Stanley Joel, *Medicine and the Reign of Technology*, London, Cambridge University Press, 1979.

Reiss, Timothy, 'Denying the body: Memory and the dilemmas of history in Descartes', *Journal of the History of Ideas*, vol. 57, no. 4, 1996, pp. 537-607.

Reines, B. P., 'On the role of clinical anomaly in Harvey's discovery of the mechanism of the pulse', *Perspectives in Biology and Medicine*, vol. 34, 1990, pp. 128-33.

Richardson, Ruth, *Death, Dissection and the Destitute*, London, Routledge & Kegan Paul, 1987.

Riese, Walther, 'Descartes as a psychotherapist: the uses of rational philosophy in the treatment of discomfort and disease, its limitations', *Medical History*, vol. 10, no. 3, 1966, pp. 237-44.

Righini-Bonelli, M. L. and Shea, W. R. (eds), *Reason, Experiment and Mysticism in the Scientific Revolution*, London, MacMillan, 1975.

Roberts, R. S., 'The personnel and practice of medicine in Tudor and Stuart England', *Medical History*, vol. 6, 1962 & vol. 8, 1964.

Roger, Jacques, 'The Mechanistic Conception of Life', in D. C. Lindberg and R. Numbers (eds), *God and Nature: Historical Essays on the Encounter between Christianity and Science*, Berkeley, University of California Press, 1986, pp. 277-95.

Rolleston, Humphrey, 'The early history of the teaching of human anatomy in London and morbid anatomy and pathology in Great Britain', *Annals of Medical History*, vol. 1, 1939, pp. 203-38.

Rolleston, Humphrey, 'Walter Charleton, DM, FRCP, FRS', *Bulletin of the History of Medicine*, vol. 8, 1940, pp. 403-16.

Rorty, Richard, *Contingency, Irony and Solidarity*, Cambridge, Cambridge University Press, 1989.

Rossi, Paolo, 'Hermeticism, rationality and the scientific revolution', in M. L. Righini Bonelli & W. R. Shea (eds), *Reason, mysticism and Experiment in the Scientific Revolution*, New York, Macmillan, 1975, pp. 247-74.

Rouse, Joseph, 'The narrative reconstruction of science', *Inquiry*, vol. 33, 1990, pp. 179-96.

Russell, Kenneth F., 'John Browne, 1642-1702: A seventeenth century surgeon, anatomist and plagiarist', *Bulletin of the History of Medicine*, vol. 33, 1959, pp. 393-414; 505-25.

Russell, Kenneth F., 'Anatomy and the Barber Surgeons', *Medical Journal of Australia*, vol. 1, 1973, pp. 1109-15.

Russell, Kenneth F., *British Anatomy: 1525-1800: A Bibliography*, Melbourne, Melbourne University Press, 1987.

Salman, Phillips, 'William Harvey and art misplaced', *Annals of Science*, vol. 49, no. 1, 1992, pp. 3-19.

Salmon, Vivian, *The Study of Language in Seventeenth Century England*, Amsterdam, Benjamins, 1979.

Sarasohn, Lisa T., 'A science turned upside-down: Feminism and the natural philosophy of Margaret Cavendish', *Huntingdon Library Quarterly*, vol. 47, 1984, pp. 299-307.

Sarasohn, Lisa T., 'Motion and morality: Pierre Gassendi, Thomas Hobbes and the mechanical world view', *Journal of the History of Ideas*, vol. 46, 1985, pp. 363-79.

Sargent, Rose-Mary, *The Diffident Naturalist: Robert Boyle and the Philosophy of Experiment*, Chicago, Chicago University Press, 1995.

Sawday, Jonathan, 'The Mint at Segovia: Digby, Hobbes, Charleton and the body as a machine in the seventeenth century', *Prose Studies*, vol. 6, no. 1, 1983, pp. 21-36.

Sawday, Jonathan, 'The Fate of Marsyas: Dissecting the renaissance body', in Lucy Gent and Nigel Llewellyn (eds), *Renaissance Bodies: The Human Figure in English Culture, c1540-1660*, London, Reaktion, 1990, pp. 111-35.

Sawday, Jonathan, 'Re-writing a Revolution: History, symbol and text in the Restoration', *Seventeenth Century*, vol. 7, no. 2, 1992, pp. 171-99.

Sawday, Jonathan, *The Body Emblazoned: Dissection and the Human Body in Renaissance Culture*, London, Routledge, 1996.

Schaffer, Simon, 'Godly men and mechanical philosophers: Souls and spirits in restoration natural philosophy', *Science in Context*, vol. 1, no. 1, 1987, pp. 55-85.

Schaffer, Simon, 'Making Certain', *Social Studies of Science*, vol. 14, 1984, pp. 137-52.

Schaffer, Simon, 'The Glorious Revolution and medicine in Britain and the Netherlands', *Notes and Records of the Royal Society of London*, vol. 43, no. 2, 1989, pp. 167-90.

Schein, Oliver D., 'Michael Servetus and Giordano Bruno: Philosophical trends with physiological implications', *Synthesis*, vol. 3, no. 3, 1976, pp. 8-24.

Schiebinger, Londa, 'Margaret Cavendish: Natural Philosopher', in M. E. White (ed.), *A History of Women Philosophers*, Dordrecht, Kluwer, 1988.

Schneider, Ulrich Johannes, 'Eclecticism rediscovered', *Journal of the History of Ideas*, vol. 59, no. 1, 1998, pp. 173-82.

Schuster, John, 'The scientific revolution', in R. C. Olby, G. N. Cantor, J. R. R. Christie and M. J. S. Hodge (eds), *Companion to the History of Modern Science*, London, Routledge, 1990, pp. 217-42.

Sennett, Richard, *Flesh and Stone: the body and the city in western civilization*, London, Faber and Faber, 1994.

Shank, Michael H., 'From Galen's Ureters to Harvey's Veins', *Journal of the History of Biology*, vol. 18, no. 3, 1985, pp. 331-55.

Shapere, Dudley, 'Plausibility and justification in the development of science', *Journal of Philosophy*, vol. 3, 1966, pp. 611-27.

Shapin, Steven, 'Pump and Circumstance: Robert Boyle's Literary Technology', *Social Studies of Science*, vol. 14, 1984, pp. 481-519.

Shapin, Steven, 'Talking history: reflections on discourse analysis', *Isis*, vol. 75, 1984, pp. 125-30.

Shapin, Steven and Schaffer, Simon, *Leviathan and the Air-Pump: Hobbes, Boyle and the Experimental Life*, Princeton, Princeton University Press, 1985.

Shapin, Steven, 'The house of experiment in seventeenth century England', *Isis*, vol. 79, 1988, pp. 373-404.

Shapin, Steven, 'Robert Boyle and mathematics: reality, representation and experimental practice', *Science in Context*, vol. 2, no. 1, 1988, pp. 23-58.

Shapin, Steve, 'Closure and credibility in seventeenth century science', in Ron L. Numbers and J. V. Pickstone (eds), 'Program, papers and abstracts for the Joint conference of the British society for the History of science, and the History of science society', Manchester, 11-15 July, 1988, p. 147-54.

Shapin, Steven, ' "The mind is its own place": Science and solitude in seventeenth-century England', *Science in Context*, vol. 4, no. 1, 1990, pp. 191-218.

Shapin, Steven, ' "A scholar and a Gentleman": The problematic identity of the scientific practitioner in early modern England', *History of Science*, vol. 29, 1991, pp. 279-327.

Shapin, Steven, *A Social History of Truth: Civility and Science in Seventeenth Century England*, Chicago, University of Chicago Press, 1994.

Shapin, Steven, *The Scientific Revolution*, Chicago, Chicago University Press, 1997.

Shapiro, Barbara, *John Wilkins, 1614-1672; An Intellectual Biography*, London, University of California Press, 1969.

Shapiro, Barbara, *Probability & Certainty in Seventeenth Century England: A study of the relations between Natural Science, Religion, History, Law, and Literature*, Princeton, Princeton University Press, 1983.

Sharp, Lindsay, 'Walter Charleton's Early Life, 1620-1659, and Relationship to Natural Philosophy in Mid-Seventeenth Century England', *Annals of Science*, vol. 30, 1973, pp. 311-40.

Sharp, Lindsay, 'The RCP and Interregnum Politics', *Medical History*, vol. 19, no. 2, 1975, pp. 107-28.

Sheets-Johnstone, Maxine (ed.), *Giving the Body its Due*, Albany, State University of New York Press, 1992.

Sherman, Nancy, *The Fabric of Character: Aristotle's Theory of Virtue*, Oxford, Clarendon Press, 1989.

Shortland, Michael, 'Bodies of history: some problems and perspectives', *History of Science*, vol. 24, 1986, pp. 303-26.

Siraisi, Nancy and McVaugh, Michael R. (eds), *Renaissance Medical Learning: Evolution of a Tradition*, Philadelphia, History of Science Society, 1990.

Siraisi, Nancy, *The Clock and the Mirror: Giordano Cardano and Renaissance Medicine*, Princeton, Princeton University Press, 1997.

Skinner, Quentin R. D., 'Meaning and understanding in the history of ideas', *History and Theory*, vol. 8, 1969, pp. 3-53.

Skulsky, Harold, *Language Recreated: Seventeenth Century Metaphorists and the Act of Metaphor*, Athens, Georgia, University of Georgia Press, 1992.

Slaughter, Mary M., *Universal Language and Scientific Taxonomy in the Seventeenth Century*, Cambridge, Cambridge University Press, 1982.

Slawinski, Maurice, 'Rhetoric and science/rhetoric of science/rhetoric as science', in M. Slawinski, S. Pumfrey and P. L. Rossi (eds), *Science, Culture and Popular Belief in Renaissance Europe*, New York, Manchester University Press, 1991, pp. 71-99.

Sloan, R., 'Descartes, the sceptics, and the rejection of Vitalism in seventeenth-century physiology', *Studies in the History & Philosophy of Science*, vol. 8, no. 1, 1977, pp. 1-28.

Small, Carola and Small, Alastair, 'John Evelyn and the Garden of Epicurus', *Journal of the Warburg and Courtauld Institutes*, vol. 60, 1997, pp. 192-214.

Smith, Robert William Innes, *English Speaking Students of Medicine at the University of Leyden*, Edinburgh, Oliver and Boyd, 1932.

Sontag, Susan, *Illness As Metaphor*, London, Penguin, 1991.

Southgate, Beverly, 'The Power of Imagination: Psychological Explanations in Mid-Seventeenth Century England', *History of Science*, vol. 30, no. 89, 1992, pp. 281-95.

Stafford, Barbara M., *Body Criticism: Imaging the Unseen in Enlightenment Art and Medicine*, Massachusetts, MIT Press, 1991.

Stearns, R. P., 'The relations between science and society in the later seventeenth century', in *The Restoration of the Stuarts, Blessing or Disaster?*, Washington, Folger Shakespeare Library, 1960.

Stevenson, Lloyd G., ' "New diseases" in the seventeenth century', *Bulletin of the History of Medicine*, vol. 39, 1965, pp. 1-21.

Stewart, Larry, *The Rise of Public Science: Rhetoric, Technology and Natural Philosophy in Newtonian Britain 1660-1750*, New York, Cambridge University Press, 1992.

Stillman, Robert E., 'Assessing the revolution: Ideology, language and rhetoric in the new philosophy of early modern England', *The Eighteenth Century: Theory and Interpretation*, vol. 35, 1994, pp. 99-118.

Stimson, Dorothy, 'Dr Willis and the Royal Society', *Journal of Modern History*, vol. 3, 1931, pp. 539-63.

Sutton, Clive, ' "Nullius in Verba and Nihilis in Verbis": Public understanding of the role of language in science', *British Journal of the History of Science*, vol. 27, 1994, pp. 55-64.

Suzuki, Akihito, 'Dualism and the transformation of psychiatric language in the seventeenth and eighteenth centuries', *History of Science*, vol. 33, 1995, pp. 417-47.

Suzuki, Akihito, ' "A Duumvirate of Rulers Within Us": Politics and medical pneumatology in Restoration England', in W. G. Marshall (ed.), *The Restoration Mind*, Newark, University of Delaware Press, 1997.

Synott, Anthony, *The Body Social: Symbolism, Self and Society*, London, Routledge, 1993.

Tansey, E. M., 'The physiological tradition', in W. F. Bynum and R. Porter (eds), *The Companion Encyclopedia of the History of Medicine*, London, Routledge, 1993, pp. 120-152.

Temkin, Owsei, 'Metaphors of Human Biology', in R. C. Stauffer (ed.), *Science and Civilization*, New York, University of Wisconsin Press, 1949, pp. 169-96.

Temkin, Owsei, 'The classical roots of Glisson's doctrine of irritation', *Bulletin of the History of Medicine*, vol. 38, 1964, pp. 297-328.

Temkin, Owsei, *Galenism: Rise and Decline of a Medical Philosophy*, Ithaca, Cornell University Press, 1973.

Temkin, Owsei, *The Double Face of Janus, and other essays in the history of medicine*, Baltimore, Johns Hopkins University Press, 1977.

Temkin, Owsei, 'The scientific approach to disease: specific entity and individual sickness', in A. Caplan, H. T. Engelhardt and J. J. McCartney (eds), *Concepts of Health and Disease: Interdisciplinary Perspectives*, Massachusetts, Addison Wesley, 1981, pp. 247-64.

Thomas, Henry, 'The Society of Chemical Physicians; An echo of the Great Plague of London, 1665', in E. A. Underwood (ed.), *Science, Medicine and History; Essays on the Evolution of Scientific Thought and Medical Practice Written in Honour of Charles Singer*, London, Oxford University Press, 1953, Vol. II, pp. 56-71.

Thomas, Keith, *Man and the Natural World: Changing Attitudes in England 1500-1800*, London, Allen Lane, 1983.

Thomas, Keith, *Religion and the Decline of Magic: Studies in Popular Beliefs in Sixteenth- and Seventeenth-Century England*, London, Penguin, 1991.

Thorpe, Clarence De Witt, 'Two disciples of Hobbes', in *The Aesthetic Theory of Thomas Hobbes*, Ann Arbor, University of Michigan Press, 1940, pp. 176-88.

Thurston, A. J., 'Giovanni Borelli and the study of human movement: An historical review', *Australian and New Zealand Journal of Surgery*, vol. 69, no. 4, 1998.

Tiggert, William D., 'Noah Biggs, Author, fl. 1651', *Journal of the History of Medicine and the Allied Sciences*, vol. 38, no. 4, 1983, pp. 452-55.

Tomaselli, Sylvana, 'The First Person: Descartes, Locke and mind-body dualism', *History of Science*, vol. 22, no. 56, 1984, pp. 185-202.

Tomlinson, Howard & Gregg, David, *Politics, Religion and Society in Revolutionary England, 1640-1660*, London, Macmillan, 1989.

Tumbleson, Raymond D., '"Reason and Religion": The Science of Anglicanism', *Journal of the History of Ideas*, vol. 57, no. 1, 1996, pp. 131-56.

Turner, Bryan S., 'Recent developments in the theory of the body', in B. S. Turner (ed.) *The Body: social process and cultural theory*, London, SAGE, 1991, pp. 1-35.

Turner, Bryan S., 'The Anatomy Lesson: A Note on the Merton Thesis', *Sociological Review*, vol. 38, no. 1, 1990, pp. 1-18.

Turner, Bryan S., *Regulating Bodies; Essays in Medical Sociology*, London, Routledge, 1992.

Turner, Bryan S., *The Body and Society: Explorations in Social Theory*, Oxford, Basil Blackwell, 1984.

Tyacke, Nicholas (ed.), *The History of the University of Oxford*, Oxford, Clarendon Press, 1997.

Valadez, Frances Margaret, *Anatomical Studies at Oxford During the Seventeenth Century*, Ann Arbor, University Microfilms, 1981.

Van Leeuwen, Henry, *The Problem of Certainty in English Thought, 1630-1690*, The Hague, Martinus Nijhoff, 1970.

Vickers, Brian, 'On the Practicalities of Renaissance Rhetoric', in B. Vickers (ed.), *Rhetoric Revalued: Papers From the International Society for the History of Rhetoric*, New York, Medieval and Renaissance Texts and Studies, 1982, pp. 133-42.

Vickers, Brian, 'Analogy versus identity: the rejection of occult symbolism, 1580-1680', in B. Vickers (ed.), *Occult and Scientific Mentalities in the Renaissance*, New York, Cambridge University Press, 1984, pp. 95-163.

Vickers, Brian and Struever, Nancy, *Rhetoric and The Pursuit of Truth: Language Change in the Seventeenth and Eighteenth Centuries*, California, William Andrews Clark Memorial Library, 1985.

Vigne, R., 'Mayerne and his successors: some Huguenot physicians under the Stuarts', *Journal of the Royal College of Physicians*, London, vol. 20, no. 3, 1986, pp. 222-6.

Von Staden, Heinrich, 'Anatomy as rhetoric: Galen on dissection and persuasion', *Journal of the History of Medicine and the Allied Sciences*, vol. 50, 1995, pp. 47-66.

Walker, D. P., The Astral Body in Renaissance Medicine, *Journal of the Warburg and Courtauld Institutes*, vol. 21, 1958, pp. 119-33.

Walker Bynum, Carol, 'Why all the fuss about the body? A medievalist's perspective', *Critical Inquiry*, vol. 22, 1995, pp. 1-33.

Wasserman, E. R., 'Dryden's Epistle to Charleton', *Journal of English and German Philology*, vol. 55, 1956.

Wear, Andrew, 'William Harvey and the "way of the anatomists"', *History of Science*, vol. 21 no. 53, 1983, pp. 223-49.

Wear, Andrew, 'Medical practice in late seventeenth- and early eighteenth-century England: continuity and union', in A. Wear and R. French (eds),

The Medical Revolution of the Seventeenth Century, Cambridge, Cambridge University Press, 1989.

Wear, Andrew (ed.), *Medicine in Society; Historical Essays*, New York, Cambridge University Press, 1992.

Wear, Andrew (ed.), *Health and healing in early modern England: studies in social and intellectual history*, Aldershot, Ashgate, 1998.

Webster, Charles, 'The College of Physicians: "Solomon's House" in Commonwealth England', *Bulletin of the History of Medicine*, vol. 41, 1967, pp. 393-412.

Webster, Charles, 'English Medical Reformers of the Puritan Revolution: A Background to the Society of Chymical Physitians', *Ambix*, vol. 14, 1967, pp. 16-41.

Webster, Charles, 'The Helmontian George Thomson & William Harvey: The revival & application of splenectomy to physiological research', *Medical History*, vol. 15, 1971.

Webster, Charles, 'Science and the challenge to the scholastic curriculum, 1640–1660', in *The Changing Curriculum*, London, Methuen, 1971.

Webster, Charles, 'The Helmontian George Thomson and William Harvey', *Medical History*, vol. 15, 1971, pp. 154-68.

Webster, Charles (ed.), *The intellectual revolution of the seventeenth century*, London, Routledge and Kegan Paul, 1974.

Webster, Charles, *The Great Instauration: Science, Medicine and Reform, 1626-1660*, London, Duckworth, 1975.

Webster, Charles (ed.), *Health, medicine, and mortality in the sixteenth century*, New York, Cambridge University Press, 1979.

Webster, Charles, *From Paracelsus to Newton: Magic and the making of modern science*, New York, Cambridge University Press, 1982.

Westfall, Richard S., *Science and Religion in Seventeenth century England*, Hamden, Archon, 1970.

Westfall, Richard S., *The Construction of Modern Science: mechanisms and mechanics*, New York, John Wiley and Sons, 1971.

Westman, Robert S., and McGuire, J. F., *Hermeticism and the scientific revolution*, Los Angeles, William Andrews Clark Memorial Library, 1977.

White, Hayden, *The Content of the Form: Narrative Discourse and Historical Representation*, Baltimore, Johns Hopkins University Press, 1987.

White, John S., 'William Harvey and the primacy of the blood', *Annals of Science*, vol. 43, no. 3, 1986, pp. 239-55.

Whitteridge, Gweneth, 'William Harvey: A Royalist and No Parliamentarian', in C. Webster (ed.) *The Intellectual Revolution of the Seventeenth Century*, London, Routledge & Kegan Paul, 1974.

Whitteridge, Gweneth, *William Harvey and the circulation of blood*, New York, American Elsevier, 1971.

Wightman, William D., 'Myth and method in seventeenth century biological thought', *Journal of the History of Biology*, vol. 2, 1969, pp. 321-36.

Wightman, William D., *The Growth of Scientific Ideas*, London, Oliver & Boyd, 1950.

Williamson, George, 'Milton and the Mortalist heresy', *Studies in Philology*, vol. 32, 1935, pp. 553-79.

Williamson, George, 'The Ephesian Matron versus the Platonic Lady', *Review of English Studies*, vol. 12, 1936, pp. 445-49.

Wilson, Catherine, 'Visual surface and visual symbol: The microscope and the occult in early modern science', *Journal of the History of Ideas*, vol. 49, no. 3, 1988.

Wilson, Catherine, 'Enthusiasm and its critics: Historical and modern perspectives', *History of European Ideas*, vol. 17, 1993, pp. 461-78.

Wilson, Catherine, *The Invisible World: Early Modern Philosophy and the Invention of the Microscope*, New Jersey, Princeton University Press, 1995.

Wilson, L. G., 'William Croune's theory of muscular contraction', *Notes and Records*, vol. 16, 1961, pp. 158-78.

Wilson, Luke, 'William Harvey's *Prelectiones*: the performance of the body in the Renaissance theatre of anatomy', *Representations*, vol. 17, 1987, pp. 62-95.

Winn, James Auderson, *John Dryden and His World*, New Haven, Yale University Press, 1987.

Wittgenstein, Ludwig, *Philosophical Investigations*, transl. G. E. M. Anscombe, Oxford, Basil Blackwell, 1953.

Wittgenstein, Ludwig, *Tractatus Logico-Philosophicus*, London, Routledge & Kegan Paul, 1962.

Wolfe, David E., 'Sydenham and Locke on the limits of anatomy', *Bulletin of the History of Medicine*, vol. 35, 1961, pp. 193-220.

Wright, J. P., 'Hysteria and Mechanical Man', *Journal of the History of Ideas*, vol. 41, 1980, pp. 233-49.

Wright, P. and Treacher, A. (eds), *The Problem of Medical Knowledge: Examining the Social Construction of Medicine*, Edinburgh, Edinburgh University Press, 1982.

Yeo, Richard and Schuster, John (eds), *The Politics and Rhetoric of Scientific Method: historical studies*, Boston, Reidel Publishing Co., 1986.

STUDIES IN HISTORY AND PHILOSOPHY OF SCIENCE

1. R. McLaughlin (ed.): *What? Where? When? Why?* Essays on Induction, Space and Time, Explanation. Inspired by the Work of Wesley C. Salmon. 1982 ISBN 90-277-1337-5
2. D. Oldroyd and I. Langham (eds.): *The Wider Domain of Evolutionary Thought*. 1983 ISBN 90-277-1477-0
3. R.W. Home (ed.): *Science under Scrutinity*. The Place of History and Philosophy of Science. 1983 ISBN 90-277-1602-1
4. J.A. Schuster and R.R. Yeo (eds.): *The Politics and Rhetoric of Scientific Method*. Historical Studies. 1986 ISBN 90-277-2152-1
5. J. Forge (ed.): *Measurement, Realism and Objectivity*. Essays on Measurement in the Social and Physical Science. 1987 ISBN 90-277-2542-X
6. R. Nola (ed.): *Relativism and Realism in Science*. 1988 ISBN 90-277-2647-7
7. P. Slezak and W.R. Albury (eds.): *Computers, Brains and Minds*. Essays in Cognitive Science. 1989 ISBN 90-277-2759-7
8. H.E. Le Grand (ed.): *Experimental Inquiries*. Historical, Philosophical and Social Studies of Experimentation in Science. 1990 ISBN 0-7923-0790-9
9. R.W. Home and S.G. Kohlstedt (eds.): *International Science and National Scientific Identity*. Australia between Britain and America. 1991 ISBN 0-7923-0938-3
10. S. Gaukroger (ed.): *The Uses of Antiquity*. The Scientific Revolution and the Classical Tradition. 1991 ISBN 0-7923-1130-2
11. P. Griffiths (ed.): *Trees of Life*. Essays in Philosophy of Biology. 1992 ISBN 0-7923-1709-2
12. P.J. Riggs (ed.): *Natural Kinds, Laws of Nature and Scientific Methodology*. 1996 ISBN 0-7923-4225-9
13. G. Freeland and A. Corones (eds.): *1543 and All That*. Image and Word, Change and Continuity in the Proto-Scientific Revolution. 1999 ISBN 0-7923-5913-5
14. H. Sankey (ed.): *Causation and Laws of Nature*. 1999 ISBN 0-7923-5914-3
15. R. Nola and H. Sankey (eds.): *After Popper, Kuhn and Feyerabend*. Recent Issues in Theories of Scientific Method. 2000 ISBN 0-7923-6032-X
16. K. Neal: *From Discrete to Continuous*. The Broadening of Number Concepts in Early Modern England. 2002 ISBN 1-4020-0565-2
17. S. Clarke and T.D. Lyons (eds.): *Recent Themes in the Philosophy of Science*. Scientific Realism and Commonsense. 2002 ISBN 1-4020-0831-7
18. E. Booth: *'A Subtle and Mysterious Machine'*. 2005 ISBN 1-4020-3377-X
19. P.R. Anstey and J.A. Schuster (eds.): *The Science of Nature in the Seventeenth Century*. Patterns of Change in Early Modern Natural Philosophy. 2005 ISBN 1-4020-3603-5

springeronline.com